Symmetry and Quantum Systems

An Introduction to Group Representations

THE MODERN UNIVERSITY PHYSICS SERIES

This series is intended for readers whose main interest is in physics, or who need the methods of physics in the study of science and technology. Some of the books will provide a sound treatment of topics essential in any physics training, while other, more advanced volumes will be suitable as preliminary reading for research in the field covered. New titles will be added from time to time.

Bagguley: *Electromagnetism and Linear Circuits*
Clark: *A First Course in Quantum Mechanics*
Littlefield and Thorley: *Atomic and Nuclear Physics* (2nd edition)
Lothian: *Optics and its Uses*
Lovell, Avery and Vernon: *Physical Properties of Materials*
Perina: *Coherence of Light*
Tilley and Tilley: *Superfluidity and Superconductivity*
Wolbarst: *Symmetry and Quantum Systems*

Symmetry and Quantum Systems

ANTHONY B. WOLBARST
Department of Physics, University of the Witwatersrand

VAN NOSTRAND REINHOLD COMPANY
New York Cincinnati Toronto London Melbourne

©Anthony Wolbarst 1977

ISBN: 0 442 30163 4 (cloth)
ISBN: 0 442 30180 4 (paper)

Library of Congress Catalog Card Number 76-18216

All rights reserved. No part of this work covered by the copyright hereon may be reproduced or used in any form or by any means —graphic, electronic, or mechanical, including photocopying, recording taping, or information storage or retrieval systems—without written permission of the publishers.

**Published by Van Nostrand Reinhold Company Limited
Molly Millars Lane, Wokingham, Berkshire, England**

*Published in 1977 by Van Nostrand Reinhold Company
A division of Litton Educational Publishing, Inc.,
450 West 33rd Street, New York, NY 10001, U.S.A.*

*Van Nostrand Reinhold Limited
1410 Birchmount Road, Scarborough, Ontario, M1P 2E7, Canada*

*Van Nostrand Reinhold Australia Pty. Limited
17 Queen Street, Mitcham, Victoria 3132, Australia*

Library of Congress Cataloging in Publication Data

Wolbarst, Anthony B
  Symmetry and quantum systems.

  (The Modern university physics series)
  Bibliography: p.
  Includes index.
  1. Symmetry (Physics) 2. Quantum theory.
3. Schrödinger equation. I. Title.
QC174.17.S9W64   539.7'21'0151222   76-18216
ISBN 0-442-30163-4
ISBN 0-442-30180-4 pbk.

*Printed in Great Britain at The Spottiswoode Ballantyne Press
by William Clowes and Sons Limited,
London, Colchester and Beccles*

Since receiving his B.A. in 1966 from Trinity (Conn.) and Ph.D. from Dartmouth in 1971, Dr Wolbarst has been involved with research projects at SUNY Buffalo, Harvard, Massachusetts General Hospital, and the University of the Witwatersrand. He is currently Senior Research Scientist in the High Pressure Physics Division of the National Physical Research Laboratory, C.S.I.R., Pretoria. His research interests involve EPR and ENDOR spectroscopy, the properties of point defects in solids, ferroelectricity, and the effects of high pressures upon materials.

to
Wm. T. Doyle, Dartmouth
Charles R. Miller, Trinity
Jon F. Reichert, SUNYAB

Sir Thomas More:... Why not be a teacher? You'd
   be a fine teacher. Perhaps, a great one.
Rich: And if I was, who would know it?
More: You, your pupils, your friends, God. Not a bad public,
   that... Oh, and a *quiet* life.

A Man for All Seasons
Robert Bolt

Preface

'Symmetry, as wide or as narrow as you may define its meaning, is one idea by which man through the ages has tried to comprehend and create order, beauty and perfection.'   H. Weyl†

'I like to recall (M. von Laue's) question as to which results derived in the present volume I considered most important. My answer was that the explanation of Laporte's rule (the concept of parity) and the quantum theory of the vector addition model appeared to be most significant. Since that time, I have come to agree with his answer that the recognition that almost all rules of spectroscopy follow from the symmetry of the problem is the most remarkable result.'   E. P. Wigner‡

Symmetry arguments can be of great value in solving numerous sorts of physical problems. For example, with a particle constrained to a circular orbit, the use of polar co-ordinates leads to a much simpler analysis than would the use of Cartesian co-ordinates. Here we choose a mathematical description which reflects the basic symmetry of the system, and we end up with one variable to handle rather than two. Similarly, we know that the Fourier sum representing an even periodic function $f_{\text{even}}(x)$ contains only cosine (even) terms: the coefficients of the sine terms, of the form $\int_{-a}^{a} f_{\text{even}}(x) \sin(nx) \, dx$, vanish because the product of an even and an odd function is odd, and the integral of this odd product is zero. Again, a helpful symmetry argument.

On perhaps a more fundamental level, an awareness of the relevant symmetries may be necessary for any deep understanding of observed physical phenomena. An apple falls with the same acceleration in San Francisco and in Los Angeles *because* the

†H. Weyl[40].
‡From author's preface, E. P. Wigner[41].

basic laws of nature are invariant to a symmetry transformation —
in this case a spatial displacement in which the force of gravity
(and, presumably, the other relevant factors) does not change.
And the energy eigenfunctions of a hydrogen atom may be
selected to be even or odd *because* the system possesses a center of
inversion.

But to take full advantage of the presence of symmetries in more
complex situations, either in simplifying specific problems or in
shedding light on general processes, one may have to make use of
group theory. This branch of mathematics involves an extension
of the elementary arguments employed above, and many of the
ideas should be no more difficult to grasp. But the entities which
one juggles in calculations, such as the traces (characters)
of certain somewhat arbitrary-seeming matrices, suggest little
apparent connection with the original physical problem. And because the link between the physical symmetry and the mathematical methods is not obvious, one may find oneself using
character tables to predict vanishing dipole moments and the
degeneracies of perturbed quantum levels with no true comprehension of what is really going on. The calculations in the
technique are straightforward, but the underlying mathematical
foundations quite abstract, and so this application without understanding may easily be the case. The dilemma is that mathematical
rigor and completeness seldom provide illumination to one first
learning a broad and complex subject like quantum mechanical
group representation theory, and yet there is little point in considering the subtle implications of the theory without a firm grasp
on the underlying basic concepts.

In *Symmetry and Quantum Systems* I have attempted to
steer a middle course, concentrating on the central issues: the
nature of the physical symmetries displayed by quantum mechanical systems, the symmetry properties of the mathematical functions used in describing such systems, and the way in which the
two are related. The primary aim is to help the reader find an
intuitive comfortableness with these issues. The book is more
concerned with the questions 'what does it mean, and why does it
work' than with 'how do you do it for this particular complicated
system, and how do you prove it?' There are several excellent
texts, listed in the Bibliography, which clearly explain and demonstrate the versatility of group theoretic methods in handling
physical problems, and there is no need to repeat what is in these.
Rather, it is hoped that the material presented herein will allow the
reader to attack more rigorous or applications-oriented books with
an understanding of the fundamental ideas already under his belt.
When approaching a diverse and extensive new area, such an
introduction can, in the long run, serve to advantage.

Symmetry and Quantum Systems was written with the third or fourth year undergraduate student of applied mathematics, physics, or chemistry in mind. It is assumed only that the reader has finished a one-semester course of quantum mechanics or quantum chemistry, and has acquired some facility in manipulating vectors, matrices, and simple differential equations. Hopefully the book will also be useful to graduate students or practising experimentalists who would like to learn, as quickly and as painlessly as possible, what is meant by an E mode of vibration, an A_1 electronic state, the Jahn-Teller effect, or an irreducible tensor operator. Although the reader may not be able to perform complex manipulations after working through this introduction, he should at least be able to follow the gist of most such calculations he encounters.

Our approach is based upon the view that the concept of paramount importance, one whose meaning is easily grasped, is that of a subspace invariant with respect to a single geometric symmetry operation. The nature of such an 'invariant subspace' is examined in detail first through the example of rotations in ordinary three-space. Here we employ a three-dimensional closure relation to generate the familiar three-by-three rotation matrix; the matrix is thus seen to be a representation of a rotation operator, established through the use of a particular basis vector set. We show that the block-diagonal form of the matrix for rotations about the 3-, or z-, axis of three-space indicates that three-space can be broken into two subspaces (with inherently different symmetry properties) invariant relative to the rotation operator. This is true not only for a single operator, but even for a group of symmetry operations such as C_{3v}, the set of rotations and reflections which leave invariant the appearance of an ammonia molecule. Irreducible representations are first introduced in this fashion.

Then after a link is established between geometric transformation of the vector $\bar{r}$ of real space and transformations of the vector $f(\bar{r})$ or $|f(\bar{r})\rangle$ of function space, the already familiar ideas of closure, representation, and subspace invariant relative to an operator are generalized so as to apply in function space. Here and later, we employ the well-known hydrogen atomic orbitals as basis vectors, and exploit the orthogonality condition $\langle\psi_{nl}(\bar{r})|\psi_{n'l'}(\bar{r})\rangle = \delta_{nn'}\delta_{ll'}$ in the construction of representations. This makes it especially easy to demonstrate the central point of the entire development: the physical symmetry of a quantum mechanical system predetermines an essentially unique set of mutually orthogonal invariant subspaces which may describe the system. And all the useful results of group theory are consequences of the symmetry and orthogonality properties of these invariant subspaces of the Hilbert space. Characters are therefore used as little as possible in this book, and are presented as computationally, rather than conceptually, useful constructs.

We have not analysed, or even listed, the various point and space groups of molecules and solids; instead, C_{3v} is employed in nearly all examples. We have chosen this path because the approach of group representations should be sufficiently novel, for the uninitiated, to warrant the repetitive use of a familiar touchstone. Once the basic ideas are clearly understood, it is a relatively easy matter to generalize to more complex geometric structures later.

Finally, concerning the exercises: they are one hundred per cent an integral part of the presentation, and must be examined. Many arise from condensations, made in the interest of brevity, of the longer original manuscript. Problems of especial importance to the development are marked with an asterisk (*); these usually introduce, easily and in terms of what has already been discussed, ideas which will be considered in detail later. Others demand, for a satisfying answer, more background material than is provided by the text—hopefully they raise the sort of nagging questions which will lead to further study. Problems and chapter sections which may be left aside without loss of continuity in a first reading are noted with a dagger (†).

Acknowledgements

Johannesburg is full of vegetarians, and the ones I've run into all tell me the same thing: you are what you eat. It seems also true enough that you are what you learn. And the author of an elementary textbook, in particular, often only synthesizes the ideas of others, trying to present them in a novel and perhaps more palatable fashion.

I began, and am continuing, to learn physics from the excellent books of Feynman, Merzbacher, Dirac, Tinkham, and others, and from personal contact with inspiring teachers and colleagues such as Wm. T. Doyle, C. R. Miller, and J. F. Reichert. And whatever successes this book may achieve are largely due to those from whom, directly or indirectly, I have learned.

Symmetry and Quantum Systems was written while I was a research fellow at the University of the Witwatersrand in Johannesburg and its evolution was greatly influenced by discussions with members of the Physics Department. I am especially indebted to Yuda Tuval, Ted Lowther, Jochen Urban, Bob Zwicker, and Arthur Every, each of whom read one version or another of the entire manuscript, made many suggestions, and pointed out a number of unclear areas and outright inaccuracies.

Full responsibility for the end product, of course, rests with me; and I would be most appreciative if any reader would offer further comments or suggestions for improvements.

I wish to thank Prof. F. R. N. Nabarro, F.R.S., and Prof. J. H. N. Loubser for arranging for the University to cover a part of the typing and copying costs of the manuscript.

Richard Janse van Rensburg kindly prepared the program used in sorting the index, and very carefully proof-read the galleys.

I am extremely grateful to David Winsor, Georgina Malcolm, and Lesley Ward of Van Nostrand Reinhold for their unfailing patience and support. They could not have been more helpful or reasonable.

And finally, had it not been for the constant encouragement and support of Adele Strauss, this task would never have been begun, much less completed. Words cannot express my sense of appreciation and gratitude for what she has done and for what she is.

Department of Physics Anthony B. Wolbarst
University of the Witwatersrand December 1975
Johannesburg

Contents

Preface	ix
Acknowledgments	xiii
Chapter 1—Introduction	1
1.1 Symmetry and Symmetry Operators	1
1.2 Symmetry Operations of the Equilateral Triangle	4
1.3 Products of Operators, Groups	6
1.4 Group Theory and Quantum Systems	9
1.5 Summary	11
Chapter 2—Vectors and Dirac Notation	12
2.1 Basis Sets and Vector Components	12
2.2 Vectors	14
2.3 Scalar Product	17
2.4 Closure Relation	19
2.5 Dirac Notation	19
2.6 Summary	23
Chapter 3—Operators and Function Spaces	25
3.1 Operators	25
3.2 Invariant Spaces and Inverse Operators	29
3.3 Vector Functions of Vectors	30
3.4 Hermitian Conjugate	31
3.5 Hermitian and Unitary Operators	32
3.6 Eigenequations	33
3.7 Vectors in Function Space	34
3.8 Continuous Basis Sets	36
3.9 A Weighted String	39
3.10 Transformation to Normal Modes	41
3.11 Summary	42
Chapter 4—Representations of an Operator in Three-Space	44
4.1 A Representation of the Rotation Operator	45
4.2 Usefulness of the Matrix Representation of the Rotation Operation	48

4.3	Subspaces in Real Space Invariant with Respect to Rotations about an Axis	51
4.4	Generalizing to Function Space	53
4.5	A Representation of the Group C_{3v}	54
4.6	Summary	57

Chapter 5—Review of Some Quantum Mechanics   59

5.1	States and Observables	60
5.2	Compatible Operators Commute: Pure States	63
5.3	Measured Values are Eigenvalues	64
5.4	Statistical Interpretation	66
5.5	The Schroedinger Equation	67
5.6	Diagonalizing the Hamiltonian	70
5.7	Diagonalizing the Hamiltonian as a Principal Axis Transformation	72
5.8	Symmetry Operators Commute with the Hamiltonian	74
5.9	Use of Symmetry Arguments in Diagonalizing a Hamiltonian	77
5.10	Degenerate Solutions to the Schroedinger Equation Define a Subspace of Hilbert Space	78
5.11	Atomic Orbitals	80
5.12	Construction and Diagonalization of a Hamiltonian—a Simple Example	84
5.13	Time-Independent Perturbation Theory	89
5.14	Summary	92

Chapter 6—Representations of an Operator in Function Space   93

6.1	The Link between Real Space and Function Space: the State $\psi(\bar{r})$	94
6.2	The Link between Real Space and Function Space: the Operator $\hat{P}_R$	95
6.3	Linear Displacements	96
6.4	The Operators $\{\hat{P}_R\}$ form a Group Isomorphic to $\{\hat{R}\}$	100
6.5	Representations of $\hat{P}_R$	102
6.6	The Link between Real Space and Function Space: $\hat{R}$ and $\hat{P}_R$	103
6.7	Representation of $\hat{P}_R$ by d-States: $\Gamma^{(d)}(\hat{R})$	106
6.8	Block-Diagonal Form of $\Gamma^{(d)}(\hat{R}_\phi)$	109
6.9	How can an Incomplete Basis Set Generate a Meaningful Representation?	110
6.10	Summary	111

Chapter 7—Representations of Groups of Operators   113

7.1	More about Groups	113
7.2	Permutation and Colour Groups	118
7.3	Classes	120

7.4	Representation of a Group: $\Gamma^{(p)}$ of C_{3v}	122
7.5	Homomorphic Representations: $\Gamma^{(s)}$ of C_{3v}	124
7.6	Equivalent Representations	126
7.7	Irreducible Representations	127
7.8	Summary	130

Chapter 8—Irreducible Group Representations   132
8.1	Irreducible Group Representations: A Review	133
8.2	The Number of Non-Equivalent Irreducible Representations Equals the Number of Distinct Classes	134
8.3	Labelling Irreducible Representations and Symmetry Types	137
8.4	Regular Representation	139
8.5	The Great Orthogonality Theorem	141
8.6	Characters	142
8.7	Which Irreducible Representations Lie Hidden within a Given Reducible Representation?	143
8.8	Reduction of a Representation: A C_{3v} Example	144
8.9	Summary	146

Chapter 9—Quantum Mechanical States: How Symmetry in Real Space Predetermines Symmetry in Hilbert Space   148
9.1	The Wigner Theorem	149
9.2	Symmetries of State Functions	153
9.3	Good Quantum Numbers	155
9.4	Projection Operators	157
9.5	Direct Product Spaces	159
9.6	Exchange Symmetry	162
9.7	Time Reversal Symmetry	164
9.8	Molecular Orbitals	165
9.9	Molecular Vibrations	172
9.10	Classical Applications	177
9.11	Summary	180

Chapter 10—More Examples of the Use of the Wigner Theorem: Continuous Groups   182
10.1	Bloch Functions	182
10.2	Systems of Axial Symmetry	185
10.3	Full Rotational Group	188
10.4	Spin and Double Groups	191
10.5	Irreducible Spherical Tensor Operators	194
10.6	Clebsch–Gordan Coefficients and the Wigner–Eckart Theorem	199
10.7	Summary	201

Chapter 11—Perturbations Static and Dynamic   203
11.1   Why Integrals Vanish   203
11.2   Perturbations   205
11.3   Reduction of Symmetry   209
11.4   Level Crossing Theorem   213
11.5   Born–Oppenheimer Approximation   215
11.6   The Jahn-Teller Theorem   218
11.7   A C_{3v} Jahn-Teller Example   220
11.8   Pseudo-Jahn-Teller Effect   223
11.9   What Group Theory Cannot Do   226
11.10   Summary   226

Chapter 12—Symmetry and Conservation Laws   227
12.1   Conservation of Linear Momentum   228
12.2   Broken Symmetry: Parity   229
12.3   Charge Conjugation   232
12.4   Other Symmetries   233
12.5   Summary   234

Chapter 13—Epilogue   235
Bibliography, and Where to Go from Here   238
List of Symbols   241
Important Relationships   244
Index   246

CHAPTER 1

Introduction

An unmarked equilateral triangle has no especial 'up' vertex; consequently one can rotate and reflect it in certain ways so that its appearance is left unchanged by the transformation. A sequence of several such symmetry operations also leaves the appearance unchanged; this observation leads to a natural definition of 'composition of operations', and then to the group concept. The symmetry group C_{3v} of the ammonia molecule, or of the equilateral triangle when we neglect 'horizontal' reflections, serves as our standard example in this chapter and throughout the rest of the book.

The physical characteristics of a quantum system depend as much upon the spatial arrangement (geometric symmetry properties) of its components as upon the chemical makeup of these components. It is for this reason that group theory plays an integral role in the study of quantum mechanics.

1.1 Symmetry and Symmetry Operators

Symmetry is something which catches our attention hundreds of times a day: the streets of our city are laid down as a rectilinear grid; an earthworm looks pretty much the same forwards or backwards or upside down; a snowflake has unmistakable six-fold symmetry; and a ball-bearing seems unchanged after any rotation; your reflection in the bathroom mirror resembles you closely (with the notable difference that the hair is parted on the wrong side)—and so on.

The essential, and self-evident, point is illustrated in Fig. 1.1. Here we have two pictures of the same equilateral triangle before

Figure 1.1.

and after it is rotated through some multiple of 120 degrees. But by which multiple? Clearly, there is no way of telling by inspection; a rotational transformation through $0°$, $\pm 120°$, $\pm 240°$, $\pm 360°$, etc., will leave the triangle's appearance unchanged. This, of course, is equivalent to saying that its absolute orientation was indeterminate in the first place; our inability to measure the initial condition of the system with assurance (which of the three vertices should be 'up'?) means that it can be rotated into new but, for all intents and purposes, equivalent conditions indistinguishable from the original.

A geometric operation which leaves unchanged the appearance of a symmetrical physical object is called a *symmetry operation* for the object. The triangle of Fig. 1.1 has a three-fold axis of rotation, and appears invariant to the action of the $0°$, $120°$, $240°$, $360°$, $-480°, \ldots$ rotational symmetry operations.

Figure 1.1 displays other symmetry properties as well. As demonstrated in Fig. 1.2(a), the mirror image is completely indistinguishable from the original triangle, if its base lies normal to the mirror. Alternatively, consider what happens when we bisect the triangle evenly, and bring one half up to the mirror, as in Fig. 1.2(b). The whole figure reappears, a consequence again of its mirror symmetry. The equilateral triangle (but not, of course, an

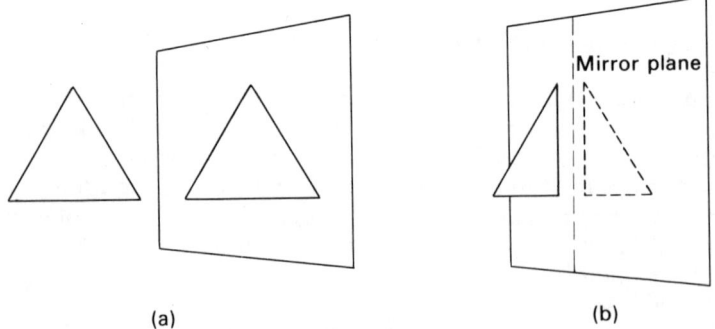

Figure 1.2.

arbitrary triangle) is said to be invariant under reflections across each of four mirror planes (the fourth being the plane in which the triangle lies).

A third important type of geometric symmetry is illustrated in Fig. 1.3. The outline of an object can be presented in the form of a particular set of points, designated $\{\bar{r}_i\}$, where the $\bar{r}$'s are position vectors drawn from a suitable origin. If for every $\bar{r}_k$ on the locus, the point $-\bar{r}_k$ also is in the set, then the origin of the figure is a *centre of inversion*.

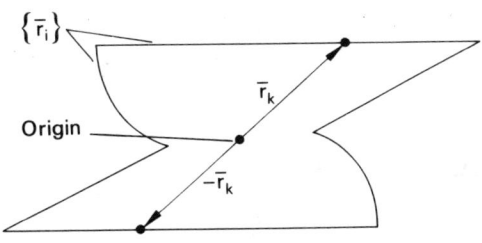

Figure 1.3.

One can speak of the symmetry types not only of geometrical objects, but also of mathematical functions; this is just a slight generalization of the ideas already considered. Figure 1.4(a) for example, is the graph of a function of the form

$$f_+(x) = c_0 + c_2 x^2 + c_4 x^4 + \cdots c_{2n} x^{2n} \qquad (1.1)$$

The left-hand side of the graph is a mirror image of the right-hand side, and since $f_+(x)$ contains only even powers of x, the function is said to be *even*. The graph of an *odd* function,

$$f_-(x) = c_1 x + c_3 x^3 + c_5 x^5 + \cdots c_{2m+1} x^{2m+1} \qquad (1.2)$$

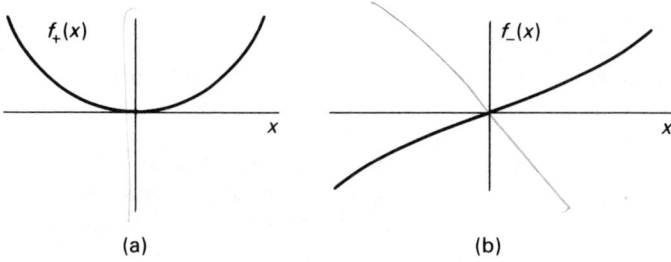

Figure 1.4.

on the other hand, displays a centre of inversion at the origin, as in Fig. 1.4(b).

The term 'symmetry', therefore, applies meaningfully both to objects and to functions. To illustrate this point further, consider the real-space *parity operator*, $\hat{R}_p$, defined through

$$\hat{R}_p(\bar{r}) = -\bar{r} \tag{1.3a}$$

From Fig. 1.3, we see that $\hat{R}_p$ is a *symmetry* operator for the locus of points outlining an object with a centre of inversion. That is, if each $\bar{r}_k$ is replaced by $-\bar{r}_k$, and yet everything looks the same, then the object is invariant to the parity operation. Figure 1.4 suggests that by analogy, we can also define a related parity operator $\hat{P}_p$ which acts upon *functions* according to

$$\hat{P}_p g(\bar{r}) = g(-\bar{r}) \tag{1.3b}$$

Then of all the types of functions which exist, even and odd are the two special, highly symmetric kinds for which

$$\hat{P}_p f_\pm(\bar{r}) = f_\pm(-\bar{r}) = \pm f_\pm(\bar{r}) \tag{1.4}$$

$\hat{P}_p$ leaves $f_+(\bar{r})$ totally invariant, and its effect upon an odd function is only to multiply it by (-1). Although the factor of (-1) cannot be ignored, the basic forms of $f_+(\bar{r})$ and $f_-(\bar{r})$ are not changed under the action of $\hat{P}_p$. Parity is thus, in a sense, a symmetry operator not only for some objects, but also for some functions. $\hat{P}_p(\hat{R}_p)$ is still defined as an *operator*, of course, for arbitrarily shaped functions (objects), but not as a *symmetry* operator.

We shall later have much to say about the eigenequations Eq. (1.4), which describe one particular symmetry operator and two kinds of functions which appear to be intimately associated with it. Here we only wish to suggest briefly what is meant by a symmetry operator, and by functions of well-defined (such as even or odd) symmetry.

1.2 Symmetry Operations of the Equilateral Triangle

Let us consider the symmetry properties of a plane, equilateral triangle. For convenience, we shall examine one without thickness, and *ignore*† the 'horizontal' reflection through the plane of the object itself, which is illustrated in Fig. 1.5. Alternatively, we could think of a solid whose vertices coincide with the positions of the atoms of an ammonia molecule NH_3, Fig. 1.6. The base of such a solid is an equilateral triangle, and the three sides are mutually congruent isoselese triangles. A bit of trial and error

---

†For our standard example of the equilateral triangle, we shall nearly always ignore $\hat{M}_h$ and yet still loosely refer to the group as C_{3v}. But see problem 7.10.

INTRODUCTION

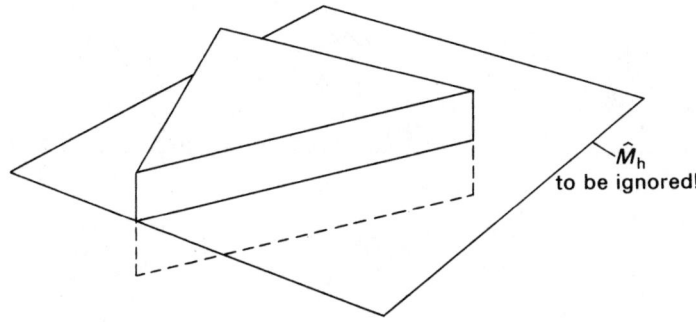

Figure 1.5.

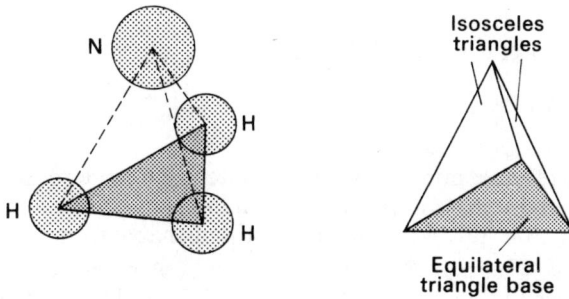

Figure 1.6.

will reveal that the only geometric symmetry operations of the NH$_3$ molecule are those of its base; and so here, too, it is sufficient to consider the reflection and rotation properties of a two-dimensional equilateral triangle.

The triangles of Figs 1.1, 1.5, and 1.6 are unmarked, as are the background regions. It is helpful to turn to the closely related situation in which we have a fully marked triangle resting on a marked field. We shall permanently label the three vertices of the triangle as x, y, and z, and three fixed background positions with Roman numerals I, II, and III. There are six ways of aligning the triangle absolutely in space, with a vertex at each Roman numeral, as demonstrated in Fig. 1.7, and we can think of the triangle as having six possible orientational states, which we name ψ_1 through ψ_6.

We shall now define $\hat{R}_{120}$ to be an operator which rotates any triangle of Fig. 1.7 through 120 degrees counter-clockwise. $\hat{R}_{120}$ acting on a triangle in the state ψ_1, then, leaves it in ψ_3. This

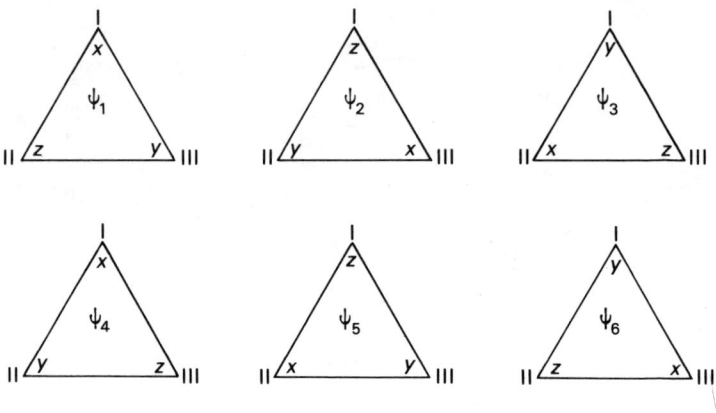

Figure 1.7.

transformation is written as

$$\hat{R}_{120}\psi_1 = \psi_3 \tag{1.5}$$

and the process is illustrated in Fig. 1.8(a).

There are three non-equivalent counter-clockwise rotation operators of interest (notice that we did *not* say 'symmetry rotation operators'; the state ψ_1, is clearly distinguishable from the state $\hat{R}_{120}\psi_1 = \psi_3$), namely $\hat{R}_{120}$, $\hat{R}_{240}$, and $\hat{R}_{360}$. A more extensive rotation has the same net effect as one of these three; for example $\hat{R}_{600}\psi_2 = \hat{R}_{240}\psi_2 = \psi_3$.

In addition to the three rotation operations, we can define mirror-like operations. Let $\hat{M}_I$ be that operator which causes an interchange of the vertices sitting in the fixed II and III positions, or across the mirror plane which passes through I. For example Fig. 1.8(b),

$$\hat{M}_I \psi_2 = \psi_5 \tag{1.6}$$

There are three reflection operators for our equilateral triangle, $\hat{M}_I$, $\hat{M}_{II}$, and $\hat{M}_{III}$.

1.3 Products of Operators, Groups

We can perform any combination of these six operations in sequence. Acting upon ψ_1, say, with $\hat{R}_{240}$ and then with $\hat{M}_I$, the process goes

$$\hat{M}_I(\hat{R}_{240}\psi_1) = \hat{M}_I \psi_2 = \psi_5 \tag{1.7}$$

or

$$(\hat{M}_I \hat{R}_{240})\psi_1 = \psi_5$$

INTRODUCTION

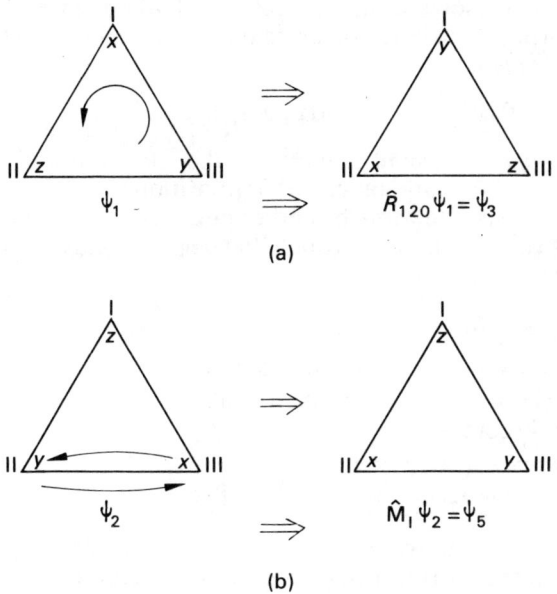

Figure 1.8.

$(\hat{M}_I\hat{R}_{240})$ is called the *composition* or *product* of the two operators $\hat{M}_I$ and $\hat{R}_{240}$. Note the importance of the ordering of the terms in the product.

PROBLEM 1.1 Does $\hat{M}_I\hat{R}_{240}\psi_1 = \hat{R}_{240}\hat{M}_I\psi_1$?

$\hat{R}_{120}\hat{R}_{240}\psi_3 = \hat{R}_{240}\hat{R}_{120}\psi_3$?

We might suspect that the product of two geometrical operators chosen from the set $\{\hat{R}_{120}, \hat{R}_{240}, \hat{R}_{360}, \hat{M}_I, \hat{M}_{II}, \hat{M}_{III}\}$ will itself reside within the set. A quick perusal of Fig. 1.7 shows that just as $\hat{M}_I\hat{R}_{240}\psi_1 = \psi_5$, so also $\hat{M}_{III}\psi_1 = \psi_5$, or

$$(\hat{M}_I\hat{R}_{240})\psi_1 = \hat{M}_{III}\psi_1 = \psi_5 \tag{1.8}$$

In fact, comparison of $\hat{M}_I\hat{R}_{240}\psi_i$ and $\hat{M}_{III}\psi_i$ for all six states $\{\psi_i\}$, $i = 1, \ldots, 6$, reveals that the results are always the same:

$$\hat{M}_I\hat{R}_{240}\psi_i = \hat{M}_{III}\psi_i \qquad i = 1, \ldots, 6 \tag{1.9}$$

We can therefore say that

$$\hat{M}_I\hat{R}_{240} = \hat{M}_{III} \tag{1.10}$$

where operation upon all possible states $\{\psi_i\}$ is implied. Equation (1.10) is not an isolated case. Inspection of all possible combinations of products will demonstrate that *the composition of any two elements in the set*

$$\{\hat{R}_{120}, \hat{R}_{240}, \hat{R}_{360}, \hat{M}_I, \hat{M}_{II}, \hat{M}_{III}\}$$

yields a unique element also in the set; the set is thus said to exhibit the property of *closure* under the composition.

This set of six operators has other interesting properties as well. A second is that rotation through 360 degrees yields the same state as no rotation at all

$$\hat{R}_{360}\psi_i = \hat{R}_0\psi_i = \psi_i \qquad i = 1, \ldots, 6 \tag{1.11}$$

and consequently $\hat{R}_{360} = \hat{R}_0$ is called the *identity operator*. Because of its special significance, we shall relabel it $\hat{I}$, and our set of operators becomes

$$\boxed{\{\hat{I}, \hat{R}_{120}, \hat{R}_{240}, \hat{M}_I, \hat{M}_{II}, \hat{M}_{III}\}} \tag{1.12}$$

It is easily seen that the effect of the product of $\hat{I}$ and any other operator in the set is just that of the other operator alone. For example $\hat{I}\hat{M}_{II} = \hat{M}_{II}\hat{I} = \hat{M}_{II}$.

Thirdly, every element of the set has a unique *inverse*, such that the product of an operator and its inverse has no effect upon any state. The inverse of an operator, in essence, undoes what the operator did in the first place. The observation that $\hat{R}_{120}\hat{R}_{240}\psi_i = \hat{R}_{240}\hat{R}_{120}\psi_i = \hat{I}\psi_i = \psi_i$, $i = 1, \ldots, 6$, tells us that $\hat{R}_{240}$ and $\hat{R}_{120}$ are inverses of one another. Similarly, $\hat{M}_{II}^2 \equiv \hat{M}_{II}\hat{M}_{II} = I$ says that $\hat{M}_{II}$ is its own inverse.

Finally, and we leave the demonstration of this as an exercise, the operators of the set obey a law of associativity. For example:

$$(\hat{M}_{III}\hat{R}_{120})\hat{M}_I = \hat{M}_{III}(\hat{R}_{120}\hat{M}_I) \tag{1.13}$$

In this section we have been discussing a set of six operators which cause transformations among the states of Fig. 1.7. But the operators are just as meaningful when describing six distinct *symmetry operations* which leave an *unmarked* object visibly unchanged. We shall now consider the elements in the set $\{\hat{I}, \hat{R}_{120}, \hat{R}_{240}, \hat{M}_I, \hat{M}_{II}, \hat{M}_{III}\}$ as being the symmetry operators of an unmarked equilateral triangle. Clearly this set of six symmetry operators still displays the properties

$$\boxed{\begin{array}{l}\text{(i) closure} \\ \text{(ii) there exists an identity} \\ \text{(iii) every element has an inverse} \\ \text{(iv) associativity}\end{array}} \tag{1.14}$$

INTRODUCTION

Any set of elements for which a law of binary composition (a product) is defined, and to which these four properties apply, is called a *group*. If the elements happen to be the symmetry operators of some physical object, the group is called a *symmetry group*.

Examples of groups are ubiquitous. The set of all rotations $\{\hat{R}_\phi\}$ about an axis forms a group, and the 'product' of $\hat{R}_\alpha$ and $\hat{R}_\beta$ is the net rotation $\hat{R}_{\alpha+\beta}$. The integers under the operation of addition, where the 'inverse' of n is $-n$, the reals excluding 0 under division, the complex numbers excluding 0 under multiplication, the rationals under subtraction, the set of all n-by-n matrices with non-vanishing (why?) determinants under matrix multiplication, the set of solutions to an ordinary homogeneous linear differential equation under addition, and the list goes on.

*PROBLEM 1.2 Some nomenclature: the ammonia molecule exhibits a three-fold axis of rotational symmetry, and also three 'vertical' mirror planes which contain the axis of rotation. A system with an n-fold rotational axis is designated C_n; if, in addition, there exist n vertical mirror planes, a subscript v is added: C_{nv}.

Draw objects which exhibit C_3, C_{3v}, and C_{4v} symmetries. What are the symmetry operators for these systems?

*PROBLEM 1.3 Consider the group of elements $\{X_1, X_2, \ldots, X_n\}$. $[X_i, X_j] \equiv X_i X_j - X_j X_i$ is called the *commutator* of X_i and X_j. If every pair of elements in the group *commutes*, that is, if $[X_i, X_j] = 0$, or $X_i X_j = X_j X_i$, for all X_i, X_j in the set, then the group is *commuting* or *abelian*. Is C_{3v} abelian? C_3? C_4. When will a symmetry group be abelian? See Problem 1.1.

*PROBLEM 1.4 Show that an inversion is equivalent to a reflection combined with a rotation through 180 degrees about an axis normal to the reflection plane. Do a reflection and an inversion always commute?

1.4 Group Theory and Quantum Systems

Different objects display different symmetry group structures; or in other words, different spatial arrangements are invariant under different sets of symmetry operations. The physical properties of a quantum mechanical system depend, of course, upon the types of atoms out of which the system is built, but *just as critically* upon the overall geometrical way in which the various constituents are put together. In fact, different types of molecules with the same general three-dimensional (symmetry group) structure may easily display more chemical and physical similarities than do isomers (molecules of the same chemical formula, but put together in different geometrical arrangements).

Surprisingly, one finds that for all the millions of known molecules and solids, the number of essentially non-equivalent symmetry types is quite small. And therein lies the power of the group theoretic approach. Once we have learned the symmetry properties of a system, a glance at a ridiculously short character table (just a few pages of data comprehensively cover all the molecular and solid structures one ever runs into) reveals a wealth of information on its quantum mechanical behaviour. Having identified the system's group structure, we can state, by symmetry considerations alone:

(a) much, and sometimes everything, about the mathematical symmetry properties (i.e. angular dependences) of its electronic and vibrational state functions. This is the most basic piece of information we acquire, and it is closely linked to the remaining items in this list.
(b) the possible allowed degeneracies of energy levels of the system.
(c) the functional forms of all the degenerate quantum states in a level, given only one of them.
(d) whether or not various types of static perturbations (arising from applied stresses, external fields, etc.) will split any of these degenerate states.
(e) whether various dynamic perturbations are allowed or forbidden to cause transitions among the different levels (selection rules).
(f) how to put a complete set of state functions into a form most helpful for diagonalizing a given Hamiltonian.
(g) what sorts of states must be considered at the various stages of an LCAO (linear combination of atomic orbitals) calculation.
(h) how the angular momenta of several electrons on an ion should be added together.
(i) the possible normal modes of vibration of the system.
(j) and much more.

Group theory can tell us a great deal about the behaviour of a physical system even before we write down the equations of motion. And the amount of human and/or computer labour saved therein may even make or break a calculation.

Group theory, as applied to the study of atoms, molecules, and solids, thus provides the connection between the symmetry of a system in real, three-dimensional space, and the symmetries of the functions in Hilbert space which describe it mathematically. The application of the theory rests upon two pillars: 1. the group structure of the system's set of symmetry operations, and 2. the nature of the vector spaces inhabited by the solutions to the system's Schroedinger equation. Soon we shall see how the two pillars blend into one in group representation theory; but we shall begin

INTRODUCTION 11

in more familiar surroundings by reviewing some of the properties of vectors and functions.

1.5 Summary

1. A plane equilateral triangle has no distinguishing features which allow one to describe its absolute orientation in space. Consequently it can be rotated through any integral multiple of 120 degrees, or reflected across any of its three mirror planes, without change of appearance. A geometric operation, such as a rotation, reflection, inversion, translation, etc. which leaves the appearance of a system invariant is called a *symmetry operation*.
2. A sequence of two symmetry operations upon an object is itself a symmetry operation; moreover, there is an identity operation, and to every symmetry operation an inverse. The set of symmetry operations for an object thus displays the properties of a *group*. The symmetry operations for the unmarked equilateral triangle, for example, constitute the elements of the group C_{3v}; this group is not *abelian*.
3. Group theory is important to the study of quantum mechanics for two basic reasons; first, it often greatly simplifies complex problems; second, and perhaps more important, it offers one an understanding of the way in which degeneracies and mathematical properties of wave functions of a quantum system depend upon the physical symmetry of the system in real space.

CHAPTER 2

Vectors and Dirac Notation

We begin with standard but non-rigorous definitions of the terms vector, linear independence, basis system, and scalar product. It is shown that a vector can be represented by its components relative to a given basis system, and that this decomposition leads to the so-called closure relationship. The Dirac bra and ket symbols are introduced as a shorthand notation for manipulating vectors. Curly brackets, { }, will be used to denote a set.

Throughout the entire book, the set

$$\{|i\rangle\} \equiv \{|i\rangle, |j\rangle, |k\rangle, \ldots\} \equiv \{|1\rangle, |2\rangle, |3\rangle \ldots\}$$

will stand for orthonormal basis vectors; as such, they may be anything from the $\bar{x}, \bar{y}, \bar{z}$ unit vectors of three-space to a complete set of eigenvectors of a Hermitian operator suitable for spanning some Hilbert space.

This chapter is quite elementary, but has been included to allow the introduction of a perhaps unfamiliar approach to vector representations, one which will be used throughout.

2.1 Basis Sets and Vector Components

Consider a vector $\bar{v}$. We can think, for example, of the directed line running from Grand Central Station to the Museum of Modern Art in New York City as a vector, and it has a physical meaning regardless of how we (or even *if* we) choose to set up co-ordinate axes, Fig. 2.1(a).

VECTORS AND DIRAC NOTATION

If we wish to explain to a visitor from London how to go from the station to the museum, it helps immensely that the streets in the environ of the two points form a rectilinear grid, Fig. 2.1(b). Then we simply say: go east from the station for v_1 blocks (where v_1 may be a negative number), then turn and go v_2 blocks north. What has happened, of course, is that we have established a basis-vector system defining a set of co-ordinate axes, with the station as the origin, and this allows us to specify uniquely directions and distances (position vectors) in terms of components along the axes. That is, if '$|1\rangle$' is *defined* to be a vector one block long pointing east, and '$|2\rangle$' is a north-heading unit vector, then

$$\bar{v} = v_1|1\rangle + v_2|2\rangle \tag{2.1a}$$

Once $|1\rangle$ and $|2\rangle$ are stipulated, then we can specify *any* vector $\bar{v}$ simply by stating the *ordered set* of *components* (v_1, v_2) relative to this basis set. The ordered pair of components (v_1, v_2) is called the *representation* of the vector $\bar{v}$ in the $\{|1\rangle, |2\rangle\}$ basis system.

But we could, if we so choose, also describe the physically meaningful vector $\bar{v}$ of Fig. 2.1(a) in terms of another basis system, $\{|1'\rangle, |2'\rangle\}$, aligned at some angle to the first, as in Fig. 2.1(c). $\bar{v}$, with the *new* components v'_1 and v'_2, now looks like

$$\bar{v} = v'_1|1'\rangle + v'_2|2'\rangle \tag{2.1b}$$

(v'_1, v'_2) is the representation of the same vector $\bar{v}$, but here in the $\{|1'\rangle, |2'\rangle\}$ basis system. The physical meaning of $\bar{v}$ does not depend upon the nature of the coordinate system used; the numbers in the ordered sets (v_1, v_2), (v'_1, v'_2), etc., on the other hand, are fully dependent upon the choice of basis set.†

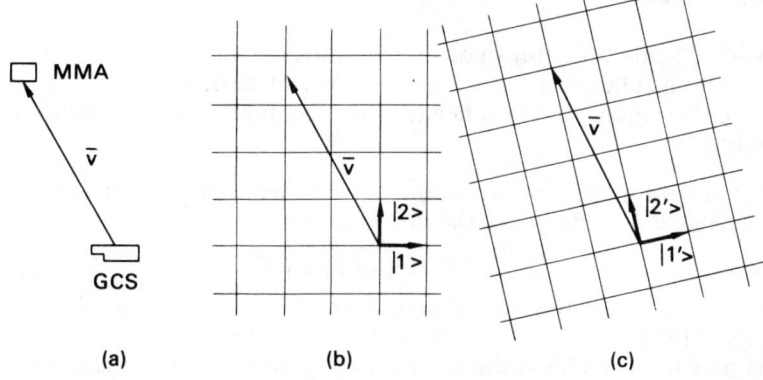

Figure 2.1.

†To avoid confusion at times it will be necessary to denote with a superscript the basis set involved in generating the representation of a vector or operator; hence $v_1^{(i)}$ and $v_1^{(i')}$, or $\bar{\bar{\Gamma}}^{(i)}(R)$ for a matrix representation of the operator $\hat{P}_R$ later on.

We shall soon find it of interest to determine what happens to the representation of an arbitrary vector if we somehow alter the coordinate system. The way in which the ordered pair (v_1, v_2) becomes (v'_1, v'_2) as $\{|1\rangle, |2\rangle\}$ is rotated into $\{|1'\rangle, |2'\rangle\}$, for example

$$\boxed{\text{as } \begin{array}{c} (v_1, v_2) \to (v'_1, v'_2) \\ \{|1\rangle, |2\rangle\} \to \{|1'\rangle, |2'\rangle\} \end{array}} \quad (2.1c)$$

is essentially the same for all vectors, and for all sorts of rotations; therefore it is possible to define a vector as an entity which (or, an entity whose representation) varies according to a precise rule under spatial rotations. We shall pursue this approach in detail in Chapter 4.

By contrast to the above, the temperature T inside the museum is 25°C, whatever basis system we use. Because the temperature is simply a number, and independent of how we align the basis vectors, it is said to be *invariant* to rotations of the co-ordinate axes, or a *scalar*. Its transformational properties are simpler to handle than are those of a vector: as the basis system rotates, T, the number describing the temperature, does not change at all!

We can measure a temperature $T(\bar{r})$ at every physical point $\bar{r}$ in the city; we have then established a *scalar field*. Similarly, if a vector such as wind velocity is defined at every point $\bar{r}$, we have a *vector field*.

PROBLEM 2.1 The scalar T is oblivious to any alteration of the coordinate axes of real space. Is $T(\bar{r})$?

2.2 Vectors†

Now to some very non-rigorous definitions, doubtless well familiar, but included just for the sake of completeness. For a start, a *vector* is something which behaves like the directed line segments of Fig. 2.1.

DEFINITION 2.1 *Vector, vector space*. Consider a set V, whose elements $\bar{u}, \bar{v}, \bar{w}$ etc., have the properties:

(a) $\quad\quad\quad \bar{u}, \bar{v} \in V \Rightarrow (\bar{u} + \bar{v}) = \bar{w} \in V \quad\quad\quad (2.2a)$

which is a shorthand way of saying that if $\bar{u}$ and $\bar{v}$ are both elements of the set V, then the sum $\bar{u} + \bar{v}$ is also some element, namely $\bar{w}$, of V. This is the same closure property as the one found in Chapter 1; here the law of binary composition is that of simple vector addition.

†After Section 2.5, *all* (not just basis) vectors, and the scalar product, will normally be presented in the standard Dirac bra-ket notation.

VECTORS AND DIRAC NOTATION

(b) there is a null or identity element $0 \in V$ such that for any $\bar{v} \in V$,

$$\bar{0} + \bar{v} = \bar{v} + \bar{0} = \bar{v} \tag{2.2b}$$

(c) for every $\bar{v} \in V$ there is a unique inverse $\overline{-v} \in V$ such that

$$\bar{v} + \overline{-v} = \overline{-v} + \bar{v} = \bar{0} \tag{2.2c}$$

(d) the set is associative under addition: $(\bar{u} + \bar{v}) + \bar{w} = \bar{u} + (\bar{v} + \bar{w})$ (2.2d)

(e) the set is commutative: $\bar{u} + \bar{v} = \bar{v} + \bar{u}$ (2.2e)

(f) if c is a number (scalar), then

$$\bar{u} \in V \Rightarrow c\bar{u} \in V \tag{2.2f}$$

If V and $\bar{u}$ obey (a) through (f), V is called a *linear vector space*, and $\bar{u}$ is a *vector* in that space. By Eq. (2.2a) through (2.2e), incidentally, V is also an abelian group under addition.

As an important example, the set of all ordered m-tuples such as $(v_1, v_2, \ldots, v_m)$ constitutes a vector space if we define the equivalence of these vectors, their addition, and the multiplication by a scalar constant (which may be a complex number) as

$$(u_1, u_2, \ldots, u_m) = (w_1, w_2, \ldots, w_m) \quad \text{if} \quad u_1 = w_1, \ldots, u_m = w_m \tag{2.3a}$$

$$(u_1, u_2, \ldots, u_m) + (v_1, v_2, \ldots, v_m) = (u_1 + v_1, u_2 + v_2, \ldots, u_m + v_m) \tag{2.3b}$$

$$c(u_1, u_2, \ldots, u_m) = (cu_1, cu_2, \ldots, cu_m) \tag{2.3c}$$

Note the distinction between the single vector $(v_1, v_2, \ldots, v_m)$ and the set of m vectors $\{\bar{v}_1, \bar{v}_2, \ldots, \bar{v}_m\}$.

DEFINITION 2.2 *Linear independence.* The vectors in the set $\{\bar{v}_1, \bar{v}_2, \ldots, \bar{v}_m\}$, none of which is the null vector, are called linearly independent if and only if

$$c_1 \bar{v}_1 + c_2 \bar{v}_2 + \cdots + c_m \bar{v}_m = \bar{0} \tag{2.4}$$

implies

$$c_1 = c_2 = \cdots = c_m = 0, \text{ scalar } c_1$$

This says that there is no possible way of taking linear combinations of the set $\{\bar{v}_1, \ldots, \bar{v}_{m-1}\}$ so as to end up with $\bar{v}_m$.

Figures 2.2(a), (b), and (c) show, respectively, two two-dimensional vectors which are linearly independent, three which are linearly dependent, and two which are linearly dependent.

DEFINITION 2.3 *Dimensionality.* The dimensionality of a

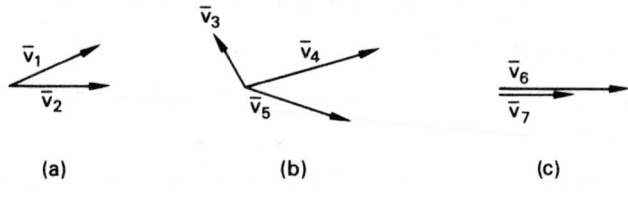

Figure 2.2.

vector space is the maximum possible number of linearly independent vectors which can coexist in the space.

The vectors of Eq. (2.3), for example, are m-dimensional.

DEFINITION 2.4 *Complete Basis set, span.* The set of m different vectors $\{|i\rangle\} \equiv \{|1\rangle, |2\rangle, \ldots |m\rangle\}$ is said to span the m-dimensional vector space V, or to form a complete basis set for V, if the $\{|i\rangle\}$ are linearly independent, and if *any* vector $\bar{v}$ in V can be written as a linear combination of the $\{|i\rangle\}$

$$\bar{v} = v_1|1\rangle + v_2|2\rangle + \cdots + v_m|m\rangle \equiv \sum_{i=1}^{m} v_i|i\rangle \qquad \text{for all } \bar{v} \text{ in } V \tag{2.5a}$$

The two vectors on Fig. 2.2(a) form a complete basis set for a two-dimensional vector space.

†PROBLEM 2.2 A vector in real space can be represented in terms of polar coordinates r, θ, ϕ. Do $\{|r\rangle, |\theta\rangle, |\phi\rangle\}$ form a complete basis?

From Eq. (2.5a), in an m-dimensional space we can define a basis system such as $\{|i\rangle\}$ and give the basis vectors a unique fixed ordering. Then the arbitrary vector $\bar{v}$ in V

$$\boxed{\bar{v} = \sum_{i=1}^{m} v_i|i\rangle} \tag{2.5b}$$

can be *represented* by the ordered m-tuple $(v_1, v_2, \ldots, v_m)$. The v_i are the components of $\bar{v}$ along the various $|i\rangle$ co-ordinate axis directions; the particular numbers actually to be found in $(v_1, v_2, \ldots, v_m)$ depend upon the initial choice of basis system; if we somehow change the basis system to $\{|i'\rangle\}$, for example by rotating the co-ordinate axis system, the m-tuple $(v_1, v_2, \ldots, v_m)$ representing $\bar{v}$ changes into $(v'_1, v'_2, \ldots, v'_m)$. But as long as the basis set is specified, the representation $(v_1, v_2, \ldots, v_m)$ has the same information content as does the vector $\bar{v}$ itself.

Equation (2.3) says that a set of ordered m-tuples $\{(v_1, v_2, \ldots, v_m)\}$ itself forms a vector space. If we have chosen a basis system $\{|i\rangle\}$ such that each m-tuple $(v_1, v_2, \ldots, v_m)$ corresponds to a particular $\bar{v}$ in V, as indicated by Eq. (2.5), then there will clearly be a close

VECTORS AND DIRAC NOTATION

link between the elements of V and those of m-tuple space:

$$\bar{v} \leftrightarrow (v_1, v_2, \ldots, v_m) \qquad (2.6)$$

V and m-tuple space are not identical, but intimately related; m-tuple space is said to be a *representation-vector space* associated with V.

The m-tuple representation in Eq. (2.6) is called a row vector; if we have reason to order the components vertically, we have the closely related column vector. Different row (or column) vectors may of course, represent the same physical vector $\bar{v}$, depending upon the choice of basis system. m-tuple column vectors inhabit the *dual space* of the space of m-tuple row vectors, and vice versa; we shall say more about dual spaces later.

2.3 Scalar Product

The scalar product offers us a way of formally incorporating the concept of length into a vector space. If you go back over Definitions 2.1 through 2.4, you will see that we have at no time spelled out how one actually determines the length of a vector. We shall generalize the notion of the standard dot product of three-space

$$\bar{r}_1 \cdot \bar{r}_2 = |\bar{r}_1| |\bar{r}_2| \cos \theta \qquad (2.7)$$

which describes the projection of $\bar{r}_1$ onto $\bar{r}_2$, or vice versa.

DEFINITION 2.5 *Scalar product.* The scalar product of two m-dimensional vectors $\bar{v}_1$ and $\bar{v}_2$ is written $\bar{v}_1 \cdot \bar{v}_2$ and is required to have the properties that

$$\bar{v}_1 \cdot \bar{v}_2 = (\bar{v}_2 \cdot \bar{v}_1)^* \qquad (2.8a)$$

$$\bar{v}_3 \cdot (c_1 \bar{v}_1 + c_2 \bar{v}_2) = c_1 \bar{v}_3 \cdot \bar{v}_1 + c_2 \bar{v}_3 \cdot \bar{v}_2 \qquad (2.8b)$$

$$\bar{v}_1 \cdot \bar{v}_1 \geq 0 \text{ and } \bar{v}_1 \cdot \bar{v}_1 = 0 \text{ if and only if } \bar{v}_1 = \bar{0} \qquad (2.8c)$$

The complex conjugate is used in Eq. (2.8a) to avoid inconsistencies when dealing with vectors whose components are complex-valued numbers. The dot product Eq. (2.7) for three-space is seen to obey all three requirements; and as in three-space, $(\bar{v} \cdot \bar{v})^{1/2} = |\bar{v}|$ gives us the length of the vector $\bar{v}$. The vector $\bar{v}/|\bar{v}|$ is of unit length, and is said to be *normalized*; two vectors (neither of which is the null vector) are *orthogonal* when their scalar product is 0. From Eq. (2.7) two vectors in real space are orthogonal when the angle between them is 90 degrees.

We shall make particular use of basis vectors which are both normalized and orthogonal. If $|i\rangle$ and $|j\rangle$ are two such basis vectors from the set spanning a space of interest, then

$$|i\rangle \cdot |j\rangle = \delta_{ij} \tag{2.9}$$

where the *Kroneker delta* δ_{ij} is defined as

$$\delta_{ij} \equiv \begin{cases} 1 & i=j \\ 0 & i \neq j \end{cases} \tag{2.10}$$

Except for three-dimensional real space, Eq. (2.7), we have still not shown how to evaluate the scalar product of two vectors. But we can easily build upon the real-space example. If the three-vectors $\bar{u}$ and $\bar{v}$ are expanded as

$$\bar{u} = u_1|1\rangle + u_2|2\rangle + u_3|3\rangle$$

$$\bar{v} = \sum_j^3 v_j |j\rangle$$

then

$$\bar{u} \cdot \bar{v} = u_1 v_1 |1\rangle \cdot |1\rangle + u_1 v_2 |1\rangle \cdot |2\rangle + \cdots + u_3 v_3 |3\rangle \cdot |3\rangle$$

$$= \sum_{i,j}^3 u_i v_j |i\rangle \cdot |j\rangle$$

$$= \sum_{i,j}^3 u_i v_j \delta_{ij}$$

$$= \sum_i^3 u_i v_i \tag{2.11}$$

Following Eqs. (2.11) and (2.8), we suggest that for an m-dimensional complex (as opposed to real-valued) vector space spanned by the orthonormal basis set $\{|i\rangle, i=1, 2, \ldots, m\}$, it is useful and consistent to employ the more general form

$$\bar{u} \cdot \bar{v} = \sum_{i=1}^m u_i^* v_i \tag{2.12}$$

as a definition of the scalar product. With some types of vector space, m may tend to infinity, or the sum may be replaced by an integral over a continuous parameter; more about that later. In any case, $\bar{u} \cdot \bar{v}$ will always be defined by some sort of extension of Eq. (2.12).

It is important to bear in mind that the scalar product of two vectors is a pure (though perhaps complex) number, a scalar. $\bar{u} \cdot \bar{v}$, moreover, is meaningful, as in Eq. (2.7), whether or not a basis set

VECTORS AND DIRAC NOTATION

has been defined; but it is usually much simpler to work in terms of representations, as in Eq. (2.12).

2.4 Closure Relation

Let us represent the vector $\bar{v}$ in some m-dimensional space in terms of its components along the elements of the complete orthonormal basis set $\{|i\rangle\}$

$$\bar{v} = \sum_{i=1}^{m} v_i |i\rangle \qquad (2.13)$$

v_i is the component of $\bar{v}$ in the ith direction, or the projection of $\bar{v}$ along $|i\rangle$ in the sense of Eq. (2.7):

$$v_i = |i\rangle \cdot \bar{v} \qquad (2.14a)$$

Then we can write Eq. (2.13) as

$$\bar{v} = \sum_{i=1}^{m} |i\rangle [|i\rangle \cdot \bar{v}] \qquad (2.14b)$$

This leads us to propose the symbol '$\hat{1}(\ )$':

$$\hat{1}(\ ) \equiv \sum_{i=1}^{m} |i\rangle [|i\rangle \cdot (\ )] \qquad (2.14c)$$

whose meaning is defined by the three preceding equations:

$$\hat{1}(\bar{v}) = \sum_{i=1}^{m} |i\rangle [|i\rangle \cdot \bar{v}] = \bar{v} \qquad (2.14d)$$

The operator $\hat{1}(\ )$ acts upon a vector simply by multiplying it by unity. But Eqs (2.14c) and (2.14d) actually say much more than that. $\hat{1}(\bar{v})$ means that a decomposition of $\bar{v}$ is occurring relative to some given complete orthonormal basis set; $|i\rangle \cdot \bar{v} = v_i$ is the projection of $\bar{v}$ in the ith direction, and summing all the $|i\rangle [|i\rangle \cdot \bar{v}]$ terms gives us back the vector $\bar{v}$ again. Such a separation of $\bar{v}$ into constituent parts relative to a basis system is called a *fourier decomposition*, and Eq. (2.14c) is called a *closure relation*. This relation has the same name as the closure property of groups, but a somewhat different meaning. The closure relation is of vital importance, and will be presented again soon in more manageable form using the Dirac notation.

*PROBLEM 2.3 Can one define a closure relation only with respect to a complete basis set? Need the set be orthonormal?

2.5 Dirac Notation

Now to Dirac's bra and ket notation. Define the ket '$|\ \rangle$' as a symbol which allows the designation of a vector by the particular index, parameter, or label which resides within the ket; that is,

$$\boxed{|v\rangle \equiv \bar{v}} \tag{2.15}$$

says that $|v\rangle$ is the vector $\bar{v}$, or the vth vector. The unit vectors $\bar{x}$, $\bar{y}$, and $\bar{z}$ of three-space could be labeled $|x\rangle$, $|y\rangle$, $|z\rangle$; we have, however, already been calling these unit vectors $|i\rangle$, $i = 1, 2, 3$, corresponding to the first (x), second (y), and third (z) co-ordinate axis, respectively, and we shall continue to do so.

We must now decide how to express the scalar product. Instead of carrying on with $\bar{u} \cdot \bar{v} = |u\rangle \cdot |v\rangle$, we shall take the Dirac approach; Dirac's formalism offers a concise, consistent vector shorthand ideally suited to the needs of quantum mechanics.

Recall, by Eq. (2.8), that

$$|u\rangle \cdot |v\rangle = (|v\rangle \cdot |u\rangle)^* \tag{2.16a}$$

$$|u\rangle \cdot (c|v\rangle) = c|u\rangle \cdot |v\rangle \tag{2.16b}$$

from which it immediately follows that

$$(d|u\rangle) \cdot |v\rangle = d^*|u\rangle \cdot |v\rangle \tag{2.16c}$$

The problem is that while the scalar product is linear in the post-factor, as noted in Eq. (2.16b), it is *anti-linear* in the pre-factor, Eq. (2.16c)[†]. If we are called upon to juggle numerous vectors and form a number of scalar products in a calculation, then a bookkeeping device which distinguishes unambiguously between pre- and post-factors would aid in avoiding confusion.

Let us therefore write $\bar{u}$ as a 'bra' $\langle u|$ when the physics dictates that it is to act as a pre-factor, and as the 'ket' $|u\rangle$ otherwise. The scalar product of $\langle u|$ and $|v\rangle$ becomes $\langle u| \cdot |v\rangle$ or $\langle u|v\rangle$, and by Eq. (2.16a),

$$\langle u|v\rangle = \langle v|u\rangle^* \tag{2.17}$$

One can think of the bra's and ket's as inhabiting two totally separate vector spaces, distinct but linked intimately through Eq. (2.17). We demand that for every ket $|w\rangle$ in ket space there exist a unique bra-vector $\langle w|$, and postulate that *the two correspond to one another in a one-to-one fashion*:

$$c^*\langle w| \leftrightarrow c|w\rangle \tag{2.18}$$

[†] Eqs (2.16b) and (2.16c) are basic ingredients in the formal definitions of linearity and antilinearity, as we shall show in Section 3.1. Among the best discussions of the need and justification for Dirac notation is that of Dirac[10] himself, p. 18. See also Dennery and Krzywicki[9], Chapters 2 and 3.

VECTORS AND DIRAC NOTATION   21

The complex conjugation of c is necessary by Eq. (2.16). Whatever the vector $|w\rangle$ stands for, the same information content is held within $\langle w|$, and for discussion of a physical problem, statements concerning $|w\rangle$ or $\langle w|$ are equally informative. Ket space and bra space are called the *dual spaces* of one another, and neither is to be considered more important or fundamental than the other.

† PROBLEM 2.4 Show that if $|u\rangle$ is represented by $\begin{pmatrix} u_1 \\ u_2 \\ \vdots \\ u_m \end{pmatrix}$, then $\langle u|$ is represented by $(u_1^*, u_2^*, \ldots, u_m^*)$. Thus if $|u\rangle$ and $\langle u|$ are dual to one another, so also are their representations. Note that the matrix product of the row and column vectors is a scalar product.

Returning to Eq. (2.14): given the complete orthonormal basis set $\{|i\rangle\}$, we can decompose the vector $|v\rangle$ as follows:

$$|v\rangle = \sum_{i=1}^{m} v_i |i\rangle \qquad (2.19a)$$

$$v_i = \langle i|v\rangle \qquad (2.19b)$$

$$|v\rangle = \sum_{i=1}^{m} |i\rangle\langle i|v\rangle \qquad (2.19c)$$

which leads immediately to the *closure relation* in Dirac notation

$$\boxed{\hat{1} = \sum_{i=1}^{m} |i\rangle\langle i|} \qquad (2.20)$$

To reiterate what was said in the last section, but in the language of bra's and ket's:

$$|i\rangle\langle i| \qquad (2.21)$$

is a *projection operator* which projects out, or extracts, from an arbitrary vector $|v\rangle$ that portion which lies along the ith direction

$$|i\rangle\langle i|v\rangle = v_i |i\rangle \qquad (2.22)$$

The sum of all such parts gives back the original vector $|v\rangle$, if the basis set $\{|i\rangle\}$ fully spans the space of interest.

The closure relation, a most useful construct, can multiply ket's from the left and bra's from the right

$$\hat{1}|v\rangle = |v\rangle$$

$$\langle v|\hat{1} = \langle v|$$

$$\langle u|\hat{1}|w\rangle = \langle u|w\rangle \qquad (2.23)$$

and it will be employed liberally as a unit multiplicative factor. But once again, it is vitally important to realize that a closure relation applies if and only if our basis set is complete.

Two significant examples may help to illustrate the workings of the closure relation. We shall first derive Eq. (2.12); then we shall consider how two different basis sets spanning the same space, $\{|i\rangle\}$ and $\{|i'\rangle\}$, might be related to one another.

(a) The scalar product of $\bar{u}$ and $\bar{v}$, $\langle u|v\rangle$, can be brought into the form of Eq. (2.12) by inserting $\hat{1} = \sum_i |i\rangle\langle i|$ between $\langle u|$ and $|v\rangle$

$$\langle u|v\rangle = (\langle u|)\hat{1}(|v\rangle) = \sum_i \langle u|i\rangle \langle i|v\rangle$$
$$= \sum_i \langle i|u\rangle^* \langle i|v\rangle$$
$$= \sum_i u_i^* v_i \tag{2.12'}$$

(b) Assume that we have two distinct basis sets, $\{|i\rangle\}$ and $\{|i'\rangle\}$, spanning the same vector space. One might obtain the $\{|i'\rangle\}$, for example, by rotating all the $\{|i\rangle\}$ vectors about some axis in the space; it is perfectly legitimate here to think of the three-dimensional case as an illustration. An arbitrary basis vector from one set, say $|i'\rangle$ in $\{|i'\rangle\}$, may be expressed in terms of the members of the other set as

$$\boxed{|i'\rangle = \sum_j |j\rangle\langle j|i'\rangle = \sum_j C_{ji}|j\rangle} \tag{2.24a}$$

where the $\langle j|i'\rangle \equiv C_{ji}$ are scalars. Similarly

$$|i\rangle = \sum_{j'} |j'\rangle\langle j'|i\rangle = \sum_j C'_{ji}|j'\rangle \tag{2.24b}$$

We shall have much to say concerning transformations such as these in the coming chapters.

*PROBLEM 2.5 Let the set $\{|1\rangle, |2\rangle, |3\rangle\}$ span real space. If we rotate these three unit vectors counter clockwise by the angle ϕ about an axis along $|3\rangle$, Fig. 2.3, then they become, respectively, $|1'\rangle$, $|2'\rangle$, and $|3'\rangle$. Show that $|1'\rangle$ can be described in terms of the original basis set as

$$|1'\rangle = \cos\phi\,|1\rangle + \sin\phi\,|2\rangle \tag{2.25}$$

Find $|2'\rangle$ and $|3'\rangle$ also in this fashion. Compare with Eq. (2.24).

*PROBLEM 2.6 Show that if $|u\rangle$ can be decomposed as $|u\rangle = \sum_i u_i|i\rangle = \sum_{i'} u'_i|i'\rangle$, and if $\{|i\rangle\}$ and $\{|i'\rangle\}$ obey Eq. (2.24), then the

VECTORS AND DIRAC NOTATION

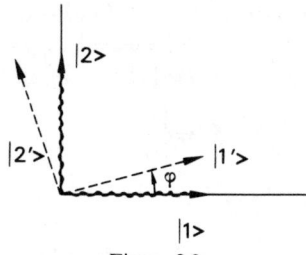

Figure 2.3.

components of $|u\rangle$ are related by

$$u_i = \sum_j C_{ij} u'_j \quad \text{and} \quad u_j = \sum_i C_{ji} u'_i \tag{2.26a}$$

$$u'_i = \sum_j C'_{ij} u_j \quad \text{and} \quad u'_j = \sum_i C'_{ji} u_i \tag{2.26b}$$

Show that $\sum_j C_{ij} C'_{jk} = \delta_{ik}$, or in matrix notation,

$$\bar{\bar{C}}' = \bar{\bar{C}}^{-1} \tag{2.26c}$$

*PROBLEM 2.7 If a subset of a vector space itself forms a vector space, then this subset is termed a *subspace*. Do $\{|1\rangle, |2\rangle\}$ span a subspace of real-space?

†PROBLEM 2.8 Do the m-dimensional vectors $\{(1, 0, 0, \ldots, 0), (0, 1, 0, \ldots, 0), \ldots, (0, 0, 0, \ldots, 1)\}$ form an orthonormal basis set for the vector space of Eq. (2.3)?

2.6 Summary

1. A set of elements $\{|u\rangle, |v\rangle, |w\rangle, \ldots\}$ constitutes a *vector space* V if

   (a) $|u\rangle, |v\rangle \in V \Rightarrow (|u\rangle + |v\rangle) \in V$

   (b) $|0\rangle \in V$

   (c) $|u\rangle \in V \Rightarrow |-u\rangle \in V$

   (d) the set is associative
   (e) the set is commutative

   (f) $|u\rangle \in V \Rightarrow c|u\rangle \in V,$ scalar c

In some circumstances, a vector may be defined as an entity which has a particular type of transformational behaviour under spatial rotations and reflections.

2. Given a vector space $V = \{|v\rangle\}$, we can define the *dual space* $\{\langle v|\}$ such that for a complex scalar c and any index u, $c^*\langle u| \leftrightarrow c|u\rangle$ in a unique, one-to-one fashion. Then the *scalar product* of the uth

vector with the vth vector is written $\langle u|v\rangle = \langle v|u\rangle^*$. If $\langle u|v\rangle = 0$ and neither $|u\rangle$ nor $|v\rangle$ is $|0\rangle$, then $|u\rangle$ and $|v\rangle$ are said to be orthogonal.

3. A set of m vectors $\{|i\rangle\}$, $i = 1, 2, \ldots m$, in V which displays

(a) *orthonormality:* $\langle i|j\rangle = \delta_{ij} = \begin{cases} 1 & i=j \\ 0 & i \neq j \end{cases}$

(b) *closure:* $\sum_{i=1}^{m} |i\rangle\langle i| = \hat{1}$

is a *complete m-dimensional* orthonormal *basis set* which *spans V*. Then any vector $|v\rangle \in V$ can undergo the *fourier decomposition*

$$|v\rangle = \sum_{i=1}^{m} |i\rangle\langle i|v\rangle = \sum_{i=1}^{m} v_i |i\rangle$$

4. We can *represent* the vector $|v\rangle$ as the ordered m-tuple $\begin{pmatrix} v_1 \\ v_2 \\ \vdots \\ v_m \end{pmatrix}$,

called a column vector. The entries in this representative matrix, unlike $\bar{v}$ itself, depend upon the choice of basis set. With the complete orthonormal basis set $\{|i\rangle\}$, the scalar product of $|u\rangle$ and $|v\rangle$ is

$$\langle u|v\rangle = \sum_{j=1}^{m} \sum_{i=1}^{m} u_j^* v_i \langle j|i\rangle = \sum_{i=1}^{m} u_i^* v_i$$

In some situations, to be discussed in the next chapter, $m \to \infty$, or the sum may even be replaced by an integral. In this last case, we may think of the representations of $|u\rangle$ and $|v\rangle$ as being functions parameterized by some variable ξ, rather than by the discrete index i, and $\langle u|v\rangle = \sum_i u_i^* v_i$ becomes $\int u(\xi)^* v(\xi)\, d\xi$.

5. If two suitable basis sets $\{|i\rangle\}$ and $\{|i)\}$ span the same vector space, a unit vector in one ($|k\rangle$, for example) can be expressed in terms of those in the other as

$$|k\rangle = \sum_{i'} |i'\rangle\langle i'|k\rangle = \sum_{i} C_{ik}|i'\rangle$$

CHAPTER 3

Operators and Function Spaces

In this chapter, linear operators are introduced, and the Hermitian conjugate and the eigenequation of an operator are examined in some detail. Function spaces are then presented through the example of the Fourier sum, and generalized to the case of continuous basis vectors via the Fourier integral. The chapter closes with discussion of the representation of the abstract vector $|f\rangle$ by its 'components' relative to the $|x\rangle$-basis system, $\langle x|f\rangle = f(x)$.

The operator $\hat{A}$ has eigenvalues which constitute the set $\{a_n\}$, and eigenvectors $\{|a_n\rangle\}$. The Hermitian conjugate of the operator $\hat{A}$ is $\hat{A}^+$, and the matrix representation of $\hat{A}$ in the $\{|i\rangle\}$ basis system is $\bar{\bar{A}}$ or $\bar{\bar{A}}^{(i)}$; the matrix element $\langle i|\hat{A}|j\rangle$ is written A_{ij}. The vectors and basis vectors of Hilbert space may be functions of a continuous parameter such as x; thus $|f(x)\rangle$, $|i(x)\rangle$.

3.1 Operators

In the last chapter we defined two kinds of products. The first is the scalar product of Eq. (2.17) and (2.8), $\langle u|v\rangle$, and the other is the product $|i\rangle\langle i|$ found in the closure relation $\sum\limits_{i=1}^{m}|i\rangle\langle i| = \hat{1}$, Eq. (2.20).

The latter entity is not a scalar number, but rather constitutes an *operator* which projects out of a vector $|v\rangle$ its ith component part: $|i\rangle\langle i|v\rangle = v_i|i\rangle$.

This brings us to the topic of operators. One frequently finds situations in which a vector is affected by an operator, say $\hat{A}$:

$$\boxed{\hat{A}|u\rangle = |v\rangle} \tag{3.1}$$

If $|u\rangle$ and $|v\rangle$ are in real space, the operator might cause a rotation, reflection, or displacement of $|u\rangle$ into the new vector $|v\rangle$; in generalized vector spaces, the operation may be of a more abstract nature.

Two operators $\hat{A}$ and $\hat{B}$ are said to be *equivalent* if for every $|u\rangle$ in our space,

$$\hat{A}|u\rangle = \hat{B}|u\rangle \tag{3.2a}$$

Similarly, the *sum* and *product* of $\hat{C}$ and $\hat{D}$ are defined through

$$\hat{E} = (\hat{C} + \hat{D}) \text{ if } \hat{E}|u\rangle = \hat{C}|u\rangle + \hat{D}|u\rangle \tag{3.2b}$$

$$\hat{F} = (\hat{C}\hat{D}) \text{ if } \hat{F}|u\rangle = \hat{C}(\hat{D}|u\rangle) \tag{3.2c}$$

for all $|u\rangle$ in the space.

All the operators with which we shall be concerned have the simplifying property that they are *linear*:

DEFINITION 3.1: *Linear operator*. The operator $\hat{A}$ is *linear* if for all $|u_1\rangle$ and $|u_2\rangle$ in the space within which $\hat{A}$ operates,

$$\hat{A}(c_1|u_1\rangle + c_2|u_2\rangle) = c_1\hat{A}|u_1\rangle + c_2\hat{A}|u_2\rangle \tag{3.3}$$

where c_1 and c_2 are complex scalars.

The operators of quantum mechanics must be linear (or of the closely related anti-linear type) if the fundamental Principle of Superposition of States, discussed in Chapter 5, is to obtain.

Equation (3.1) says that with *every* vector $|u\rangle$ in the space V, the operator $\hat{A}$ associates in a unique fashion a *new* vector $|v\rangle = \hat{A}|u\rangle$. Thus we might try to present the meaning of $\hat{A}$ by listing each $|u\rangle$ of V, and then the corresponding $|v\rangle = \hat{A}|u\rangle$, in a table. Even in a finite dimensional vector space, this could be a rather tedious undertaking.

But we can fully, and much more conveniently, describe the affect of $\hat{A}$ upon any *arbitrary* vector in terms only of what $\hat{A}$ does to m *basis vectors* $\{|i\rangle\}$ spanning the space in question. Equation (3.1) becomes, in the particular case where $\hat{A}$ operates on a basis vector such as $|i\rangle$,

$$\hat{A}|i\rangle = |g_i\rangle \tag{3.4a}$$

With the aid of the closure relation on $\{|i\rangle\}$, and calling

$$\{|i\rangle\} = \{|j\rangle\} = \ldots, \text{ so that } \sum_i^m |i\rangle\langle i| = \sum_j^m |j\rangle\langle j| = \ldots, \text{ etc.,}$$

$$\hat{A}|i\rangle = \sum_{j=1}^m |j\rangle\langle j|(\hat{A}|i\rangle) = \sum_{j=1}^m |j\rangle\langle j|g_i\rangle = \sum_{j=1}^m |j\rangle A_{ji} \tag{3.4b}$$

This says that $\hat{A}$ transforms $|i\rangle$ into some other vector, namely

OPERATORS AND FUNCTION SPACES

$|g_i\rangle$, which can then itself be expanded in the $\{|i\rangle\}$ basis system. We surely can take the scalar product of $|g_i\rangle$ with the bra $\langle j|$, and so $\langle j|g_i\rangle = \langle j|(\hat{A}|i\rangle)$ is simply a scalar number, which we shall call A_{ji}.

We can completely define the workings of $\hat{A}$ by recording what it does, via Eq. (3.4), to each of the m basis vectors in $\{|i\rangle\}$. And this information can then be stored within a matrix or table of the numbers A_{ji}. In an m-dimensional space, there will be only m^2 such numbers A_{ji}.

And yet, as emphasized in the last chapter, especial importance must not be ascribed either to bra or to ket space. We might, therefore, be inclined to think that the related symbol $(\langle j|\hat{A})|i\rangle$ also has a meaning; it looks like the scalar product of $|i\rangle$ with the bra vector $(\langle j|\hat{A})$, which in turn results from the action of $\hat{A}$ upon the bra $\langle j|$. The problem is that we have *defined* the operator $\hat{A}$ as something which transforms the *ket* basis vectors according to some clear prescription, manifest in the particular set of numbers $\{A_{ji}\}$—but we have said nothing about what $\hat{A}$ might do to a bra!

To escape this dilemma, we must first agree that the entity $\langle j|\hat{A}$ *does* exist as a bra—which is to say, $\hat{A}$ can operate in some fashion on a bra so as to produce a new bra vector. We then run into no inconsistencies if (and only if!) we construct or define $(\langle j|\hat{A})$ to be that unique bra vector whose *components* are determined by

$$(\langle j|\hat{A})|i\rangle \equiv \langle j|(\hat{A}|i\rangle) = \langle j|\hat{A}|i\rangle = A_{ji} \qquad (3.5)$$

This is the meaning of $\hat{A}$ operating to the left on a bra. By the nature of the definition, the parentheses in Eq. (3.5) are no longer needed, and we may think of $\hat{A}$ as operating either to the left or to the right, as the need arises.

A_{ji} is a shorthand notation for $\langle j|\hat{A}|i\rangle$, and may be considered as the $j-i$th entry into a *matrix representation* $\bar{\bar{A}}$ of $\hat{A}$ in the particular $\{|i\rangle\}$ basis system. Equation (3.4) now reads

$$\boxed{\hat{A}|i\rangle = |g_i\rangle = \sum_{j=1}^{m} |j\rangle\langle j|\hat{A}|i\rangle = \sum_{j=1}^{m} |j\rangle A_{ji}} \qquad (3.6)$$

We can return to Eq. (3.1) and examine the effect of $\hat{A}$ upon an *arbitrary* vector $|u\rangle$. Employing the closure relation in several places,

$$\hat{A}|u\rangle = |v\rangle \qquad (3.1)$$

$$\left(\sum_h |h\rangle\langle h|\right)\hat{A}\left(\sum_j |j\rangle\langle j|\right)|u\rangle = \left(\sum_k |k\rangle\langle k|\right)|v\rangle$$

$$\sum_{hj} |h\rangle\langle h|\hat{A}|j\rangle\langle j|u\rangle = \sum_k |k\rangle\langle k|v\rangle$$

$$\sum_{hj} |h\rangle A_{hj}u_j = \sum_k |k\rangle v_k \tag{3.7}$$

At this stage all is still equivalent to the original vector equation $\hat{A}|u\rangle = |v\rangle$; but we can project out the ith component of Eq. (3.7) by taking its scalar product with the basis vector $\langle i|$:

$$\sum_{hj} \langle i|h\rangle A_{hj}u_j = \sum_k \langle i|k\rangle v_k$$

$$\boxed{\sum_j A_{ij}u_j = v_i, \quad i = 1, 2, 3, \ldots, m} \tag{3.8a}$$

where the last step follows from the orthonormality of the $\{|i\rangle\}$: $\langle i|k\rangle = \delta_{ik}$. Equation (3.8a) is the immensely useful relationship which tells us how a *representation* of any ket vector is affected if $\hat{A}$ operates on the vector.

Equation (3.8a) is one of the sources of the familiar rules of matrix multiplication. If we make column vectors (m–by–1 matrices) out of the components u_i and v_i, then in matrix language, the n distinct equations of Eq. (3.8a) can be summarized as

$$\begin{pmatrix} A_{11} & A_{12} & \cdots & A_{1m} \\ \vdots & \vdots & & \vdots \\ A_{m1} & A_{m2} & \cdots & A_{mm} \end{pmatrix} \begin{pmatrix} u_1 \\ \vdots \\ u_m \end{pmatrix} = \begin{pmatrix} v_1 \\ \vdots \\ v_m \end{pmatrix} \tag{3.8b}$$

PROBLEM 3.1 Derive the rule for the multiplication of $m \times m$ matrices:

$$(AB)_{ij} = \sum_k^m A_{ik}B_{kj} \tag{3.9}$$

Interpret this relationship in light of the above discussion.

*PROBLEM 3.2 Equation (2.25) describes the rotation of basis vectors in three-space. What happens to the $\{|1\rangle, |2\rangle, |3\rangle\}$ representation of an arbitrary three-vector $\bar{r}$ under the same transformation?

It is important to note that Eq. (2.24a) is closely related to Eq. (3.4b). In the last chapter, we employed a closure relation to expand a unit vector $|i'\rangle$ from $\{|i'\rangle\}$ in terms of another basis set $\{|i\rangle\}$, as

$$|i'\rangle = \sum_j C_{ji}|j\rangle \tag{2.24a}$$

from which it followed that the *representations* of an *arbitrary* vector $|u\rangle$ generated by the two sets are related by

$$|u\rangle = \sum_i u_i |i\rangle = \sum_{i'} u'_{i'}|i'\rangle$$

and

$$u_i = \sum_j C_{ij} u'_j \qquad (2.26a)$$

Equation (3.6), on the other hand, tells us explicitly how to get the $\{|i'\rangle\}$ from the $\{|i\rangle\}$, given the right operator $\hat{A}$:

$$\hat{A}|i\rangle = |g_i\rangle \equiv |i'\rangle = \sum_j A_{ji}|j\rangle \qquad (3.6)$$

PROBLEM 3.3 What is the connection between Eq. (2.26b) and (3.8)? Note that Eq. (2.26b) describes the change in a representation of $|u\rangle$ with a transformation (due, say, to an operator $\hat{C}$) of the basis vectors; Eq. (3.8) is concerned with the direct influence of some $\hat{A}$ upon $|u\rangle$ itself.

*PROBLEM 3.4 What happens to the $\{|1\rangle, |2\rangle, |3\rangle\}$ representation of an operator $\hat{A}$ under the transformation of Eq. (2.25), where $\hat{A}$ acts upon vectors in three-space?

*PROBLEM 3.5 Is it worrysome that in Eq. (3.6), we sum over the first subscript in the representation of $\hat{A}$, while in Eq. (3.8), we sum over the second?

3.2 Invariant Spaces and Inverse Operators

We have so far been making the tacit assumption that the $|u\rangle$ and $|v\rangle$ vectors of Eq. (3.1) inhabit the same space. This would be the case if, say, $\hat{A}$ caused a rotation or reflection of the $\{|u\rangle\}$, perhaps compounded with a change of length. If $\hat{A}|u\rangle = |v\rangle$, and if the $\{|v\rangle\}$ do inhabit the same space as the $\{|u\rangle\}$, then this space is said to be *invariant relative to the operator* $\hat{A}$. Of course, a space invariant relative to one operation is not necessarily invariant to another.

If to every $|u\rangle$ in our space there corresponds a unique $|v\rangle$ determined by $\hat{A}|u\rangle = |v\rangle$, and if similarly to every $|v\rangle$ there corresponds one and only one $|u\rangle$, then $\hat{A}$ is called a one-to-one mapping of the space onto itself; this will be the case only if the space is invariant relative to $\hat{A}$. Given a one-to-one mapping $\hat{A}$, we can define its inverse operator $\hat{A}^{-1}$ by $\hat{A}\hat{A}^{-1} = \hat{A}^{-1}\hat{A} = \hat{1}$. It follows that

$$\hat{A}|u\rangle = |v\rangle \leftrightarrow \hat{A}^{-1}|v\rangle = |u\rangle \qquad (3.10)$$

If it should happen that a set of operators of interest is to form a group, then of course every operator must have a unique inverse.

PROBLEM 3.6 Is three-space invariant relative to the projection operator $(|1\rangle\langle 1|+|2\rangle\langle 2|)$? to $(|1\rangle\langle 1|+|2\rangle\langle 2|+|3\rangle\langle 3|)$?

†3.3 Vector Functions of Vectors

The above is not the only approach to the transformation properties of vectors. One could, alternatively, consider linear *vector functions of vectors*. Thus the linear function $\bar{f}(\ )$ associates with every vector $|u\rangle$ a new vector $|v\rangle$

$$\bar{f}(|u\rangle)=|v\rangle \tag{3.11}$$

where we are mixing standard and Dirac notations. $\bar{f}(\ )$ is defined as a *linear* function if

$$\bar{f}(c_1|u_1\rangle + c_2|u_2\rangle)=c_1\bar{f}(|u_1\rangle)+c_2\bar{f}(|u_2\rangle) \tag{3.12}$$

The $|v\rangle$ of Eq. (3.11) may be, but are not necessarily, in the same space as the $|u\rangle$. If the two spaces do coincide, and are spanned by $\{|i\rangle\}$, than any $\bar{f}(|u\rangle)$ can be expressed as a linear combination of the $\{|i\rangle\}$. In particular, the function $\bar{f}$ associates with every basis vector $|i\rangle$ a new vector $\bar{f}(|i\rangle)$, which must be expressible as

$$|g_i\rangle \equiv \bar{f}(|i\rangle)=\sum_j |j\rangle f_{ji} \tag{3.13}$$

The matrix of numbers f_{ji} then fully specifies the nature of the function $\bar{f}$, since it reveals what $\bar{f}$ does to *any* vector $|u\rangle$:

$$\begin{aligned}|v\rangle &= \bar{f}(|u\rangle) \\ &= \bar{f}\left(\sum_j u_j|j\rangle\right) \\ &= \sum_j u_j \bar{f}(|j\rangle) \quad \text{by Eq. (3.12)} \\ &= \sum_{i,j} u_j f_{ij}|i\rangle\end{aligned}$$

therefore

$$v_i = \sum_j f_{ij} u_j \tag{3.14}$$

Eqs (3.11), (3.13), and (3.14) correspond, respectively, to Eq. (3.1), (3.6), and (3.8).

An advantage of the concept of a linear vector function is that it can be readily generalized mathematically. For example, the idea of a functional, or *scalar* function of a vector, $\phi(|v\rangle)$, leads directly into formal study of dual spaces. This topic is discussed in books on linear algebra.

3.4 Hermitian Conjugate

Let $\hat{A}|u\rangle = |v\rangle$. Because bra and ket spaces are equally important, we should be able to define an operator, designated $\hat{A}^+$, which operates in *bra* space in such a manner as to preserve our relationship between $|u\rangle$ and $|v\rangle$ of ket space. If $\hat{A}|u\rangle = |v\rangle$, and if we associate $\langle u|$ and $\langle v|$ with $|u\rangle$ and $|v\rangle$, respectively, then define a new operator $\hat{A}^+$ which works in *bra* space such that $\langle u|\hat{A}^+ = \langle v|$ for all u, v. Thus

$$\boxed{\langle u|\hat{A}^+ = \langle v| \Leftrightarrow \hat{A}|u\rangle = |v\rangle} \tag{3.15}$$

Clearly, given the one-to-one correspondences between $|u\rangle$ and $\langle u|$ and $|v\rangle$ and $\langle v|$, there should also be an intimate relationship between $\hat{A}$ and $\hat{A}^+$.

$\hat{A}^+$ which (unlike $\hat{A}$) is defined initially as operating in bra space and to the left, is called the Hermitian conjugate of $\hat{A}$. We can record the effect of $\hat{A}^+$ in bra space in an equation directly analogous to Eq. (3.4):

$$\langle i|\hat{A}^+ = \sum_{j=1}^{m} \langle i|\hat{A}^+|j\rangle \langle j| = \sum_{j=1}^{m} A_{ij}^+ \langle j| \tag{3.16}$$

where again we drop the parentheses, by the argument preceeding Eq. (3.5).

Given $\hat{A}$ and a basis set $\{|i\rangle\}$, or the A_{ji} of Eq. (3.4), we can easily find the entries in the matrix representation A^+ of the Hermitian conjugate of $\hat{A}$. $\langle u|\hat{A}^+ = \langle v|$ is to correspond to $\hat{A}|u\rangle = |v\rangle$. Select an arbitrary state $|w\rangle$ and its dual $\langle w|$ and form the scalar products $\langle u|\hat{A}^+|w\rangle = \langle v|w\rangle$ and $\langle w|\hat{A}|u\rangle = \langle w|v\rangle$. But $\langle w|v\rangle = \langle v|w\rangle^*$, so $\langle u|\hat{A}^+|w\rangle = \langle w|\hat{A}|u\rangle^*$. For the case $u=i$, $w=j$

$$\boxed{A_{ji}^+ = A_{ij}^*} \tag{3.17}$$

Thus taking the complex conjugate of the transpose of $\bar{\bar{A}}$ gives us A^+.

In the development of this section, we have found it convenient to associate the superscript '+' with bra space. But consideration of the operator $\hat{B} \equiv \hat{A}^+$ shows immediately that $\hat{A}$ and $\hat{A}^+$ are meaningful only *relative* to one another, and that the superscript '+' is not tied preferentially to either kind of space.

PROBLEM 3.7 Show that $(\hat{A}^+)^+ = \hat{A}$.

†PROBLEM 3.8 Eq. (3.8) or (2.26) provides the rule governing the transformation properties of the column vector representation of a ket vector. Find the corresponding rule for transforming the representation of an arbitrary bra vector. See Problem 2.4.

3.5 Hermitian and Unitary Operators

In fact, for one special and very important type of operation, no distinction at all need be made between an operator and its Hermitian conjugate.

DEFINITION 3.2 *Hermitian operator*. An operator $\hat{H}$ is *Hermitian* if it is equal to its own Hermitian conjugate

$$\hat{H} = \hat{H}^+ \tag{3.18a}$$

It follows from Eq. (3.17) and $\bar{\bar{H}} = \bar{\bar{H}}^+$ that

$$\langle i|\hat{H}|j\rangle = \langle j|\hat{H}|i\rangle^*, \quad \text{or} \quad H_{ij} = H_{ji}^* \tag{3.18b}$$

The matrix representation of a Hermitian operator is called a Hermitian matrix.

The requirement of being Hermitian is quite constraining upon an operator, but at times necessary for a vitally significant reason: the eigenvalues of a Hermitian operator are always real numbers. If we wish (as we shall) to associate the possible results of a physical measurement in the laboratory with the possible eigenvalues of a corresponding mathematical operator, it is essential that the operator be Hermitian: the eigenvalues of Hermitian operators are, like the results of physical measurements, real numbers. It is also true, and of great importance to the formalism, that the eigenvectors of a Hermitian operator satisfy a closure relation (can be used as a complete basis set) and can be orthonormalized.

A second type of operator central to the development of quantum mechanics is the *unitary* operator. This is a generalization of the familiar rotation operator of three-space, which changes the alignment of any vector without altering its length.

DEFINITION 3.3 *Unitary operator*. An operator $\hat{U}$ is *unitary* if its inverse is equal to its Hermitian conjugate;

$$\hat{U}^{-1} = \hat{U}^+ \tag{3.19}$$

The essential characteristic of $\hat{U}$ is that it must leave the length of an arbitrary vector, such as $|v\rangle$, invariant. Since by Eq. (3.15) the vector $\hat{U}|v\rangle$ in ket space is associated with $\langle v|\hat{U}^+$ of bra space, this constraint is expressed as

$$\langle v|v\rangle = (\langle v|\hat{U}^+)(\hat{U}|v\rangle) = \langle v|\hat{U}^+\hat{U}|v\rangle \tag{3.20}$$

which can be valid for all $|v\rangle$ only if $\hat{U}^+\hat{U} = \hat{1}$. Eq. (3.19) follows from the definition of an operator's inverse.

Finding the inverse of an arbitrary matrix may involve a fair amount of computation. It is fortunate that for quantum mechanical purposes one usually does not need the inverse of the

OPERATORS AND FUNCTION SPACES   33

matrix representation of a Hermitian operator, and that the inverses of unitary matrices, which are frequently used, are easy to come by:

$$U_{ij}^{-1} = U_{ij}^{+} = U_{ji}^{*} \qquad (3.21)$$

We shall see in Chapter 5 that operators associated with dynamical quantum mechanical entities, such as momentum, energy, or position, are Hermitian; the operators reflecting upon the system's geometric symmetry ($\hat{M}_1$, $\hat{R}_{120}$, etc. for a C_{3v} molecule) are unitary. And operators or matrices describing a change of basis system are also usually unitary.

†PROBLEM 3.9 One sometimes finds the definitions: $\hat{H}$ is Hermitian if

$$\langle \hat{H}\psi | \phi \rangle = \langle \psi | \hat{H}\phi \rangle, \text{ all } \psi, \phi; \qquad (3.22a)$$

$$\hat{U} \text{ is unitary if } \langle \hat{U}\psi | \hat{U}\phi \rangle = \langle \psi | \phi \rangle \text{ all } \psi, \phi \qquad (3.22b)$$

Explain the notation. For arbitrary $\hat{A}$, does $\langle \psi | \hat{A} | \phi \rangle$ equal $\langle \psi | \hat{A}\phi \rangle$, $\langle \hat{A}\psi | \phi \rangle$, both, or neither?

3.6 Eigenequations

We have not explicitly shown what any particular $\hat{A}$ 'looks' like, but only how it works:

$$\boxed{\hat{A}|i\rangle = \sum_{j}^{m} A_{ji}|j\rangle, \quad A_{ji} = \langle j | \hat{A} | i \rangle} \qquad (3.6)$$

The possible effects of $\hat{A}$ can be fully described in terms of the matrix $\bar{\bar{A}}$ of numbers A_{ji}, which are the coefficients in the set of linear equations

$$\boxed{\begin{aligned} \hat{A}|1\rangle &= A_{11}|1\rangle + A_{21}|2\rangle + \cdots + A_{m1}|m\rangle \\ \hat{A}|2\rangle &= A_{12}|1\rangle + A_{22}|2\rangle + \cdots + A_{m2}|m\rangle \\ &\cdots\cdots\cdots\cdots\cdots\cdots\cdots\cdots\cdots \\ \hat{A}|k\rangle &= A_{1k}|1\rangle + A_{2k}|2\rangle + \cdots + A_{mk}|m\rangle \\ &\cdots\cdots\cdots\cdots\cdots\cdots\cdots\cdots\cdots \\ \hat{A}|m\rangle &= A_{1m}|1\rangle + A_{2m}|2\rangle + \cdots + A_{mm}|m\rangle \end{aligned}} \qquad (3.23)$$

As such, we can think of the $\{A_{ji}\}$ as the *admixture coefficients* which tell us how great an amount of each of the various basis vectors in $\{|i\rangle\}$ is mixed into, for example, the vector $|k\rangle$ when $|k\rangle$ is affected by the operation $\hat{A}$. Just as with the representation of vectors discussed in the last chapter, an important point is that the matrix elements depend fully upon the choice of basis vectors used. Had we employed a totally different (but complete and orthonormal) basis set $\{|\alpha\rangle\}$ instead of $\{|i\rangle\}$, we would have found

$\hat{A}|\alpha\rangle = \sum_\beta A_{\beta\alpha}|\beta\rangle$, where $A_{ji} \neq A_{\beta\alpha}$.

If it should happen, for some reason, that $\hat{A}$ has but a slight effect upon all the $|i\rangle$'s, such that $\langle i|A|j\rangle$ is small except in the case $i = j$, then it might be possible to use approximation schemes in analysing Eq. (3.6) or (3.23), as is done in standard quantum mechanical perturbation theory. But it may even be that there exist certain vectors in vector space which are *not at all* disturbed or altered by $\hat{A}$—that is, there may be certain vectors $\{|a_n\rangle\}$ such that

$$\boxed{\hat{A}|a_n\rangle = a_n|a_n\rangle} \qquad (3.24)$$

Thus when $\hat{A}$ operates on one of these vectors, the resultant involves no admixture with other vectors at all; we start off with $|a_n\rangle$ alone, and $\hat{A}|a_n\rangle$ leaves us in the same condition. The 'length' of the vector may be altered by the complex or real number a_n, but the 'direction' in Hilbert space remains the same. The vectors $\{|a_n\rangle\}$ to which Eq. (3.24) applies are known as the *eigenvectors* of A, and the corresponding $\{a_n\}$ are the eigenvalues. There is a *unique set of eigenvalues and eigenvectors* associated with, and characteristic of, any operator of the kind with which we shall deal. In Chapter 5 we shall offer demonstrations of the process involved in finding the eigenvectors and eigenvalues of an operator.

Eigenequations play a singularly important role in all branches of the physical sciences; and to a large extent, the study of quantum mechanical systems amounts to solving the eigenequations of certain particular Hermitian operators.

†PROBLEM 3.10 Prove that the eigenvalues and any expectation value of a Hermitian operator are real numbers, and that eigenvectors corresponding to different eigenvalues are orthogonal.
*PROBLEM 3.11 Consider the operator $\hat{A}$ with eigenvectors $\{|a_n\rangle\}$ and eigenvalues $\{a_n\}$. Prove that a representation of $\hat{A}$ generated by the basis set of its own eigenvectors $\{|a_n\rangle\}$ is a diagonal matrix, $A_{ij} = a_i \delta_{ij}$. Under what conditions might off-diagonal elements of $\bar{A}$ be small but non-vanishing?
PROBLEM 3.12 Is Eq. (3.23) a necessary and/or sufficient condition for the operator $\hat{A}$ to be linear?

3.7 Vectors in Function Space

The vectors of the preceding sections have, for the most part, behaved like the familiar three-dimensional vectors of real spaces. The ideas developed, however, could have been modelled upon vectors of a much more general nature.

OPERATORS AND FUNCTION SPACES

The sum of two functions continuous over some domain, for example, is itself a continuous function, as is a continuous function multiplied by a scalar (even 0 or -1). Thus the set of all such continuous functions forms a vector space; and the space consists of functions, not directed line segments or ordered n-tuples. It is, moreover, generally not too difficult to define a scalar product for this space, and then to choose a subset of vectors which are mutually orthogonal, normalized, and which span the space. Such a set can serve as a set of basis vectors for our function space.

As an important example, the set of solutions to a homogeneous linear differential equation forms a vector space. All the standard sets of orthonormal functions used extensively in quantum mechanics and classical physics are the basis sets which span spaces of solutions to specific differential equations. For example, a common equation of physics is the one-dimensional Helmholtz equation

$$\frac{d^2}{dx^2}f(x) + k^2 f(x) = 0$$

$$f(x) = 0 \text{ at } x = 0, x = L \tag{3.25a}$$

where both $f(x)$ and k are to be determined. Such eigenequations and boundary conditions might arise, for example, in solving Laplace's equation for the electrostatic fields inside a box of dimension L whose walls are held at specified potentials. Normalized solutions to Eq. (3.25a) are of the form

$$f_m(x) = (2/L)^{1/2} \sin kx,$$

$$k = \frac{\pi m}{L} \quad \text{integer } m \tag{3.25b}$$

as can be verified easily by direct substitution. One can use $\{f_m(x)\}$ of Eq. (3.25b) as a basis set to decompose any odd, periodic function of x as a Fourier series. Spherical harmonics, Bessel functions, etc., come about in a like manner.

As a second but closely related example, consider a basis set in which the mth basis vector is denoted

$$|m, x\rangle = L^{-1/2} \exp(2\pi i m x/L), \text{ integer } m \tag{3.26}$$

$\{|m, x\rangle\}$ forms a complete set of orthonormal functions, each labeled by the index 'm', but each of which (as with Eq. (3.25b)) happens to be a function of the parameter x. A periodic function $f(x)$ can thus be considered as a vector in an infinite dimensional vector space, and can be decomposed into linearly independent parts

$$f(x) = \sum_{m=-\infty}^{\infty} c_m \exp(2\pi i m x/L) \tag{3.27a}$$

with the 'component in the mth direction' given by

$$c_m = \int_0^L f(x) L^{-1/2} \exp(-2\pi i m x/L)\, dx \tag{3.27b}$$

These two transcribe to the Dirac notation as

$$|f(x)\rangle = \sum_{-\infty}^{\infty} c_m |m, x\rangle \tag{3.28a}$$

$$c_m = \langle m, x | f(x) \rangle \tag{3.28b}$$

where the scalar product in Eq. (2.28b) now involves integration over x; as a generalization of Eq. (2.12), and comparing Eqs (3.27b) and (3.28b), we define

$$\boxed{\langle f(x) | g(x) \rangle = \int f(x)^* g(x)\, dx} \tag{3.29}$$

†PROBLEM 3.13 How is Eq. (3.28b) related to Eq. (2.12)? Does the scalar product of Eq. (3.29) obey the rules of Eq. (2.8)? What is the measure of length?

†PROBLEM 3.14 Eq. (3.25a) is an eigenequation; is the operator Hermitian? What are the orthonormality and closure conditions for the basis sets of Eq. (3.26) and (3.28)?

3.8 Continuous Basis Sets

We began with the finite, discrete basis set $\{|m\rangle, m=1, 2, 3\}$; this was extended by allowing m to become an integer larger than three, eventually with $m \to \infty$. In the last section we discussed a set $\{|m, x\rangle\}$ again parameterized by integer m where $m \to \infty$, but in which every basis vector happened to be a continuous function of x.

The final topic of this chapter is a somewhat subtle extension of the one just covered. Here the set $\{|m, x\rangle\}$ is generalized so that even the index m may vary continuously over the real numbers, rather than being constrained to integer values.

We shall introduce this idea through the example of the Fourier integral. A non-periodic function $f(x)$, such as a wave packet or square pulse, can be expressed as a linear superposition:

$$f(x) = \left(\frac{1}{2\pi}\right)^{1/2} \int_{-\infty}^{\infty} g(y) \exp(ixy)\, dy \tag{3.30a}$$

where

$$g(y) = \left(\frac{1}{2\pi}\right)^{1/2} \int_{-\infty}^{\infty} f(x) \exp(-ixy)\, dx \tag{3.30b}$$

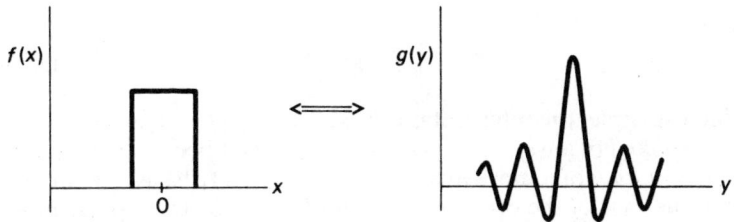

Figure 3.1.

Comparison with Eqs (3.27) and (3.28) suggests that we consider Eq. (3.30a) as the expansion of a function $f(x)$ in terms of the continuously varying basis set of elements $|y, x\rangle = (2\pi)^{-1/2} \exp(ixy)$, where the basis vector-identifying index y is a real *continuous* parameter. Then we can write

$$f(x) = \int_{-\infty}^{\infty} g(y) |y, x\rangle \, dy \qquad (3.31\text{a})$$

The magnitude of the component of $f(x)$ in the particular 'direction' y' in function space would be $g(y')$. One must not think of y as having any significance as an axis direction in real space; it is simply a label which, rather than assuming the values 1, 2, and 3 (as in real space), or 1, 2, 3, ... (as with a Fourier Series), varies continuously. In analogy to Eq. (3.28b) then, we set

$$g(y) = \langle y, x | f(x) \rangle \qquad (3.31\text{b})$$

where an integration over x is implied. But there is an apparent difficulty with Eq. (3.31); the obvious symmetry between (3.30a) and (3.30b) is not preserved. The reason for this should become clear if we rise to a slightly higher level of abstraction.

Consider $f(x)$ which lies in some function space. We can say that the 'real' vector of interest should be called $|f\rangle$, and that the particular value $f(x')$ for some x' is merely the 'component' of $|f\rangle$ in the x'-direction, or

$$\boxed{f(x') = \langle x' | f \rangle} \qquad (3.32)$$

Recall decomposing a vector $|v\rangle$ as $|v\rangle = \sum_i |i\rangle \langle i|v\rangle$ in the discrete basis system $\{|i\rangle\}$; if instead of a discrete basis set we had needed a continuous set parameterized by x, for which the closure relation would have been

$$\boxed{\int |x\rangle \, dx \langle x| = \hat{1}} \qquad (3.33\text{a})$$

then

$$|f\rangle = \int |x\rangle\langle x|f\rangle \, dx = \int f(x)|x\rangle \, dx \qquad (3.33b)$$

which suggests the interpretation of $f(x)$ as $\langle x|f\rangle$; thus when we have spoken of a 'vector' $|f(x)\rangle$, we have in fact been referring to the *representation* of a more abstract $|f\rangle$ in the $\{|x\rangle\}$ basis system. But, as in the case of the ordered n-tuples of Eq. (2.6), the representation $\langle x|f\rangle$ contains complete information about $|f\rangle$, once the basis system is specified, and itself inhabits its own vector space.

We have not said what the elements in the set $\{|x\rangle\}$ look like when explicitly written out—we simply claim that they are labelled by the continuous parameter x, and that each value of x corresponds to a different basis vector. The orthonormality condition on the x basis set is

$$\boxed{\langle x|x'\rangle = \delta(x - x')} \qquad (3.34)$$

where the Dirac delta function $\delta(x)$ is defined by

$$\int_{-\infty}^{\infty} f(x)\delta(x - x_0) \, dx = f(x_0)$$

$$\delta(x - x_0) = 0, \quad x \neq x_0$$

$$\int_{-\infty}^{\infty} \delta(x - x_0) \, dx = 1 \qquad (3.35)$$

One can think of $\delta(x)$ as an infinitesimally narrow, infinitely high spike of unit area† centered at $x = x_0$.

It is expected that the meaning of Eq. (3.32) through (3.35) will become clearer, or at least more palatable, with experience. To delve deeply into discussion of them would take us far afield of our primary aims, but the example of the next section should remove some of their mystery.

†PROBLEM 3.15 Resolve the paradox of the asymmetry of Eq. (3.31). [HINT: is $|y, x\rangle = \langle y|x\rangle$ meaningful?]

†PROBLEM 3.16 The solutions to $(d^2y/dx^2) + y = 0$ are of the form $y = \alpha_1 \cos x + \alpha_2 \sin x$. Find a set of orthonormal basis vectors spanning the space of the solutions. Now find another basis set by rotating the first through θ in function space.

PROBLEM 3.17 Does a Fourier transform define a projection operation?

---

†$\delta(x)$, a generalization of the Kronecker delta δ_{ij}, does not behave like an ordinary function, and in fact is called a *distribution*. Must the spike be infinitely high?

OPERATORS AND FUNCTION SPACES   39

*PROBLEM 3.18 Interpret the quantum mechanical state $\psi(\vec{r}) = \langle r|\psi\rangle$.

PROBLEM 3.19 Demonstrate and interpret

$$\delta(x-x_0) = \frac{1}{2\pi} \int_{-\infty}^{\infty} \exp\{i(x-x_0)y\}\, dy$$

3.9 A Weighted String

As an illustration of some of the ideas we have just been discussing, we consider the vibrational modes of a string to which are attached some minute weights. The string, fixed at $x=0$ and $x=L$, has N equal weights distributed evenly along its length at $x=na$, where $n=1,2,\ldots, N=(L/a-1)$.

If the system is free to move only vertically, then any displacement of the masses from equilibrium can be noted in terms of components along unit vectors $\{|y_i\rangle, i=1,\ldots N\}$, Fig. 3.2; $|y_k\rangle$ describes the situation in which the kth mass has moved a distance of one unit upward from the horizon, all the other weights remaining in their original positions. A general configuration of displacements $|w\rangle$ is written, with this basis system,†

$$|w\rangle = \sum_i^N |y_i\rangle\langle y_i|w\rangle = \sum_i^N w_i^{(y)}|y_i\rangle \qquad (3.36a)$$

where we have assumed the scalar product and

$$\langle y_i|y_j\rangle = \delta_{ij} \qquad (3.36b)$$

to be meaningful.

If the masses become even lighter but their number increases indefinitely, the limiting case of Eq. (3.36a) defines an integral:

$$|w\rangle = \lim_{N\to\infty} \sum_i^N w_i^{(y)}|y_i\rangle \equiv \int w^{(y)}(x)|y(x)\rangle\, dx \qquad (3.37a)$$

where the discrete dummy index i has turned into the continuous variable x. There is no loss of information if we now replace the

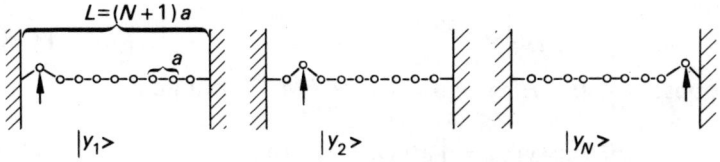

Figure 3.2.

†See footnote, Page 13.

symbol $|y(x)\rangle$ by $|x\rangle$, and Eq. (3.37a) assumes the form of Eq. (3.33b)

$$|w\rangle = \int w^{(y)}(x)|x\rangle \, dx \tag{3.37b}$$

$|w\rangle$ is an abstract vector describing a particular shape of the *entire* string; and since we know precisely what the $\{|x\rangle\}$ mean, it is sufficient to work with the *representation* $w^{(y)}(x)$ of $|w\rangle$ until such time as we have cause to make a change of basis set.

Equation (3.37b) suggests that the $\{|x\rangle\}$ obey closure and orthonormality relations

$$\hat{1} = \int |x\rangle \, dx \langle x| \tag{3.33a}$$

$$\langle x|x'\rangle = \delta(x-x') \tag{3.34}$$

Now for a rather substantial leap of faith: although we have developed the $\{|x\rangle\}$ through a model in which $|x\rangle \equiv |y(x)\rangle$, we can think of the $\{|x\rangle\}$ more generally as being something like, but not identical to simple position vectors. In our new sense, it is no longer true that $|x_1\rangle + |x_2\rangle = |x_1 + x_2\rangle$ as in a simple one-dimensional 'real' space; rather, the $\{|x\rangle\}$ are the basis vectors spanning an infinite dimensional *function* space! It is this sort of generalized basis set which leads to our interpretation of $f(x) = \langle x|f\rangle$ in Eq. (3.33). We shall examine the connection between the $\{|x\rangle\}$ and real-space vectors more carefully in Section 6.6.

*PROBLEM 3.20 In a space spanned by $\{|x\rangle\}$, we can define a 'position operator' $\hat{x}$ whose orthonormal eigenvectors are the complete set $\{|x\rangle\}$

$$\hat{x}|x\rangle = x|x\rangle$$
$$\langle x'|x\rangle = \delta(x'-x) \tag{3.38a}$$

Show that with this basis set, the matrix elements of the operator $F = F(\hat{x})$ are of the form

$$\langle x'|F(\hat{x})|x\rangle = F(x)\delta(x-x') \tag{3.38b}$$

†PROBLEM 3.21 We define $\hat{H}$ as being Hermitian if for arbitrary u and v

$$\langle u|\hat{H}|v\rangle = \langle v|\hat{H}|u\rangle^* \tag{3.18b}$$

Suppose that $\hat{H} = H(\hat{x})$; show that $H(\hat{x})$ is Hermitian if

$$\int u^*(x)H(\hat{x})v(x) \, dx = \int v^*(x)H(\hat{x})u(x) \, dx \tag{3.39}$$

Compare with Eq. (3.22a)

OPERATORS AND FUNCTION SPACES

†3.10 Transformation to Normal Modes

Let us now, in conclusion, carry our example a step further, and consider the consequences of a change of basis. Returning to the case of discrete masses, for simplicity we begin by thinking of the situation in which $N=2$. For the state $|w\rangle$, the displacements of the particles at $x=a$ and $x=2a$ are represented in the $\{|y_1\rangle, |y_2\rangle\}$ basis system by $(w_1^{(y)}, w_2^{(y)})$, or $(w^{(y)}(a), w^{(y)}(2a))$ in continuum notation. Figure 3.3 demonstrates that $|w\rangle$ can be represented just as meaningfully (since the weights are assumed to be of negligible mass) in terms of a basis derived from the string's *normal modes* of vibration. The string's fundamental and first harmonic modes are of shapes $\sin(\pi x/L)$ and $\sin(2\pi x/L)$ (not worrying about the time dependence), and the displacements of the masses may be given by

$$|w\rangle = \sum_{i=1}^{2} w_i^{(q)}|q_i\rangle \qquad (3.40)$$

where $|q_1\rangle$ and $|q_2\rangle$ are defined by the displacement arrows in Fig. 3.3. For $N>2$ evenly distributed particles, N normal coordinates $\{|q_i\rangle\}$ are required.

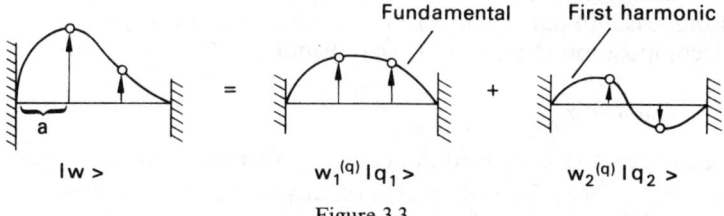

Figure 3.3.

As in Eq. (2.24a), a relationship between the *unit* vectors of one basis system and those of the other is

$$|q_i\rangle = \sum_j |y_j\rangle\langle y_j|q_i\rangle = \sum_j U_{ji}|y_j\rangle \qquad (3.41a)$$

Consequently the components of two *representations* of an *arbitrary* state $|w\rangle$ are linked by

$$w_j^{(y)} = \sum_i U_{ji} w_i^{(q)}, \quad U_{ji} \equiv \langle y_j|q_i\rangle \qquad (3.41b)$$

as in Eqs (2.26a) and (3.8). Figure 3.3 suggests that the scalar product of $\langle y_j|$ with $|q_i\rangle$, and consequently the elements of the transformation matrix U, be

$$\langle y_j|q_i\rangle \equiv U_{ji} = \sqrt{\frac{2a}{L}} \sin\frac{ji\pi a}{L} \qquad (3.41c)$$

where the normalization is obtained from

$$1 = \langle y_l | y_l \rangle = \sum_k \langle y_l | q_k \rangle \langle q_k | y_l \rangle = \sum_k U_{kl}^2$$

This points out an important general property of transformation matrices such as $\bar{\bar{U}}$. Consider the particular matrix elements U_{ji} obtained from Eq. (3.41) by fixing i; the column vector

$$\begin{pmatrix} U_{1i} \\ U_{2i} \\ \vdots \\ U_{Ni} \end{pmatrix} = \begin{pmatrix} \langle y_1 | q_i \rangle \\ \langle y_2 | q_i \rangle \\ \vdots \\ \langle y_N | q_i \rangle \end{pmatrix} \qquad (3.42)$$

constitutes the representation of the specific unit vector $|q_i\rangle$ generated by the $\{|y_j\rangle\}$ basis system. We arrive at the transformation matrix $\bar{\bar{U}}$ by stacking the N such column vectors, one for each i, side by side. We shall make use of this observation later.

PROBLEM 3.22 Is $\bar{\bar{U}}$ unitary? Can you find an operator $\hat{U}$ for which $\bar{\bar{U}}$ is a matrix representation?

†PROBLEM 3.23 Extend the analysis of this section to the continuum case. In particular, show that a Fourier transform is a decomposition in a plane-wave (normal mode) basis system.

3.11 Summary

1. The operator $\hat{A}$, defined in ket space, alters the arbitrary vector $|u\rangle$: $\hat{A}|u\rangle = |v\rangle$. The effect of $\hat{A}$ on any vector may be conveniently described by considering its effect on the various members of a basis vector set:

$$\hat{A}|i\rangle = \sum_{j=1}^m |j\rangle\langle j|\hat{A}|i\rangle = \sum_{j=1}^m |j\rangle A_{ji}$$

Then $\hat{A}|u\rangle = |v\rangle$ can be written as the representation equation

$$\sum_j A_{ij} u_j = v_i$$

2. To operator $\hat{A}$ defined to operate in ket space, there exists a corresponding Hermitian conjugate $\hat{A}^+$, such that if $\hat{A}|u\rangle = |v\rangle$, then $\langle u|\hat{A}^+ = \langle v|$. Given the matrix representation of $\hat{A}$, $\bar{\bar{A}}$, the representation of the $\hat{A}^+$ is $\bar{\bar{A}}^+$. $\hat{A}$ is Hermitian if $\hat{A} = \hat{A}^+$, and unitary if $\hat{A}^{-1} = \hat{A}^+$.

3. An eigenequation is a very important special case of $\hat{A}|u\rangle = |v\rangle$ in which $\hat{A}|a_n\rangle = a_n|a_n\rangle$. The eigenvectors of a Hermitian operator constitute a complete, orthogonizable basis system, and the eigenvalues $\{a_n\}$ are real-valued.

4. The solutions to a linear differential equation constitute a vector space. The basis vectors of this space may be discrete (in the Fourier sum, $\{|m,x\rangle\}$), or continuous, (as in the Fourier integral with $\{|y,x\rangle\}$). The formalism is greatly simplified once one determines that $f(x')$ is the component of $|f\rangle$ in the x'-direction, relative to the basis set $\{|x\rangle\}$, and

$$|f\rangle = \int |x\rangle \langle x|f\rangle \, dx = \int f(x)|x\rangle \, dx$$

$$\int |x\rangle \, dx \langle x| = \hat{1}, \qquad \langle x|x'\rangle = \delta(x-x')$$

That is, $f(x)$ is the representation of $|f\rangle$ generated by the $\{|x\rangle\}$ basis set. The argument is valid even if $|f(x)\rangle$ is itself a basis vector, such as $|y,x\rangle$, of function space.

CHAPTER 4

Representations of an Operator in Three-Space

Vectors in three-space can be rotated or reflected by a geometric transformation operator $\hat{R}$. The rotations described in this chapter will generally be about the z- or three-axis and through the angle ϕ, produced by the operator $\hat{R}_\phi$; reflections across a vertical mirror plane aligned at an angle ϕ from $|1\rangle$ are produced by $\hat{M}_\phi$. Making use of a three-dimensional closure relation

$$\sum_{i=1}^{3} |i\rangle\langle i| = \hat{1}$$

it is shown that the familiar three-by-three real-space transformation matrices are the matrix representations of $\hat{R}$ in the basis system $\{|i\rangle\} = \{|1\rangle, |2\rangle, |3\rangle\}$; $\bar{\bar{R}}$ is the unitary matrix whose elements are $R_{ji} \equiv \langle j|\hat{R}|i\rangle$.

The block-diagonal form of $\bar{\bar{R}}_\phi$, in particular,

$$\bar{\bar{R}}_\phi = \begin{pmatrix} \cos\phi & -\sin\phi & 0 \\ \sin\phi & \cos\phi & 0 \\ \hline 0 & 0 & 1 \end{pmatrix}$$

is a consequence of a fortuitous (or clever!) alignment of the coordinate axes of real space so as to reflect the physical presence of an axis of rotation. This particular choice of basis vectors results in a partition of real space into two subspaces, each of which is invariant relative to rotations about the axis; the two subspaces are composed of vectors of two inherently different symmetry types. These ideas apply not only to a representation of a single operator, but even to a group of symmetry operators. This provides us with a relatively painless first exposure to irreducible representations.

4.1 A Representation of the Rotation Operator

In Chapter 2 we discussed vectors and their representations, and in Chapter 3, operators and their representations. Here we shall solidify and extend what we have done by examining in greater detail the case of real, three-dimensional space. We shall find, for example, that the familiar three-by-three matrix used in rotating vectors in three-space is the representation of a unitary rotation operator; a particular form for this representation is determined by a specific choice of three orthonormal unit vectors, $\{|1\rangle, |2\rangle, |3\rangle\}$ defining the coordinate axis directions of real space.

We shall tentatively† assume that $\{|1\rangle, |2\rangle, |3\rangle\}$ span our space of interest; then for any vector $|v\rangle$ in the space,

$$|v\rangle = \sum_{i=1}^{3} |i\rangle \langle i|v\rangle = \sum_{i=1}^{3} v_i |i\rangle \tag{4.1}$$

where we have employed the closure relation:

$$\sum_{i=1}^{3} |i\rangle\langle i| = \hat{1} \tag{4.2}$$

For the time being, consider rotations about an axis normal to the $|1\rangle - |2\rangle$, or x–y, plane, Fig. 4.1. We define the operator $\hat{R}_\phi$ to rotate *any* three-vector $|u\rangle$ counter-clockwise through the angle ϕ about the $|3\rangle$ axis, as in Fig. 4.1a; then Figs 4.1(b) and (c) illustrate what happens to the *unit* vectors $|1\rangle$ and $|2\rangle$, in particular, for ϕ of approximately 40°. From Fig. 4.1(d), we see that

$$|1\rangle \to |1'\rangle = \hat{R}_\phi|1\rangle = |1\rangle \cos\phi + |2\rangle \sin\phi \tag{4.3a}$$

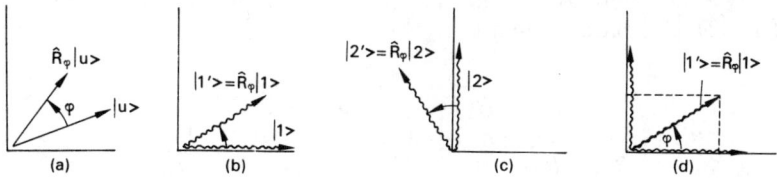

Figure 4.1.

Similarly

$$\begin{aligned} |2\rangle \to |2'\rangle &= \hat{R}_\phi|2\rangle = |1\rangle(-\sin\phi) + |2\rangle\cos\phi \\ |3\rangle \to |3'\rangle &= \hat{R}_\phi|3\rangle = |3\rangle \end{aligned} \tag{4.3b}$$

On the other hand, we may use the closure condition Eq. (4.2)

---

†In Section 8.2 we shall find that when we consider both polar and axial vectors, this assumption is not valid.

to decompose $\hat{R}_\phi$ as

$$\hat{R}_\phi|i\rangle = \sum_{j=1}^{3} |j\rangle\langle j|\hat{R}_\phi|i\rangle = \sum_{j=1}^{3} R_{ji}|j\rangle \qquad (4.4)$$

where $R_{ji} = \langle j|\hat{R}_\phi|i\rangle$, and the numbers R_{ji} constitute a matrix $\bar{\bar{R}}_\phi$; $\bar{\bar{R}}_\phi$ is the representation of the operator $\hat{R}_\phi$ generated in the $\{|i\rangle\}$ basis system.

Explicit expressions for $\bar{\bar{R}}_\phi$ can be obtained now by either of two equivalent methods:

Method 1. Obtain the $\{R_{ji}\}$, term by term, directly from Eq. (4.3) by making use of the orthonormality of the basis vectors, $\langle i|j\rangle = \delta_{ij}$:

$$\hat{R}_\phi|1\rangle = |1\rangle\cos\phi + |2\rangle\sin\phi$$

therefore

$$R_{11} \equiv \langle 1|(\hat{R}_\phi|1\rangle) = \langle 1|\hat{R}_\phi|1\rangle = \overbrace{\langle 1|1\rangle}^{1}\cos\phi + \overbrace{\langle 1|2\rangle}^{0}\sin\phi = \cos\phi$$
$$R_{21} = \langle 2|\hat{R}_\phi|1\rangle = \sin\phi$$
$$R_{31} = \langle 3|\hat{R}_\phi|1\rangle = 0, \qquad (4.5)$$
etc.

Method 2. Obtain the $\{R_{ji}\}$ through term by term comparison of Eqs (4.3) and (4.4):

$$\hat{R}_\phi|1\rangle = |1\rangle\cos\phi + |2\rangle\sin\phi + |3\rangle\,0$$
$$\hat{R}_\phi|1\rangle = |1\rangle R_{11} + |2\rangle R_{21} + |3\rangle R_{31} \qquad (4.6)$$

therefore

$$R_{11} = \cos\phi, \; R_{21} = \sin\phi, \; R_{31} = 0, \text{ etc.}$$

By either method, we find that the matrix representing $\hat{R}_\phi$ in the $\{|1\rangle, |2\rangle, |3\rangle\}$ basis system is

$$\bar{\bar{R}}_\phi = \begin{pmatrix} R_{11} & R_{12} & R_{13} \\ R_{21} & R_{22} & R_{23} \\ R_{31} & R_{32} & R_{33} \end{pmatrix} = \begin{pmatrix} \cos\phi & -\sin\phi & 0 \\ \sin\phi & \cos\phi & 0 \\ 0 & 0 & 1 \end{pmatrix} \qquad (4.7)$$

where we have followed the convention that R_{ij} is to be the entry in the ith row and jth column of $\bar{\bar{R}}_\phi$. Shortly we shall examine the usefulness of such a matrix representation.

PROBLEM 4.1 Generate the representation of $\hat{R}_\phi$ with the basis set $\{2^{-1/2}(|1\rangle \pm |2\rangle), |3\rangle\}$.

It is important to note, as in Problem 4.1, that had we started with a different basis system, we would have found a representa-

REPRESENTATIONS OF AN OPERATOR IN THREE-SPACE

tive matrix different from the right-hand side of Eq. (4.7). In particular, if we had chosen a set of three vectors none of which lies parallel to the axis of rotation, we would have generated a representation by the same procedure as above, but without the 1 and four 0's. The simplicity of Eq. (4.7) is a consequence of a fortuitously compatible choice of basis system and axis of rotation.

*PROBLEM 4.2 Consider a reflection plane which contains the $|3\rangle$ axis; thus if $|1\rangle$ and $|2\rangle$ lie in the plane of this page, $|3\rangle$ and the reflection plane extend out of it. Figures 4.2(a) and (b) show $|1\rangle$ and $|2\rangle$, respectively, and their images $\hat{M}_\phi|1\rangle$ and $\hat{M}_\phi|2\rangle$; the reflection plane has been constructed so as to make an angle ϕ with the $|1\rangle - |3\rangle$ plane. Show that the matrix representation of the operator for mirror reflection across this plane is

$$\bar{\bar{M}}_\phi = \begin{pmatrix} \cos 2\phi & \sin 2\phi & 0 \\ \sin 2\phi & -\cos 2\phi & 0 \\ 0 & 0 & 1 \end{pmatrix} \quad (4.8)$$

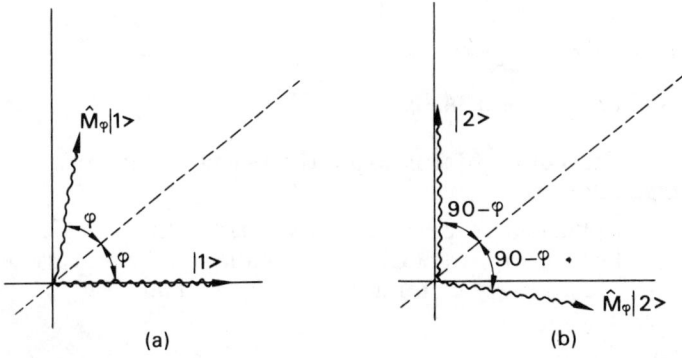

Figure 4.2.

In the above, $\hat{R}_\phi$ has rotated our basis vectors counter-clockwise through ϕ, and indeed, could likewise transform all *vectors* in three-space. It is apparent from Fig. 4.3, however, that Eq. (4.4) would still apply had we turned the *coordinate axes* by ϕ clockwise instead. The two approaches yield completely equivalent results. The former approach, which we have used throughout this section, is called the *active*, in the sense that physically meaningful vectors such as $|u\rangle$ of Fig. 4.1(a) are actually twisted into new vectors by $\hat{R}_\phi$; this picture can be easily generalized for discussion of all types of operations. The *passive* point of view, on the other

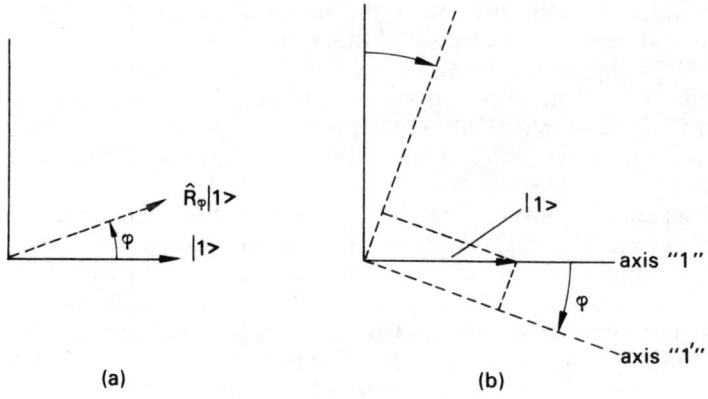

Figure 4.3.

hand, Fig. 4.3(b) is useful mainly in describing changes of basis sets.

PROBLEM 4.3 Prove that $\hat{R}_\phi$ is a unitary operator and that $\bar{\bar{R}}_\phi$ is a unitary matrix:

$$\hat{R}_\phi^{-1} = \hat{R}_\phi^{+} \tag{4.9a}$$

$$R_{ji}^{-1} = R_{ij}^{*} \tag{4.9b}$$

Is $\bar{\bar{M}}_\phi$ unitary? (See Eq. (4.8)).

4.2 Usefulness of the Matrix Representation of the Rotation Operation

The value of the matrix representation of the rotation or reflection operation becomes apparent upon examination of the process of rotating an arbitrary vector $|u\rangle$ (rather than a unit vector) in real space.

$$|v\rangle = \hat{R}_\phi |u\rangle \tag{4.10}$$

says that the vector $|u\rangle$ is rotated by the operator $\hat{R}_\phi$ through ϕ in the counter-clockwise direction about $|3\rangle$, and consequently becomes $|v\rangle$. By the closure condition applied in several places, exactly as in Eq. (3.8),

$$v_i = \sum_j^3 R_{ij} u_j \quad i = 1, 2, 3 \tag{4.11a}$$

or in matrix (shorthand) form,

$$\begin{pmatrix} v_1 \\ v_2 \\ v_3 \end{pmatrix} = \bar{\bar{R}}_\phi \begin{pmatrix} u_1 \\ u_2 \\ u_3 \end{pmatrix} \tag{4.11b}$$

In the case of rotation through an angle ϕ about $|3\rangle$ this is written out long-hand as

$$\begin{aligned} v_1 &= u_1 \cos\phi - u_2 \sin\phi \\ v_2 &= u_1 \sin\phi + u_2 \cos\phi \\ v_3 &= u_3 \end{aligned} \tag{4.11c}$$

Equation (4.11) tells us that once we have chosen a basis system and have generated a three-space representation of the operator $\hat{R}_\phi$, we can immediately determine how *any* vector in three-space is transformed by $\hat{R}_\phi$ simply by seeing how its components transform. This immensely useful result carries over, of course, to the case of a rotation about an arbitrary axis or reflection across any mirror plane; one simply must calculate the appropriate transformation matrix once and for all.

Note again, as with Eqs (3.6) and (3.8), that while we sum over the first subscript on the matrix representation of $\hat{R}_\phi$ in Eq. (4.4), we sum over the second index in Eq. (4.11a). This happens because the two equations describe different processes. In Eq. (4.4), we see how basis vectors are affected by a rotation; Eq. (4.11a), on the other hand, tells us how the *components* of an arbitrary vector are changed when the vector is rotated. These components form a representation of $|v\rangle$, but we should not expect a representation of a vector to behave exactly like the *vector* itself.

PROBLEM 4.4 Combining Eqs (4.1), (4.10), and (4.11a) gives

$$\hat{R}_\phi |u\rangle = \sum_{ij}^{3} |i\rangle R_{ij} u_j$$

Is this consistent with Eq. (4.4) for the case in which $|u\rangle$ is a unit vector? Put Eq. (4.3) into the shorthand notation of Eq. (4.11b).
*PROBLEM 4.5 What is the representation of $|1\rangle$ in the $\{|1\rangle, |2\rangle, |3\rangle\}$ basis system? What becomes of this representation under the transformation $\hat{R}_\phi |1\rangle = |1'\rangle$? See Problem 2.8.
*PROBLEM 4.6 What is the connection between Eq. (4.4) and Eq. (2.24)?

$$\hat{R}|i\rangle = |i'\rangle = \sum_{j}^{3} |j\rangle R_{ji} \tag{4.4}$$

$$|i'\rangle = \sum_{j}^{3} |j\rangle\langle j|i'\rangle = \sum_{j} C_{ji} |j\rangle \tag{2.24a}$$

See the comment following Eq. (3.8a). Use Eq. (2.24) to find $\bar{\bar{R}}_\phi$.

If we adopt the passive point of view, Eq. (4.11) tells us how a vector representation changes under a clockwise rotation of the coordinate axes by ϕ, Fig. (4.3b). An operator $\hat{R}_\phi$ which could describe such a re-alignment of the axes must, of course, be unitary. It is instructive now to examine the effect of such a change of basis system upon the matrix representation of an arbitrary operator $\hat{A}$, which need not be unitary.

We shall re-write Eq. (4.11a) as

$$u'_i = \sum_j R_{ij} u_j \tag{4.12a}$$

to emphasize our taking the passive approach. This says that as the coordinate axes are rotated, or equivalently, as the $\{|i\rangle\}$ basis system is replaced by the $\{|i'\rangle\}$, then the representation of the ket vector $|u\rangle$ changes from $\begin{pmatrix} u_1 \\ u_2 \\ u_3 \end{pmatrix}$ to $\begin{pmatrix} u'_1 \\ u'_2 \\ u'_3 \end{pmatrix}$

In simpler notation

$$\tilde{u}' = \bar{\bar{R}} \tilde{u} \tag{4.12b}$$

where the curly bar here denotes a column vector. Similarly for the representation of another vector $|w\rangle$

$$\tilde{w}' = \bar{\bar{R}} \tilde{w} \tag{4.12c}$$

Consider now an operator $\hat{A}$ which relates $|u\rangle$ to $|w\rangle$; $\hat{A}|u\rangle = |w\rangle$. In the $\{|i\rangle\}$ basis system, this relationship is represented as

$$w_i = \sum_j A_{ij} u_j, \text{ or } \tilde{w} = \bar{\bar{A}} \tilde{u} \tag{4.13a}$$

Since a change of basis should have no effect upon the information content of the original equation $\hat{A}|u\rangle = |w\rangle$, in the $\{|i'\rangle\}$ system Eq. (4.13a) becomes

$$\tilde{w}' = \bar{\bar{A}}' \tilde{u}' \tag{4.13b}$$

We wish to find the relationship between the matrix representations $\bar{\bar{A}}'$ and $\bar{\bar{A}}$ of $\hat{A}$.

With the aid of Eq. (4.12), Eq. (4.13b) becomes

$$\bar{\bar{R}} \tilde{w} = \bar{\bar{A}}' \bar{\bar{R}} \tilde{u} \tag{4.13c}$$

Multiplying Eq. (4.13a) on the left by the matrix $\bar{\bar{R}}$, however, yields $\bar{\bar{R}} \tilde{w} = \bar{\bar{R}} \bar{\bar{A}} \tilde{u}$. It follows that

$$\boxed{\bar{\bar{A}}' = \bar{\bar{R}} \bar{\bar{A}} \bar{\bar{R}}^{-1} = \bar{\bar{R}} \bar{\bar{A}} \bar{\bar{R}}^+} \tag{4.14}$$

where the last step depends on the unitarity of $\hat{R}$. Equation (4.14) describes a *similarity transformation* of the matrix $\bar{\bar{A}}$ under the change of basis designated by $\bar{\bar{R}}$.

With our passive approach, the *operator* (as opposed to matrix)

REPRESENTATIONS OF AN OPERATOR IN THREE SPACE

equation

$$\boxed{\hat{A}' = \hat{R}\hat{A}\hat{R}^{-1}} \tag{4.15a}$$

extracted from Eq. (4.14) has no obvious significance. From the active viewpoint, however, it is meaningful. Consider

$$\hat{A}|u\rangle = \hat{A}(\hat{R}^{-1}\hat{R})|u\rangle = |w\rangle$$

therefore

$$\hat{R}\hat{A}|u\rangle = (\hat{R}\hat{A}\hat{R}^{-1})\hat{R}|u\rangle = \hat{R}|w\rangle \tag{4.15b}$$

Thus if some $\hat{R}$ influences every vector $|u\rangle$ and $|w\rangle$ of the space, $|u\rangle \to \hat{R}|u\rangle$, etc., the nature of $\hat{A}|u\rangle = |w\rangle$ is preserved only if $\hat{A} \to \hat{R}\hat{A}\hat{R}^{-1} = \hat{A}'$ also. $\hat{R}\hat{A}\hat{R}^{-1}$ is, however, not as easy to visualize as is $\hat{R}|u\rangle$.

4.3 Subspaces in Real Space Invariant with Respect to Rotations about an Axis

Recall Eq. (4.4) and (4.7)

$$\hat{R}_\phi |i\rangle = \sum_j^3 |j\rangle R_{ji}, \qquad R_{ji} = \langle j|\hat{R}_\phi|i\rangle \tag{4.4}$$

$$\bar{\bar{R}}_\phi = \begin{pmatrix} \cos\phi & -\sin\phi & 0 \\ \sin\phi & \cos\phi & 0 \\ 0 & 0 & 1 \end{pmatrix} \tag{4.7}$$

Equation (4.4) tells us that, in general, if we wish to rotate a unit vector, or equivalently, if we wish to describe the unit vector after rotating the co-ordinate system in the opposite direction, we end up with a linear combination of *all* the basis vectors. The matrix elements of Eq. (4.7) tell how much of each individual basis vector is admixed into the sum. This point is discussed in the paragraph following Eq. (3.23).

The representation Eq. (4.7) is of an operation which rotates vectors counter-clockwise about the $|3\rangle$ axis. The fact that $R_{13} = R_{31} = R_{23} = R_{32} = 0$ tells us, through Eq. (4.4), that if we operate with $\hat{R}_\phi$ upon any vector $|u\rangle$ which happens to lie entirely in the $|1\rangle - |2\rangle$ plane, we end up with a vector still in the $|1\rangle - |2\rangle$ plane; thus if $|u\rangle$ originally has no component along $|3\rangle$, then $|u'\rangle = \hat{R}_\phi |u\rangle$ will also have no component along $|3\rangle$. And $\hat{R}_\phi$ leaves any vector which is colinear with $|3\rangle$ unchanged. This all follows naturally from our original choice that $|3\rangle$ should lie along the axis of rotation (or vice versa).

The central point considered in the above paragraph can be illustrated by the case of a solid, cylindrically symmetric wheel.

For this system, there is one and only one special direction in Euclidean space, that of the axis of rotation. Radially directed vectors will transform into other radially directed vectors under rotation of the wheel, and vectors pointing along the axis are invariant. Thus by a judicious selection of the orientation in three-space of our basis system relative to the symmetry in three-space of our physical system (e.g. if there is an axis of rotation we direct one of our three unit vectors along it), we break three-space into subspaces, the vectors of which do not mix under certain spatial transformations, (i.e. rotation about the axis of rotation).

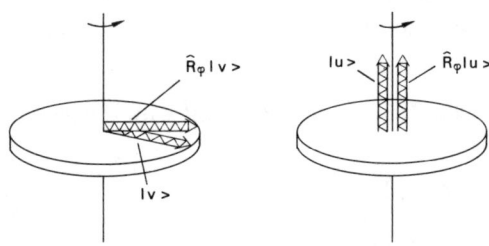

Figure 4.4.

By 'orienting' space, that is, by choosing a basis set or co-ordinate system suitable to the symmetry of the problem at hand, we can perform a useful separation of variables, and greatly simplify the analysis. The fundamental idea, then, is this: the possible invariant subspaces are determined by, and are a direct reflection upon, both the symmetry of the physical system (e.g. rotational symmetry of a wheel) and of the operations with which we are concerned (rotation about the axis of symmetry). The two, of course, are intimately related.

*PROBLEM 4.7 When is Eq. (4.4) an eigenequation?
*PROBLEM 4.8 In generating the representation Eq. (4.7), we have used the basis vectors $|1\rangle$, $|2\rangle$ and $|3\rangle$. What would we have found if we had tried to use $|1\rangle$ and $|2\rangle$ alone? $|3\rangle$ alone? $|1\rangle$ alone? $|1\rangle$, $|2\rangle$, and $(|1\rangle+|2\rangle)/\sqrt{2}$?

A final, but most fundamental issue: to say that the $|1\rangle$ and $|2\rangle$ unit vectors, and linear combinations thereof, are *mixed* under the rotation $\hat{R}_\phi$, while $|3\rangle$ remains unchanged, is equivalent to saying that $|1\rangle$ and $|2\rangle$ span one subspace invariant relative to $\hat{R}_\phi$, and $|3\rangle$ spans another. And, moreover, because the $|1\rangle$, $|2\rangle$ pair display

REPRESENTATIONS OF AN OPERATOR IN THREE-SPACE   53

behaviour under $\hat{R}_\phi$ which is clearly different from that of $|3\rangle$ (the dimensionalities of the two subspaces are not even the same!) we say that $|1\rangle$ and $|2\rangle$ are of one *symmetry type*, while $|3\rangle$ is of another. By analogy, an odd function is one which is changed in sign under the parity operator, while an even function is invariant. Similarly, $|1\rangle$ and $|2\rangle$ are of such a symmetry that they are mixed together in a specific fashion under $\hat{R}_\phi$, while $|3\rangle$ is invariant to $\hat{R}_\phi$, or left alone.

In summary, $|1\rangle$ and $|2\rangle$ span one subspace invariant relative to $\hat{R}_\phi$ and $|3\rangle$ spans another. The transformation behaviour under (symmetry-type relative to) $\hat{R}_\phi$ of any vector *lying entirely in either invariant subspace* is especially perspicuous; an inhabitant of one, moreover, is *automatically orthogonal* to any member of the other.

4.4 Generalizing to Function Space

It may seem that we have gone through a fair amount of labour here just to display some well-known results. We have done so for two reasons; first of all, we shall find that the central relationship of our later development will be a direct generalization of Eq. (4.4) to the case of a space spanned by basis *functions* $\{|i\rangle\}$ rather than the $\{|1\rangle, |2\rangle, |3\rangle\}$ directed line segments of this chapter. We shall soon define an operator $\hat{P}_R$, related to $\hat{R}$, whose representation is shown by

$$\hat{P}_R|i\rangle = \sum_j |j\rangle \Gamma_{ji}(\hat{R}), \qquad \Gamma_{ji}(\hat{R}) \equiv \langle j|\hat{P}_R|i\rangle \tag{4.16}$$

which is obviously very much like

$$\hat{R}|i\rangle = \sum_j^3 |j\rangle R_{ji} \tag{4.4}$$

$\hat{P}_R$ is an operator which somehow mirrors, *in function space*, events (such as rotations) occuring in real space.

Secondly, the matrix representation $\bar{\bar{R}}_\phi$ of $\hat{R}_\phi$ was seen to be in block-diagonal form:

$$\bar{\bar{R}}_\phi = \begin{pmatrix} R_{11} & R_{12} & 0 \\ R_{21} & R_{22} & 0 \\ 0 & 0 & R_{33} \end{pmatrix} \tag{4.17}$$

We argued that the block-diagonal form suggests that we have found a co-ordinate system appropriate for the physical symmetry of the system; that is, $\bar{\bar{R}}_\phi$ is of the form of Eq. (4.17) because, and only because, the $|3\rangle$ axis is chosen to lie along the axis of rotation. We shall see that the same sort of argument also applies in the more abstract function spaces. There the proper choice of $\{|i\rangle\}$ will cause the $\Gamma_{ji}(\hat{R})$ of Eq. (4.16) to be in block-diagonal form. This, in turn, implies that the function space spanned by $\{|i\rangle\}$ can be broken down into disjoint invariant subspaces, the members of which are not inter-mixed by $\hat{P}_R$. The nature of these invariant subspaces, and the symmetries of the functions spanning them, is the primary concern of quantum mechanical group representation theory.

4.5 A Representation of the Group C_{3v}

Before passing on to function space, and deciding what $\hat{P}_R$ is and how it should be related to $\hat{R}$, we can go one large step further with our real-space example.

Recall our discussion in Chapter 1 of the symmetry group of the ammonia molecule, or of the plane equilateral triangle. The elements of this group, C_{3v}, are the symmetry operators $\{\hat{I}, \hat{R}_{120}, \hat{R}_{240}, \hat{M}_I, \hat{M}_{II}, \hat{M}_{III}\}$. From Eq. (4.7) and (4.8), we find that the following six matrices are the representations of these particular operators generated by $\{|1\rangle, |2\rangle, |3\rangle\}$ basis set:

$$\bar{\bar{I}} = \begin{pmatrix} 1 & 0 & 0 \\ 0 & 1 & 0 \\ 0 & 0 & 1 \end{pmatrix}, \quad \bar{\bar{R}}_{120} = \tfrac{1}{2} \begin{pmatrix} -1 & -\sqrt{3} & 0 \\ \sqrt{3} & -1 & 0 \\ 0 & 0 & 2 \end{pmatrix}$$

$$\bar{\bar{R}}_{240} = \tfrac{1}{2} \begin{pmatrix} -1 & \sqrt{3} & 0 \\ -\sqrt{3} & -1 & 0 \\ 0 & 0 & 2 \end{pmatrix}, \quad \bar{\bar{M}}_I = \begin{pmatrix} -1 & 0 & 0 \\ 0 & 1 & 0 \\ 0 & 0 & 1 \end{pmatrix} \quad (4.18)$$

$$\bar{\bar{M}}_{II} = \tfrac{1}{2} \begin{pmatrix} 1 & \sqrt{3} & 0 \\ \sqrt{3} & -1 & 0 \\ 0 & 0 & 2 \end{pmatrix} \quad \bar{\bar{M}}_{III} = \tfrac{1}{2} \begin{pmatrix} 1 & -\sqrt{3} & 0 \\ -\sqrt{3} & -1 & 0 \\ 0 & 0 & 2 \end{pmatrix}$$

Two points concerning Eq. (4.18) require immediate attention. First, term by term matrix multiplication will demonstrate that this set of six matrices forms a group, and has exactly the same group structure as has $\{\hat{I}, \hat{R}_{120}, \hat{R}_{240}, \hat{M}_I, \hat{M}_{II}, \hat{M}_{III}\}$. Thus if $\hat{M}_I \cdot \hat{R}_{240} = \hat{M}_{III}$, then for the matrices, $\bar{\bar{M}}_I \cdot \bar{\bar{R}}_{240} = \bar{\bar{M}}_{III}$ also, and vice versa. The set of six matrices also constitutes a group of type C_{3v}! Instead of operators, the elements of this group are matrices; but the *group* properties are identical to those of the original set of symmetry operators.

Two groups with exactly the same internal structure (regardless of the names, labels, or form of the elements) are said to be *isomorphic* to one another. We have here, for C_{3v}, a group of

REPRESENTATIONS OF AN OPERATOR IN THREE-SPACE   55

representative matrices isomorphic to a group of symmetry operators. But beware! As we shall soon see, every set of matrices representing a group of symmetry operators will itself always form a group; but a representative group is not necessarily isomorphic to the original symmetry group!

The second point of interest concerning Eq. (4.18) is that the six matrices are all in the same block-diagonal form. What has been said in the preceding sections about the block-diagonal form can apply, then, not only to a single representative matrix, but even to groups of matrices.

Each matrix in Eq. (4.18) is of the form

$$\begin{pmatrix} R_{11} & R_{12} & 0 \\ R_{12} & R_{22} & 0 \\ \hline 0 & 0 & R_{33} \end{pmatrix} \qquad (4.19)$$

If we let Eq. (4.19) represent one operator in C_{3v}, and if a similar block-diagonal matrix $\bar{\bar{S}}$ represents a second, we find for the product $\bar{\bar{R}}\bar{\bar{S}}$

$$\begin{pmatrix} R_{11} & R_{12} & 0 \\ R_{21} & R_{22} & 0 \\ \hline 0 & 0 & R_{33} \end{pmatrix} \begin{pmatrix} S_{11} & S_{12} & 0 \\ S_{21} & S_{22} & 0 \\ \hline 0 & 0 & S_{33} \end{pmatrix} = \qquad (4.20)$$

$$\begin{pmatrix} R_{11}S_{11}+R_{12}S_{21} & R_{11}S_{12}+R_{12}S_{22} & 0 \\ R_{21}S_{11}+R_{22}S_{21} & R_{21}S_{12}+R_{22}S_{22} & 0 \\ \hline 0 & 0 & R_{33}S_{33} \end{pmatrix}$$

Thus in carrying out a matrix multiplication, the upper-left-hand two-by-two block is totally independent of the lower-right-hand one-dimensional block! No information at all is lost if we separate each of the matrices of Eq. (4.18) into two parts, and deal with the set of six two-by-two matrices, and then the set of one-dimensional matrices, independently:

$$\bar{\bar{I}} = \begin{pmatrix} 1 & 0 \\ 0 & 1 \end{pmatrix} \qquad \bar{\bar{R}}'_{120} = \tfrac{1}{2}\begin{pmatrix} -1 & -\sqrt{3} \\ \sqrt{3} & -1 \end{pmatrix}$$

$$\bar{\bar{R}}'_{240} = \tfrac{1}{2}\begin{pmatrix} -1 & \sqrt{3} \\ -\sqrt{3} & -1 \end{pmatrix} \qquad \bar{\bar{M}}'_{I} = \begin{pmatrix} -1 & 0 \\ 0 & 1 \end{pmatrix} \qquad (4.21)$$

$$\bar{\bar{M}}'_{II} = \tfrac{1}{2}\begin{pmatrix} 1 & \sqrt{3} \\ \sqrt{3} & -1 \end{pmatrix} \qquad \bar{\bar{M}}'_{III} = \tfrac{1}{2}\begin{pmatrix} 1 & -\sqrt{3} \\ -\sqrt{3} & -1 \end{pmatrix}$$

and

$$\bar{\bar{I}}''=(1) \qquad \bar{\bar{R}}''_{120}=(1)$$
$$\bar{\bar{R}}''_{240}=(1) \qquad \bar{\bar{M}}''_{I}=(1) \qquad (4.22)$$
$$\bar{\bar{M}}''_{II}=(1) \qquad \bar{\bar{M}}''_{III}=(1)$$

We say that the set of matrices, Eq. (4.18), has been *reduced* to two sets of smaller matrices; if we choose to name the set of matrices of Eq. (4.21) 'E', the set in Eq. (4.22) 'A_1', and the representation generated by $\{|1\rangle, |2\rangle, |3\rangle\}$ 'Γ', then we can describe this reduction process for C_{3v} as:

$$\Gamma \to A_1 + E \qquad (4.23)$$

The reason for the labels Γ, E, and A_1 will be explained in Section (8.3).

What has happened is this: We have employed three real-space unit vectors to generate matrix representations of the group of symmetry operators of the equilateral triangle. We have, however, been clever enough not to orient the triangle in space arbitrarily; rather we have realized that there exist optimal alignments of the object, with one of the unit vectors (in this case $|3\rangle$) lying along the axis of three-fold rotation (i.e. through 0°, 120°, or 240°); this axis also contains all three mirror planes. Such a selection of basis vectors separates three-space into two orthogonal, invariant subspaces. The basis vector sets of these two independent subspaces, namely $\{|1\rangle, |2\rangle\}$ and $\{|3\rangle\}$, generate two distinct, non-equivalent representations, called E and A_1, of C_{3v}.

Further investigation, by trial and error or otherwise, will convince one that there is no better alignment of the triangle in space, and there is no way possible to diagonalize fully all six matrices simultaneously. Therefore, while the Γ of Eq. (4.18) is a *reducible representation*, A_1 and E are *irreducible representations*.

In Section (4.3) we argued that $\{|1\rangle, |2\rangle\}$, and $\{|3\rangle\}$ display inherently different symmetry behaviour under *any* rotation or reflection about $|3\rangle$. C_{3v} consists of a small subset of the set of all these rotations and reflections, and so the separation of three-space into two invariant subspaces, of A_1 and E symmetry, remains meaningful even for a finite group of operators like C_{3v}. A_1 and E, moreover, refer to general symmetry types, and need not at all be restricted to descriptions of C_{3v}. It simply happens that two of the irreducible representations of C_{3v} are A_1 and E.

*PROBLEM 4.9 Is either an A_1 or an E matrix representation of C_{3v} isomorphic to C_{3v} itself?

REPRESENTATIONS OF AN OPERATOR IN THREE-SPACE

What we have been discussing in this chapter is central to quantum mechanical applications. Before arriving at that final goal, however, we must generalize our approach a bit, and decide how all of the above should be phrased in the language of function spaces, rather than of real space. The analysis of this section should provide helpful models for such future developments.

*PROBLEM 4.10 Which of the basis sets of Problem 4.8 generate irreducible representations of C_{3v}?

*PROBLEM 4.11 Does $\{\hat{R}_\phi, \hat{M}_\phi, \text{all } \phi\}$ form a group?

4.6 Summary

1. A rotation of an object or coordinate system can be described by the unitary operator $\hat{R}$. $\hat{R}$ influences a basis vector $|i\rangle$ as

$$\hat{R}|i\rangle = |i'\rangle = \sum_{j=1}^{3} R_{ji}|j\rangle, \quad R_{ji} = \langle j|\hat{R}|i\rangle$$

and an arbitrary vector $|u\rangle$ as $\hat{R}|u\rangle = |u'\rangle$. The representation of $|u\rangle$ in the $\{|i\rangle\}$ basis system transforms as

$$u'_i = \sum_j R_{ij} u_j \quad \text{or} \quad \vec{u}' = \overline{\overline{R}} \vec{u}$$

where $\vec{u}$ is a column vector.

2. Under a rotation of space such that all $|u\rangle \to |u'\rangle = \hat{R}|u\rangle$, the operator relation $\hat{A}|u\rangle = |w\rangle$ remains meaningful if

$$\hat{A} \to \hat{A}' = \hat{R}\hat{A}\hat{R}^{-1};$$

then $\hat{A}'|u'\rangle = |w'\rangle$.

3. The block-diagonal form $\overline{\overline{R}}_\phi$ suggests that there is an optimal choice of co-ordinate axes, determined by the nature of the operation and by the symmetry of the physical system under consideration. The use of this co-ordinate system allows us to partition vector space into invariant subspaces of different symmetry types, whose vectors behave differently under the influence of the operator.

4. One can form the representation not only of a single operator $\hat{R}$, but even of a group $\{\hat{R}\}$ such as C_{3v}. The observation that block-diagonal form is maintained under matrix multiplication suggests separating the blocks for each $\overline{\overline{R}}$ into the *direct sum* of distinct matrices:

$$\begin{pmatrix} \boxed{R^1} & & 0 \\ & \boxed{R^2} & \\ 0 & & \ddots \end{pmatrix} \to \overline{\overline{R}}^1 + \overline{\overline{R}}^2 + \cdots$$

Then the first blocks for all the $\hat{R}$'s, $\{\bar{\bar{R}}^1\}$ constitute a representation of the group, as do the second blocks $\{\bar{\bar{R}}^2\}$, etc. $\{|1\rangle, |2\rangle\}$ generate an E-type irreducible representation of C_{3v}, in particular, and A_1 is generated by a basis vector displaying fundamentally different symmetry properties under C_{3v}, $|3\rangle$).

CHAPTER 5

Review of Some Quantum Mechanics

The physical nature of a system is fully specified in terms of its state $|\psi\rangle$ in Hilbert space, and an actual measurement of the attribute A can be described mathematically by operating upon $|\psi\rangle$ with an appropriate Hermitian operator $\hat{A}$. Measurements of the physical attributes A and B of a quantum mechanical system may (or may not) be mutually incompatible. A pure state is one which is simultaneously an eigenstate of a complete set of compatible (commuting) operators $\{\hat{A}, \hat{B}, \ldots\}$.

In general, the measurement process described by $\hat{A}$ will 'interfere with' or disrupt a system in the state $|\psi\rangle$:

$$\hat{A}: |\psi\rangle \to |\psi'\rangle$$

$\hat{A}$ will leave the system undisturbed if and only if $|\psi\rangle$ happened to be an eigenstate of $\hat{A}$, such as $|a_n\rangle$, originally:

$$\hat{A}|a_n\rangle = a_n|a_n\rangle$$

In either case, a single measurement of A will yield an eigenvalue of $\hat{A}$, and after the measurement, the system will reside in the corresponding eigenstate of $\hat{A}$. For a system in the state $|\psi\rangle$, the probability that measurement of $\hat{A}$ will yield the particular eigenvalue a_n can be determined only statistically, and is

$$|\langle a_n|\psi\rangle|^2$$

Assuming the principle of the superposition of states, the Schroedinger equation

$$i\hbar \frac{d}{dt}|\psi(t)\rangle = \hat{\mathcal{H}}(t)|\psi(t)\rangle$$

is derived via the time development operator, $\hat{T}(t_2, t_1)$, and then integrated to give the time-independent Schroedinger equation

$$\mathscr{H}|\psi_n\rangle = E_n|\psi_n\rangle$$

where $\{|\psi_n\rangle\}$ are eigenstates of the Hamiltonian.

It is shown that the expressions 'diagonalizing the Hamiltonian (matrix)', 'solving the energy eigenequation', and 'solving the secular determinant for the Hamiltonian' are synonymous, and that diagonalizing the Hamiltonian is equivalent to rotating Hilbert space (i.e. the co-ordinate axes of function space) into an optimal (principal axis) alignment. Symmetry arguments or formal group theoretic techniques can, among other things, aid greatly in the diagonalization process, and indicate the presence of degeneracies in energy levels.

Finally, the principal ideas of the chapter are illustrated in an exact derivation of a simple form of the Breit–Rabi equation, and the same results are then obtained via perturbation theory.

For one who has worked through the first half of Merzbacher, Messiah, etc., this should be light fare. In any case, a flow diagram, Table 5.1 may be of some use.

The central problem of quantum mechanics is to find the Hamiltonian (total energy) operator which suitably describes a system under consideration, and then to find the energy eigenstates of the system—i.e. to diagonalize this Hamiltonian. The eigenequation of the Hamiltonian, or the 'time-independent Schroedinger equation', is an equation of motion for the system; calculation of the eigenvalues and eigenvectors yields not only the positions of the stationary energy levels, but also a means to determine the system's time development. One can employ the eigenstates, moreover, to find items such as transition energies and rates, and to examine the effects upon the system of various static perturbations.

The way in which group theory can simplify and elucidate all aspects of these problems is the main topic of this book. Here we shall go over some of the basic ideas upon which elementary quantum mechanics is built, both as a quick review, Sections 5.1–5.6, and as an introduction to some elementary symmetry-related concepts, Sections 5.7–5.10, that will be of use later.

5.1 States and Observables

The quantum mechanical view of nature differs from the classical view in one fundamental but all-important respect: quantum theory incorporates into its foundation the *experimentally* observed, inescapable fact that *some kinds of measurements on a*

REVIEW OF SOME QUANTUM MECHANICS

Table 5.1

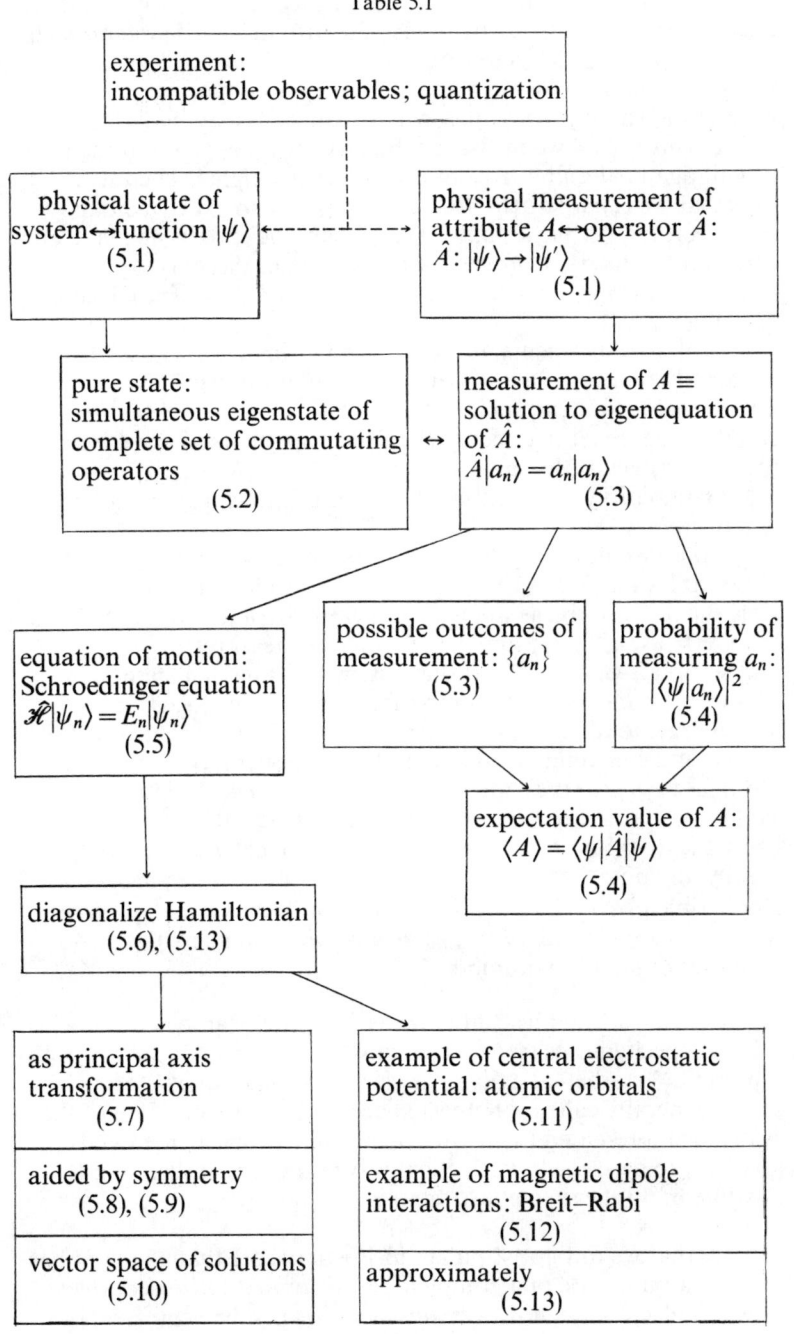

sub-microscopic system may be mutually incompatible. That is, there may be certain kinds of attributes of an atomic-sized system, such as X and Y, with the following remarkable property. We can measure X and arrive at some value x, and then measure Y finding the value y; if we measure X again, we may find some $x' \neq x$. This is as if we measure a box's weight, length, and then weight again, and find two totally different weights. The theory says that we *cannot* attribute the discrepancy to experimental error. Rather, no matter how carefully and delicately we make our observation, the very process of measurement disrupts, or 'interferes' with the system sufficiently to actually alter its initial state of being.

Not all measurements, however, are mutually incompatible. In fact, there may be a number of particular observables or attributes of a system, such as the set $\{A, B, \ldots\}$, for which even a complicated, repetitive sequence of measurements of these attributes always yields the same values, $\{a, b, \ldots\}$. A largest possible set of compatibly measurable attributes is known as a *complete set of observables*.

Suppose we have a system with a complete set of compatible observables $\{A, B, \ldots, N\}$, and suppose we know that the possible outcomes of measurement of the ath attribute, A, are the numbers $\{a_1, a_2 \ldots\}$; similarly for B, one measures $\{b_1, b_2, \ldots\}$, and so on for $C, D, \ldots N$. If the system is sitting in a state such that measurements of the compatible observables $\{A, B, \ldots N\}$ (e.g. energy, total angular momentum, and z-components of orbital and spin angular momenta for an electron on a hydrogen atom) always give the same values $\{a_j, b_k, \ldots n_p\}$, the system is said to be in the *pure state* characterized by the values $\{a_j, b_k, \ldots n_p\}$, or by the quantum numbers $\{j, k, \ldots p\}$. We can thus uniquely specify any pure state as $|jk \ldots p\rangle$ with respect to the particular compatible observables $\{A, B, \ldots N\}$; for brevity we may even write the specific state $|jk \ldots p\rangle$ as $|m\rangle$, where m refers to the entire set of quantum numbers.

POSTULATE 5.1 The physical nature of a system can be described fully by specifying its state. The ψth state, written $|\psi\rangle$, can be represented mathematically by a vector in some suitable vector space, generally called a Hilbert space. Every possible state of the system is represented by its own vector in this space. Identical systems are described by identical wave functions (except for a possible multiplicative phase factor).

Note that we did not stipulate in Postulate 5.1 that a system has to be in a pure state of a complete set of compatible observables; as we shall soon see, however, *any* state $|\psi\rangle$ can be expressed as a

REVIEW OF SOME QUANTUM MECHANICS

linear combination of pure states:

$$|\psi\rangle = \sum_m |m\rangle\langle m|\psi\rangle = \sum_m c_m |m\rangle \tag{5.1}$$

The idea, suggested by Eq. (5.1), that a linear combination of physically meaningful states corresponds to another physically meaningful state is the fundamental *Postulate of the Superposition of States*.

A space in which any state of interest can be decomposed as in Eq. (5.1) and for which a scalar product is defined is loosely called a Hilbert space.

5.2 Compatible Operators Commute; Pure States

It is usually convenient to choose a complete set of pure states to serve as a basis system for spanning Hilbert space. So we must find a way of generating pure states, which amounts to determining the conditions under which two measurements are compatible.

Consider a complete set of basis states, $\{|pq\rangle\}$, chosen to be simultaneous eigenstates of the two operators $\hat{A}$ and $\hat{B}$ which represent mathematically the observables A and B:

$$\boxed{\begin{aligned}\hat{A}|pq\rangle &= a_p|pq\rangle \\ \hat{B}|pq\rangle &= b_q|pq\rangle\end{aligned}} \tag{5.2a}$$

If the system is in a mutual eigenstate $|pq\rangle$ of $\hat{A}$ and $\hat{B}$, then it is not disturbed by either operator. From Eq. (5.2a),

$$\hat{B}\hat{A}|pq\rangle = \hat{B}a_p|pq\rangle = a_p\hat{B}|pq\rangle = a_p b_q|pq\rangle \tag{5.2b}$$

Operating on the second line of Eq. (5.2a) with $\hat{A}$ leads to a similar result, from which

$$(\hat{A}\hat{B} - \hat{B}\hat{A})|pq\rangle \equiv [\hat{A}, \hat{B}]|pq\rangle = 0$$

or, since this applies to *any* state in $\{|pq\rangle\}$

$$[\hat{A}, \hat{B}] = 0 \tag{5.3}$$

We have demonstrated

THEOREM 5.1 Pure states are simultaneously eigenvectors of a complete set of commutating operators; similarly, observables are compatible if and only if the corresponding operators commute.

If the pure states $\{|m\rangle\}$ are to serve as a basis system for describing vectors in the function spaces of quantum mechanics, as in Eq. (5.1), they should both be orthonormal and obey a closure relation:

$$\langle m|m'\rangle = \delta_{mm'} \tag{5.4a}$$

$$\sum_m |m\rangle\langle m| = \hat{1} \tag{5.4b}$$

or

$$\langle m|m'\rangle = \delta(m - m') \tag{5.4c}$$

$$\int |m\rangle\,\mathrm{d}m\langle m| = \hat{1} \tag{5.4d}$$

depending upon whether the $\{|m\rangle\}$ are discrete or continuous basis vectors.

If there are several distinct sets of possible basis vectors spanning the same space, $\{|m\rangle\}$ and $\{|\alpha\rangle\}$,† say, then the unit basis vectors can be expressed in terms of one another via the transformation relationships:

$$\boxed{\begin{aligned} |\alpha\rangle &= \sum_m |m\rangle\langle m|\alpha\rangle \\ |m\rangle &= \sum_\alpha |\alpha\rangle\langle \alpha|m\rangle \end{aligned}} \tag{5.5}$$

The numbers $\langle m|\alpha\rangle = \langle \alpha|m\rangle^*$ constitute the entries in the 'transformation matrix' which relates the two sets. Equations (5.5) and (2.24) show that, at least formally, it is a simple matter to switch from one basis system to another.

A final word of caution: there are several different ways of defining the term 'pure state'. In particular, with treatments which lead into discussions of quantum statistics (which involves extending the approach of quantum mechanics into the domain of finite temperatures), other definitions may be employed.

5.3 Measured Values are Eigenvalues

After defining pure states as the mutual eigenvectors of a complete set of commutating operators, we suggested that a physically significant general quantum state $|\psi\rangle$ can be expanded as a linear combination of such pure states, Eq. (5.1). This formalism, and the statistical interpretation of $|\psi\rangle$, are based upon important properties of the eigenvectors of Hermitian operators, which we examine in this section and the next.

Our introductory statement, that an observation may interfere

---

†The two sets of pure states might be the eigenstates of two non-equivalent sets of compatible observables; but even then, they should be obtainable from one another by a transformation in Hilbert space, Eq. (5.5).

with the state-of-being of a system, may be symbolized as

$$\hat{A}:|\psi\rangle \to |\psi'\rangle \quad (5.6)$$

Equation (5.6) looks a bit like Eq. (3.1), which we used to introduce the concept of operator; it says, however, a great deal more, by implying the fundamental physical idea that the *measurement process* can *disrupt* a *state*. Suppose a system is left in the state $|\psi\rangle$ following some observation Y. If we thereafter make a measurement of X, where X happens to be incompatible with Y, the system will be knocked out of the state $|\psi\rangle$ and into some new state $|\psi'\rangle$.

It may, however, happen that the operator $\hat{A}$ leaves the particular state $|a_n\rangle$ invariant, $\hat{A}:|a_n\rangle \to |a_n\rangle$; the measurement does not disrupt the state, and the description of the process is written

$$\hat{A}|a_n\rangle = a_n|a_n\rangle \quad (5.7)$$

Equation (5.7) is the eigenequation for the operator $\hat{A}$. The possible states $|a_n\rangle$ (and associated numbers a_n) for which Eq. (5.7) obtains usually constitute only a small subspace of the function space of all possible states of the system; but these $\{|a_n\rangle\}$ are of especial significance for the reasons expressed in the next two postulates:

POSTULATE 5.2 To every measurable attribute A associated with a system, there corresponds mathematically a linear Hermitian operator $\hat{A}$. A single measurement of the attribute A of a system will always yield a number from the set of eigenvalues $\{a_n\}$ of the eigenequation for $\hat{A}$. After the measurement, the system will reside in the corresponding eigenstate $|a_n\rangle$ of $\hat{A}$.

We stipulate that $\hat{A}$ be a Hermitian operator because we associate its eigenvalues with physically measurable quantities, and these must be real-valued; the eigenvalues of a Hermitian operator are real.

POSTULATE 5.3 The spectrum of eigenvectors of a Hermitian operator constitutes a complete, orthonormalizable basis set.

A system residing in the arbitrary state $|\psi\rangle$ can be described, as in Eq. (5.1), by

$$|\psi\rangle = \sum_m |a_m\rangle\langle a_m|\psi\rangle = \sum_m c_m|a_m\rangle \quad (5.8)$$

where $\{|a_m\rangle\}$ are the eigenvectors of some physically relevant Hermitian operator $\hat{A}$, and where the closure condition Eq. (5.4b) has been used; one particular and most important set of this sort consists of the eigenvectors $\{|\psi_n(\vec{r})\rangle\}$ of the Hamiltonian operator $\mathcal{H}$.

5.4 Statistical Interpretation

POSTULATE 5.4 $|\langle a_n|\psi\rangle|^2$ is the probability of finding the value a_n when measuring the attribute A of a system known or prepared to be in the normalized state $|\psi\rangle$. Similarly, $|\langle a_n|\psi\rangle|^2$ is the probability that a system originally in the state $|\psi\rangle$ will end up in (be scattered into) the state $|a_n\rangle$ as the consequence of a measurement of A.

Equation (5.8) and Postulates 5.3 and 5.4 lead immediately to a way of calculating the average, or expectation value, of the observable A. If a large number of identical systems are initially prepared so that all are in the state $|\psi\rangle$ (which is not necessarily a pure state!), and if A is measured, then

THEOREM 5.2 The expectation value $\langle A \rangle$ of the observable A for a system in the state $|\psi\rangle$ is

$$\langle A \rangle = \langle \psi | \hat{A} | \psi \rangle \tag{5.9}$$

The proof of this theorem is easy. Choose a basis system $\{|a_n\rangle\}$ of vectors which are eigenvectors of $\hat{A}$. Then, using these states in a closure relation, the scalar entity $\langle \psi|\hat{A}|\psi\rangle$ may be padded out as

$$\langle \psi|\hat{A}|\psi\rangle = \sum_{nn'} \langle \psi|a_{n'}\rangle \langle a_{n'}|\hat{A}|a_n\rangle \langle a_n|\psi\rangle \tag{5.10}$$

By the orthonormality of the $\{|a_n\rangle\}$, and since $|a_n\rangle$ is an eigenvector of $\hat{A}$, Postulate 5.4 leads to

$$\langle \psi|\hat{A}|\psi\rangle = \sum_n a_n |\langle \psi|a_n\rangle|^2 = \langle A \rangle$$

in agreement with Eq. (5.9).

If the system happened originally to be in the kth eigenstate of operator $\hat{A}$, then measurement of A would yield a_k with complete certainty.

PROBLEM 5.1 Explain the interpretation of $|\psi(x)|^2\,dx$ as the probability of finding a system described by $|\psi\rangle$ lying between x and $x+dx$.

PROBLEM 5.2 Heisenberg's Uncertainty Principle and the notion of pure state both involve the commutation of operators. What is the connection between the two?

REVIEW OF SOME QUANTUM MECHANICS

†PROBLEM 5.3 Consider the following situation: An ensemble of identical systems are prepared in the state $|\psi\rangle$; measurement of the attribute A is then always found to lead to the same number, a_n. Show that $|\psi\rangle$ must therefore be an eigenvector of $\hat{A}$.
[Hint: consider the expectation value of the variance of $\hat{A}$, $\langle(\hat{A}-\langle A\rangle)\rangle$].

5.5 The Schroedinger Equation

Quantum mechanical states-of-being and measurements are represented mathematically by vectors and operators in Hilbert space; the link with everyday reality is provided through the statistical interpretation. But we have yet to consider the dynamical behaviour of quantum systems.

Let $\hat{T}(t_2, t_1)$ be a linear operator which describes the time development of *any* state $|\psi(t)\rangle$ of a given physical system:

$$|\psi(t_2)\rangle = \hat{T}(t_2, t_1)|\psi(t_1)\rangle. \tag{5.11a}$$

Our state, of the form $|\psi(t_1)\rangle$ at time t_1, develops into $|\psi(t_2)\rangle$ at the later time t_2, and this process is determined and described by the operator $\hat{T}(t_2, t_1)$. $\hat{T}(t_2, t_1)$ is postulated to be a linear operator: if

$$|\psi(t_1)\rangle = c_\phi|\phi(t_1)\rangle + c_\xi|\xi(t_1)\rangle \tag{5.11b}$$

then

$$|\psi(t_2)\rangle = c_\phi \hat{T}(t_2, t_1)|\phi(t_1)\rangle + c_\xi \hat{T}(t_2, t_1)|\xi(t_1)\rangle$$

This says that if a general state is expressed as a linear combination of basis vectors, we can consider the time evolution of each part separately, and find the net result by adding up the pieces afterwards. We shall take this approach now.

Consider the operator $\hat{T}(t+\delta t, t)$ which generates the infinitesimally small change of a state occurring over the infinitesimal passage of time δt. Over δt, the state $|\psi(t)\rangle$ evolves into the slightly different state $|\psi(t+\delta t)\rangle$ in accordance with Eq. (5.11a), as

$$|\psi(t)\rangle \to |\psi(t+\delta t)\rangle = \hat{T}(t+\delta tt)|\psi(t)\rangle \tag{5.11c}$$

We can make a Taylor series expansion of the operator $\hat{T}(t+\delta t, t)$:

$$\hat{T}(t+\delta t, t) = 1 + \frac{d\hat{T}}{dt}\delta t + \cdots$$

from which

$$|\psi(t+\delta t)\rangle = \left(1 + \frac{d\hat{T}}{dt}\delta t\right)|\psi(t)\rangle \equiv \left(1 - \frac{i}{\hbar}\mathcal{H}(t)\delta t\right)|\psi(t)\rangle \quad (5.12a)$$

Equation (5.12a) defines a new operator $\mathcal{H}(t) = i\hbar(d\hat{T}/dt)$, where $\hbar = h/2\pi$, and h is Planck's constant; the factor of $(-i/\hbar)$ has been included for future convenience. $\mathcal{H}$ is seen to be the operator which tells us how $|\psi(t)\rangle$ varies in the infinitesimal span of time δt; as such, it is called the *generator of infinitesimal displacements in time*.

Similarly, we can express the change in $|\psi(t)\rangle$ over δt as

$$|\psi(t)\rangle \rightarrow |\psi(t+\delta t)\rangle = |\psi(t)\rangle + \delta t \frac{d}{dt}|\psi(t)\rangle \quad (5.12b)$$

Comparison of Eq. (5.12a) with (5.12b) reveals that the time development of $|\psi(t)\rangle$ is determined by the Schroedinger differential equation:

$$\boxed{i\hbar \frac{d|\psi(t)\rangle}{dt} = \mathcal{H}(t)|\psi(t)\rangle} \quad (5.13)$$

The equation of motion, and the physics of the problem, are therefore known once the operator $\mathcal{H}(t)$ is found.

By considering analogies with classical physics in the limiting case when $\hbar \rightarrow 0$, or by arguments based upon comparison with the special theory of relativity, we can assert that $\mathcal{H}(t)$, the generator of infinitesimal displacements in time, is none other than the Hamiltonian, or total energy operator.† Consider, for example, the one-dimensional case of a particle of momentum $\bar{p}$ and mass m in the potential $V(x)$. Classically, the total energy of the system is

$$\boxed{E = \frac{p_x^2}{2m} + V(x)} \quad (5.14a)$$

By the well-known prescriptions for converting classical entities such as p_x or x into their quantum mechanical analogues (in the 'coordinate representation'), the energy attribute Eq. (5.14a) becomes the Hamiltonian operator

$$\boxed{\mathcal{H} = -\frac{\hbar^2}{2m}\frac{d^2}{dx^2} + \hat{V}(x)} \quad (5.14b)$$

This particular Hamiltonian happens to be time independent. In such cases, it is easy to show that the differential equation Eq.

†Dirac[10], p. 110.

REVIEW OF SOME QUANTUM MECHANICS

Table 5.2

classical	quantum mechanical
x	x
p_x	$-i\hbar \dfrac{d}{dx}$
J_z	$-i\hbar \dfrac{d}{d\phi}$
$\mathcal{H}(x, p_x)$	$\mathcal{H}\left(x, -i\hbar \dfrac{d}{dx}\right)$

(5.13) reduces to the *time-independent Schroedinger equation* of the form of Eq. (5.7):

$$\boxed{\mathcal{H}|\psi_n\rangle = E_n|\psi_n\rangle} \tag{5.15a}$$

$$|\psi_n(t)\rangle = |\psi_n\rangle \exp\{-i(t-t_0)E_n/\hbar\} \tag{5.15b}$$

Equations (5.15) and (5.8) determine a system's time development; that is, a general state $|\psi\rangle$ of the system can be decomposed as a linear combination of the energy eigenvectors $\{|\psi_n\rangle\}$ obeying Eqs (5.15a) and, by the linearity of $\hat{T}(t_2, t_1)$,

$$|\psi(t)\rangle = \sum_n c_n|\psi_n\rangle \exp\{-i(t-t_0)E_n/\hbar\} \tag{5.15c}$$

All one need do is find the eigenvectors $\{|\psi_n\rangle\}$ and eigenenergies $\{E_n\}$ of $\mathcal{H}$.

*PROBLEM 5.4 Show that if the Hamiltonian is independent of time, the generator of finite displacements in time is of the form

$$\hat{T}(t, t') = \exp\{-i(t-t')\mathcal{H}/\hbar\} \tag{5.16}$$

and show that although $\mathcal{H}$ is Hermitian, $\hat{T}$ is unitary.

We close this section by asking you to prove an important theorem, which describes the time development of the expectation value of an operator.

*PROBLEM 5.5 Prove that for any Hermitian operator $\hat{A}$,

$$i\hbar \frac{d}{dt}\langle \hat{A}\rangle = \langle[\mathcal{H}, \hat{A}]\rangle + i\hbar \left\langle \frac{\partial \hat{A}}{\partial t}\right\rangle \tag{5.17}$$

If $\hat{A}$ is a dynamical operator not explicitly a function of time, then $\langle \hat{A}\rangle$ is time independent if and only if $[\mathcal{H}, \hat{A}] = 0$. In that case the observable associated with $\hat{A}$ is called a *constant of the motion*, and said to be *conserved*. At a later stage, we shall see how conservation laws are intimately connected with symmetry properties.

5.6 Diagonalizing the Hamiltonian

Our derivation of the time-independent Schroedinger equation, Eq. (5.15a), leads to the problem of finding the eigenspectrum of $\mathcal{H}$.

A unitary or Hermitian operator, such as $\mathcal{H}$, has a unique and characteristic set of eigenvectors and eigenvalues. But *any* arbitrary complete basis set spanning Hilbert space can give a matrix representation of the operator, and one may then talk about the eigenvalues and eigenvectors of the matrix, rather than of the original operator. The eigenvalues of the matrix should be independent of the basis used, and, in some sense, so also must be the eigenvectors.

As noted in Section 3.6 and in Problem 3.11, if we form a representation of the operator $\hat{A}$ by means of its *own* eigenvectors, then the resultant matrix is diagonal, and the diagonal elements are its eigenvalues. This suggests that we attack the time independent Schroedinger equation by forming a matrix representation of the Hamiltonian and diagonalizing it, by the following procedure:

1. Choose *some* complete orthonormal basis set suitable for constructing all possible states of the system. Let us assume that the basis set constitutes the eigenspectrum $\{|a_i\rangle\}$ of some physically significant operator $\hat{A}$.

2. Construct a matrix representation of the Hamiltonian $\mathcal{H}$, whose entries are the numbers $\mathcal{H}_{ij} \equiv \langle a_i|\mathcal{H}|a_j\rangle$. Since many easily manipulated basis sets $\{|a_i\rangle\}$ will be composed of vectors which are not necessarily eigenvectors of $\mathcal{H}$, the matrix representation of $\mathcal{H}$ will probably not be diagonal.

3. Perform the change of basis, indicated formally in Eq. (5.5), to another basis set in which the matrix is diagonal—i.e. transform to a co-ordinate system in which the basis vectors are eigenvectors of $\mathcal{H}$. If we can discover how to perform this transformation, we at the same time learn the eigenvectors and eigenvalues. The remainder of this section describes the method of determining the diagonalizing transformation or, equivalently, of finding the $\{|\psi_n\rangle\}$ and $\{E_n\}$ for which

$$\mathcal{H}|\psi_n\rangle = E_n|\psi_n\rangle \tag{5.18}$$

Although we have not yet found any particular $|\psi_n\rangle$ and E_n for which Eq. (5.18) works, still we can make use of its general form. We expand it in terms of an arbitrary complete orthonormal

REVIEW OF SOME QUANTUM MECHANICS

basis set $\{|a_i\rangle, i=1, 2, \ldots M\}$, and project out the kth component:

$$\sum_i^M \langle a_k|\mathcal{H}|a_i\rangle \langle a_i|\psi_n\rangle = E_n \langle a_k|\psi_n\rangle, \quad k=1, 2, \ldots M \quad (5.19a)$$

$\langle a_k|\psi_n\rangle \equiv \psi_{n_k}$ is simply the kth component of the nth eigenvector of $\mathcal{H}$, as decomposed in the $\{|a_i\rangle\}$ basis system. In matrix notation, Eq. (5.19a) becomes

$$\begin{pmatrix} \mathcal{H}_{11} & \mathcal{H}_{12} & \cdots & \mathcal{H}_{1M} \\ & & \mathcal{H}_{ki} & \cdots & \\ & & & & \\ \mathcal{H}_{M1} & & \cdots & \mathcal{H}_{MM} \end{pmatrix} \begin{pmatrix} \psi_{n_1} \\ \psi_{n_2} \\ \vdots \\ \psi_{n_M} \end{pmatrix} = E_n \begin{pmatrix} \psi_{n_1} \\ \psi_{n_2} \\ \vdots \\ \psi_{n_M} \end{pmatrix} \quad (5.19b)$$

or

$$\begin{pmatrix} \mathcal{H}_{11}-E_n & \mathcal{H}_{12} & \cdots & \mathcal{H}_{1M} \\ \mathcal{H}_{21} & \mathcal{H}_{22}-E_n & \cdots & \mathcal{H}_{2M} \\ \vdots & \vdots & & \vdots \\ \mathcal{H}_{M1} & \cdots & & \mathcal{H}_{MM}-E_n \end{pmatrix} \begin{pmatrix} \psi_{n_1} \\ \psi_{n_2} \\ \vdots \\ \psi_{n_M} \end{pmatrix} = 0 \quad (5.19c)$$

This set of simultaneous equations has a unique solution if and only if the determinant of the matrix of Eq. (5.19c) vanishes:

$$|\mathcal{H}_{ij} - E_n \delta_{ij}| = 0 \quad (5.20)$$

Equation (5.20) yields an Mth order polynomial in E_n, whose solution provides the possible energy eigenvalues. Thus to get $\{E_n\}$, we do *not* need to know $\{|\psi_n\rangle\}$ or the $\{(\psi_{n_1}, \psi_{n_2}, \ldots, \psi_{n_M})\}$. Moreover, *any* complete initial set $\{|a_i\rangle\}$ spanning Hilbert space will, by Eq. (5.20), lead to $\{E_n\}$.

But does solving Eq. (5.20) also diagonalize $\mathcal{H}$? Using the closure relation twice on Eq. (5.18) and projecting out with the energy eigenvector $\langle\psi_{n'}|$ gives

$$\sum_{ij}^M \langle\psi_{n'}|a_i\rangle \langle a_i|\mathcal{H}|a_j\rangle \langle a_j|\psi_n\rangle = E_n \delta_{n''n} \quad (5.21a)$$

The single state $|\psi_n\rangle$ can be decomposed in the $\{|a_i\rangle\}$ system as $\sum_i^M |a_i\rangle \langle a_i|\psi_n\rangle$ and represented by

$$\begin{pmatrix} \langle a_1|\psi_n\rangle \\ \langle a_2|\psi_n\rangle \\ \vdots \\ \langle a_M|\psi_n\rangle \end{pmatrix} = \begin{pmatrix} \psi_{n_1} \\ \psi_{n_2} \\ \vdots \\ \psi_{n_M} \end{pmatrix} \quad (5.21b)$$

We can now build up a matrix $\bar{\bar{N}}$ by juxtapositioning the column vector representations of all M of the eigenvectors, for $n = 1, 2, \ldots M$, (why do $\{|a_i\rangle\}$ and $\{|\psi_n\rangle\}$ have the same number of elements?) as in the process sketched in Section (3.10):

$$\bar{\bar{N}} = \begin{pmatrix} \psi_{1_1} & \psi_{2_1} & \cdots & \psi_{M_1} \\ \psi_{1_2} & \psi_{2_2} & \cdots & \psi_{M_2} \\ \vdots & \vdots & & \vdots \\ \psi_{1_M} & \psi_{2_M} & \cdots & \psi_{M_M} \end{pmatrix} \tag{5.21c}$$

and Eq. (5.21a) now reads:

$$\bar{\bar{N}}^+ \bar{\bar{\mathscr{H}}} \bar{\bar{N}} = \bar{\bar{\mathscr{H}}}_{\text{diag}} \tag{5.21d}$$

where $\bar{\bar{\mathscr{H}}}_{\text{diag}}$ is a diagonal matrix whose non-vanishing elements are the eigenvalues of $\mathscr{H}$.

PROBLEM 5.6 Show that the process of solving the secular equation for $\mathscr{H}$, Eq. (5.20), amounts to finding a matrix $\bar{\bar{N}}$ for which Eq. (5.21d) obtains, or to transforming from the $\{|a_i\rangle\}$ to a $\{|\psi_n\rangle\}$ representation in which $\mathscr{H}|\psi_n\rangle = E_n|\psi_n\rangle$.

The expressions 'diagonalize the Hamiltonian matrix', 'solve the eigenequation for the Hamiltonian', 'find the eigenvectors and/or eigenvalues of the Hamiltonian', 'solve the secular equation/determinant for the Hamiltonian', or even 'diagonalize the Hamiltonian operator' are essentially synonymous. Moreover, one can think of the transformation from the $\{|a_i\rangle\}$ to the $\{|\psi_n\rangle\}$ basis system as being a *rotation of function space* to a new 'orientation' in which the new unit vectors define a principal axis system.

†5.7 Diagonalizing the Hamiltonian as a Principal Axis Transformation

Let the set $\{|a_i\rangle\}$ form a complete orthonormal basis set spanning the function space associated with the Hamiltonian $\mathscr{H}$. Employing the closure relation twice, we perform the decomposition

$$\mathscr{H} = \sum_{ij}^{M} |a_i\rangle \langle a_i| \mathscr{H} |a_j\rangle \langle a_j| = \sum_{ij}^{M} \mathscr{H}_{ij} |a_i\rangle \langle a_j| \tag{5.22}$$

The matrix $\bar{\bar{\mathscr{H}}}$ is to be diagonalized; we could solve its secular equation, but we shall use another approach here instead. Choose an *arbitrary* fixed vector in our space, $|\phi\rangle$, and consider the scalar product of $\mathscr{H}|\phi\rangle$ with $\langle\phi|$, $\langle\phi|\mathscr{H}|\phi\rangle$. By Eq. (5.22)

$$\langle\phi|\mathscr{H}|\phi\rangle = \sum_{ij} \langle\phi|a_i\rangle \langle a_j|\phi\rangle \mathscr{H}_{ij} = \sum_{ij} \mathscr{H}_{ij} \phi_i^* \phi_j = \langle E_\phi\rangle \tag{5.23a}$$

REVIEW OF SOME QUANTUM MECHANICS

We have written the scalar product as a constant (it happens to be the expectation value of $\mathcal{H}$ for a system in the state $|\phi\rangle$) as a reminder that it should be invariant to any rotations of the co-ordinate system in function space.†

Let us assume (simply so that we can draw a picture of what we are doing) that both the matrix $\bar{\bar{\mathcal{H}}}$ and the components of $|\phi\rangle$ are real-valued:

$$\sum_{ij} \mathcal{H}_{ij}\phi_i\phi_j = \langle E_\phi \rangle \qquad (5.23b)$$

This equation represents a family of confocal ellipses, one corresponding to each value of $\langle E_\phi \rangle$, whose major axes do not necessarily lie along the axis directions a_i and a_j (we assume that $\bar{\bar{\mathcal{H}}}$ is not yet diagonal), Fig. 5.2a. Of all the co-ordinate systems in which one can write Eq. (5.23b) or draw Fig. (5.2a), there exist a few special orientations of function space (i.e. particular basis sets $\{|\psi_n\rangle\}$) for which Eq. (5.22) reduces to $\sum_n \mathcal{H}_{nn}|\psi_n\rangle\langle\psi_n|$, and Eq. (5.23b) becomes the simpler form

$$\sum_n \mathcal{H}_{nn}\phi_n'^2 = \sum_n E_n \phi_n'^2 = \langle E_\phi \rangle \qquad (5.23c)$$

$$\phi_n' = \langle \psi_n | \phi \rangle$$

In such a 'principal axis' orientation, the principal axes of the ellipse and the directions of the co-ordinate axes *do* coincide, Fig. 5.2b, and the matrix representation of $\mathcal{H}$ has only diagonal elements; $\{|\psi_n\rangle\}$ are then eigenstates of $\mathcal{H}$. Diagonalizing $\mathcal{H}$ as in Eq. (5.23c) is thus equivalent to rotating function space until a preferential orientation relative to some mathematical 'object' (such as the symmetrical ellipse of Fig. (5.2b)) is found.

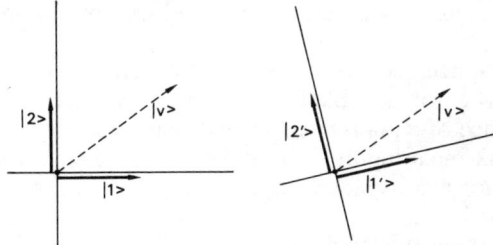

Figure 5.1.

†Figure 5.1 demonstrates an analogous example of a vector $|v\rangle$ in two dimensional space, when the $\{|1\rangle, |2\rangle\}$ basis system is rotated into the new $\{|1'\rangle, |2'\rangle\}$ system. $|v\rangle$ is oblivious to this re-alignment of space, and the scalar quantity $\langle v|v\rangle$, the square of the vector's length, is invariant to the change. The components of $|v\rangle$, namely the $v_i = \langle i|v\rangle$, assume new values, $v_i' = \langle i'|v\rangle$, and the *description* of the vector changes. But a scalar quantity associated with the space, such as $\langle v|v\rangle$, does not.

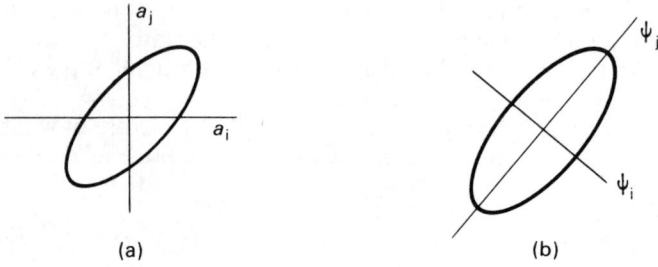

Figure 5.2.

PROBLEM 5.7. Show that the quadratic form Eq. (5.23c) is equivalent to the eigenequation

$$\mathcal{H}|\psi_n\rangle = E_n|\psi_n\rangle \tag{5.23d}$$

5.8 Symmetry Operators Commute with the Hamiltonian

In the next few sections we shall offer brief, but hopefully tantalizing, glimpses of the way in which symmetry arguments and group theory can be of use in the study of quantum mechanical systems. But first we must show that a *symmetry* operator in Hilbert space leaves the Hamiltonian operator invariant if and only if the two commute.

So far our discussion of quantum mechanics has been concerned with the Hermitian operators which correspond to dynamical variables such as position, momentum, angular momentum, energy, etc., as in Table 5.2. But just as in three-space, the operators in Hilbert space describing rotations and reflections, and therefore a system's symmetry operators, are unitary, not Hermitian.

The three-space geometric operators which we examined in the last chapter are not suitable for working in Hilbert space. Whereas the significance of $\hat{R}_{120}$ is clear in real space, it is not obvious how one can speak meaningfully about a 120 degree rotation operator appropriate for a subspace whose basis vectors are, say, the hydrogen d-orbitals.

We shall attack this problem early in the next chapter. Here it must suffice to say that to every geometric operator of real space $\hat{R}$, there corresponds a unique operator $\hat{P}_R$ that works in Hilbert space:

$$\hat{R} \leftrightarrow \hat{P}_R \tag{5.24}$$

If $\hat{R}$ describes a rotation, reflection, displacement, etc., of an object in real space, then $\hat{P}_R$ somehow conveys the same information to

REVIEW OF SOME QUANTUM MECHANICS

Hilbert space. At times one can actually write down an explicit form for $\hat{P}_R$, but often it is enough to show indirectly what $\hat{P}_R$ does to an arbitrary function.

Perhaps the simplest example of the use of such a $\hat{P}_R$ involves the inversion operator. Suppose the real-space operator $\hat{R}_p$ replaces every $\bar{r}$ in a set by $-\bar{r}$:

$$\hat{R}_p \bar{r} = -\bar{r} \tag{5.25a}$$

As noted in Section 1.2, if the sets $\{\bar{r}\}$ and $\{\hat{R}_p \bar{r}\}$ coincide, then $\{\bar{r}\}$ is said to possess a centre of inversion symmetry, and $\hat{R}_p$ is a symmetry operator for the set. But whether or not this is the case for a particular system, we can still define a parity operator $\hat{P}_p$, as in Eq. (1.3b), which works on *functions* rather than on points in three-space:

$$\hat{P}_p g(\bar{r}) = g(-\bar{r}) \tag{5.25b}$$

or equivalently,

$$\hat{P}_p g(-\bar{r}) = g(\bar{r}) \tag{5.25c}$$

Even and odd functions are those which happen to be eigenvectors of $\hat{P}_p$. If we wish, we can extend the domain of $\hat{P}_p$ a bit and allow it also to affect operators, should they be explicit or implicit functions of $\bar{r}$:

$$\hat{P}_p \hat{A}(-\bar{r}) = \hat{A}(\bar{r}) \tag{5.25d}$$

Either way, the intimate link between $\hat{R}_p$ and $\hat{P}_p$, $\hat{R}_p \leftrightarrow \hat{P}_p$, is

$$\hat{P}_p g(\hat{R}_p \bar{r}) = g(\bar{r}) \tag{5.25e}$$

Consider now the effect of $\hat{P}_p$, as in Eq. (5.25d), upon the Hamiltonian operator of a physical system which happens to exhibit inversion symmetry, i.e. for which $\hat{P}_p$ is a symmetry operator.

Figure 5.3 shows such a molecule; an electron in its vicinity will experience a coulombic potential of the form $\hat{V}(\bar{r})$, and by symmetry,

$$\hat{V}(\bar{r}) = \hat{V}(-\bar{r}) \tag{5.26a}$$

Similarly, the kinetic energy operator is invariant to inversion (prove it.). Therefore the Hamiltonian obeys

$$\hat{\mathcal{H}}(\bar{r}) = \hat{\mathcal{H}}(-\bar{r}) = \hat{\mathcal{H}}(\hat{R}_p \bar{r}) \tag{5.26b}$$

If $\psi(\bar{r})$ is an arbitrary (not necessarily odd or even) function in Hilbert space, then

$$\begin{aligned}\hat{P}_p \{\hat{\mathcal{H}}(\bar{r})\psi(\bar{r})\} &= \hat{\mathcal{H}}(-\bar{r})\psi(-\bar{r}) \\ &= \hat{\mathcal{H}}(\bar{r})\psi(-\bar{r}) \\ &= \hat{\mathcal{H}}(\bar{r})\hat{P}_p \psi(\bar{r})\end{aligned}$$

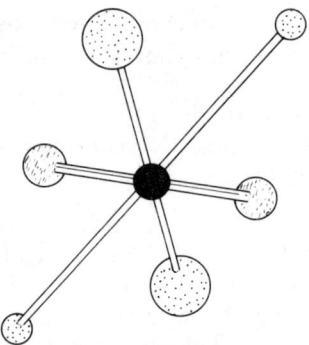

Figure 5.3.

Since $\psi(\bar{r})$ is arbitrary, this can be summarized as

$$[\hat{P}_p, \mathcal{H}] = 0. \tag{5.26c}$$

This result is generalizable: if R is *any symmetry operator* for a physical object, the corresponding Hilbert space operator $\hat{P}_R$ commutes with the system's Hamiltonian:

$$\boxed{[\hat{P}_R, \mathcal{H}] = 0} \tag{5.27a}$$

Equivalently, the Hamiltonian is left invariant by the symmetry operator:

$$\mathcal{H}(\hat{R}\bar{r}) = \mathcal{H}(\bar{r}) \tag{5.27b}$$

Consider now a C_{3v} molecule, to the base vertices of which are fixed three positive charges, and near which are seated immobile negative charges, Fig. 5.4a. The energy of the system will be altered should the molecule be twisted through, say, 43 degrees; but the energy and Hamiltonian are clearly insensitive to any of the six C_{3v} operations.

In a similar vein, the Hamiltonian for an electron on the C_{3v} molecule of Fig. 5.4b is of form

$$\mathcal{H}(\bar{r}) = \frac{-\hbar^2}{2m} \nabla^2 + e^2 \left\{ \frac{1}{|\bar{r} - \bar{r}_1|} + \frac{1}{|\bar{r} - \bar{r}_2|} + \frac{1}{|\bar{r} - \bar{r}_3|} \right\} \tag{5.28}$$

This is invariant to any interchange of $\bar{r}_1, \bar{r}_2, \bar{r}_3$, which denote the positions of the vertex atoms, brought about by the six C_{3v}

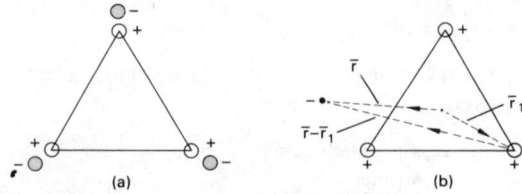

Figure 5.4.

operations; Eq. (5.27) again holds and the Hamiltonian is invariant to the system's symmetry operations!

5.9 Use of Symmetry Arguments in Diagonalizing a Hamiltonian

The procedure of Eq. (5.19), (5.20) shows that to diagonalize a Hamiltonian, any complete basis set may be used in forming the matrix representation. Some basis sets are more helpful than others, and a best choice depends upon the system's physical symmetry. Suppose the eigenvectors $\{|p_\alpha\rangle\}$ of a system's single symmetry operator $\hat{P}_R$ form a complete orthonormal set:

$$\left. \begin{array}{l} \hat{P}_R|p_\alpha\rangle = p_\alpha|p_\alpha\rangle \\ \sum_\alpha |p_\alpha\rangle \langle p_\alpha| = \hat{1} \\ \langle p_\alpha|p_{\alpha'}\rangle = \delta_{\alpha\alpha'} \end{array} \right\} \quad (5.29a)$$

We can then expand Eq. (5.27a) through closure on this set:

$$[\mathcal{H}, \hat{P}_R] = \sum_\alpha (\mathcal{H}|p_\alpha\rangle \langle p_\alpha|\hat{P}_R\rangle - (\hat{P}_R|p_\alpha\rangle \langle p_\alpha|\mathcal{H}) = 0 \quad (5.29b)$$

from which, after pre- and post-multiplying by $\langle p_{\alpha'}|$ and $|p_{\alpha''}\rangle$, respectively†,

$$\sum_\alpha (\mathcal{H}_{\alpha'\alpha}\Gamma_{\alpha\alpha''}(\hat{R}) - \Gamma_{\alpha'\alpha}(\hat{R})\mathcal{H}_{\alpha\alpha''}) = 0, \quad \Gamma_{\alpha\alpha'} = \langle p_\alpha|\hat{P}_R|p_{\alpha'}\rangle \quad (5.29c)$$

But by the assumption of Eq. (5.29a), the matrix of $\hat{P}_R$ generated by its own orthonormalized eigenvectors must be diagonal,

$$\Gamma_{\alpha\alpha'} = p_{\alpha'}\langle p_\alpha|p_{\alpha'}\rangle = p_{\alpha'}\delta_{\alpha\alpha'}$$

from which

$$\sum_\alpha (\mathcal{H}_{\alpha'\alpha}p_{\alpha''}\delta_{\alpha\alpha''} - p_{\alpha'}\delta_{\alpha'\alpha}\mathcal{H}_{\alpha\alpha''}) = 0$$

or

$$\mathcal{H}_{\alpha'\alpha''}(p_{\alpha''} - p_{\alpha'}) = 0 \quad (5.29d)$$

Thus either $p_{\alpha'} = p_{\alpha''}$, or $\langle p_{\alpha'}|\mathcal{H}|p_{\alpha''}\rangle = 0$.

This is a very useful result. The way we solve the eigenequation for a Hamiltonian, as discussed in the previous few sections, is by choosing a complete set of zero-th order functions, using them to create a matrix representation of $\mathcal{H}$, and then diagonalizing this matrix. Equation (5.29d) says that if $\hat{R}$ is a symmetry operation for the system, such that, by Eq. (5.27), $[\mathcal{H}, \hat{P}_R] = 0$, then for our zero-th order functions we should try to choose functions which

---

†In Eq. (6.19) we shall define the $\Gamma(\hat{R})$ symbol, through

$$\hat{P}_R|\psi_i\rangle = \sum_j |\psi_j\rangle \langle \psi_j|\hat{P}_R|\psi_i\rangle = \sum_j \Gamma_{ji}(\hat{R})|\psi_i\rangle$$

are already eigenvectors of $\hat{P}_R$. If we do so, many of the matrix elements $\mathcal{H}_{\alpha'\alpha''}$ will vanish, and (with proper ordering of the set $\{|p_\alpha\rangle\}$ in the construction of the matrix) the Hamiltonian matrix will be in block diagonal form, each block corresponding to a different eigenvalue of $\hat{P}_R$. The only non-vanishing terms in the matrix will be arrayed in square blocks along the diagonal:

$$\begin{pmatrix} \boxed{p_1} & & 0 \\ & \boxed{p_2} & \\ 0 & & \ddots \end{pmatrix} \tag{5.30}$$

All other elements will be zero's.

Since in the further diagonalization process each of these blocks can be attacked separately, such a situation simplifies the remaining work considerably.

PROBLEM 5.8 Consider a system which is invariant under the parity operator $\hat{P}_p$—i.e. it has a centre of inversion. Compare the matrix representations of $\mathcal{H}$ constructed out of (a) two even functions and two odd functions, (b) four functions of mixed symmetry.

5.10 Degenerate Solutions to the Schroedinger Equation Define a Subspace of Hilbert Space

A simple problem which makes a very important point:

*PROBLEM 5.9 Show that a set of degenerate solutions to the time-independent Schroedinger equation inhabit a vector subspace.

If ψ_n is an eigenvector of $\mathcal{H}$, and $\hat{R}$ is any sort of operator in three-space (not necessarily a symmetry operator), then it is perfectly legitimate to say

$$\mathcal{H}\psi_n = E_n\psi_n$$

$$\hat{P}_R\{\mathcal{H}(\hat{P}_R^{-1}\hat{P}_R)\psi_n\} = \hat{P}_R\{E_n\psi_n\}$$

or

$$\mathcal{H}'\hat{P}_R\psi_n \equiv (\hat{P}_R\mathcal{H}\hat{P}_R^{-1})\hat{P}_R\psi_n = E_n\hat{P}_R\psi_n \tag{5.31}$$

This is simply a statement of the fact that every vector and operator in Hilbert space has been affected by $\hat{P}_R$, in the sense of Eq. (4.15).

REVIEW OF SOME QUANTUM MECHANICS

If we now consider the special case in which $\hat{R}$ is a symmetry operator, Eq. (5.31) immediately leads to a most important result. By Eq. (5.27), if $\hat{R}$ is a symmetry operator for the system, then $\mathcal{H}(\hat{R}\bar{r}) = \mathcal{H}(\bar{r})$ and $\hat{P}_R$ has no effect upon $\mathcal{H}$, $[\hat{P}_R, \mathcal{H}] = 0$:

$$\mathcal{H}' \equiv \hat{P}_R \mathcal{H} \hat{P}_R^{-1} = \mathcal{H} \tag{5.32}$$

This and Eq. (5.31) tell us that if ψ_n is an eigensolution to $\mathcal{H}$, $(\hat{P}_R \psi_n)$ is also, and both have eigenvalue E_n!

$$\boxed{\mathcal{H}(\hat{P}_R \psi_n) = E_n(\hat{P}_R \psi_n)} \tag{5.33}$$

Two possible cases may occur here: (a) $\hat{P}_R$ has no effect upon ψ_n other than to multiply it by a constant; i.e. ψ_n is also an eigenvector of $\hat{P}_R$, as in the last section:

$$\boxed{\hat{P}_R \psi_n = p_n \psi_n} \tag{5.34a}$$

(b) $\hat{P}_R$ causes an admixture of linearly independent states when operating upon ψ_n:

$$\boxed{\hat{P}_R \psi_n = \sum_{n'} \Gamma_{n'n} \psi_{n'}} \tag{5.34b}$$

with admixture coefficients $\Gamma_{n'n}$.

So far we have been considering a single operator $\hat{P}_R$. The problem becomes interesting when we examine a system's entire group of symmetry operators $\{\hat{P}_R\}$. If the system's symmetry group $\{\hat{R}\}$ is abelian, and all the unitary $\hat{P}_R$ commute among themselves, then we can, as in the case of a complete set of commutating Hermitian operators, find a set $\{\psi_n\}$ which are simultaneously eigenvectors, Case (a), of all the $\{\hat{P}_R\}$. By the results of the last section, such a set of states would be most useful in the diagonalizing of $\mathcal{H}$. If $\{\hat{R}\}$ is *not* abelian, however, *no* such set $\{\psi_n\}$ can be found, and Case (b) applies. It is here that group representation theory becomes indispensable. If all symmetry groups were abelian, we wouldn't need it!

For Case (b), by Eq. (5.33), all the $\psi_{n'}$ of Eq. (5.34b) must be degenerate with ψ_n, and the $\{\psi_{n'}\}$ correspondingly constitute a function subspace of dimension greater than one. Case (a), on the other hand makes no demands with regard to the degeneracy of ψ_n. When we turn, in Chapter 9, to the connection between quantum mechanics and group theory, this observation will be our jumping-off point.

PROBLEM 5.10 Show that systems described by wave functions differing only by a constant multiplicative wave factor are physically indistinguishable. Under what conditions might $\psi_n(\bar{r})$ and $\sum_{n'} \Gamma_{n'n} \psi_{n'}(\bar{r})$ describe physically equivalent situations?

*PROBLEM 5.11 We have discussed the commutation properties of operators in five places: Problem (1.3), Theorem (5.1), Problem (5.2), Eq. (5.17), and Eq. (5.27). Which of these involve Hermitian operators, which involve unitary, and which involve both? Are there any common links among the five?

5.11 Atomic Orbitals

Until now, we have been concerned with the properties of all quantum systems, and with the general form

$$\mathcal{H}\psi_n = E_n\psi_n \tag{5.35}$$

of the time-independent Schroedinger equation. In this section we shall examine a specific example, the hydrogen atom, and determine the angular dependence of the electron's possible orbital states. In the following section we pursue this further and study the effect of a magnetic field upon the energy of its optical ground state.

An approximate Hamiltonian describing the electron of a hydrogen atom is of the form†

$$\mathcal{H} = -\frac{\hbar}{2m}\nabla^2 + \frac{e^2}{r} \tag{5.36}$$

$$\nabla^2\psi = \frac{\partial^2\psi}{\partial x^2} + \frac{\partial^2\psi}{\partial y^2} + \frac{\partial^2\psi}{\partial z^2} \qquad \text{rectangular co-ordinates} \tag{5.37a}$$

$$\nabla^2\psi = \frac{1}{r^2}\frac{\partial}{\partial r}\left(r^2\frac{\partial\psi}{\partial r}\right) + \frac{1}{r^2\sin\theta}\frac{\partial}{\partial\theta}\left(\sin\theta\frac{\partial\psi}{\partial\theta}\right) + \frac{1}{r^2\sin^2\theta}\frac{\partial^2\psi}{\partial\phi^2}$$

$$\text{spherical co-ordinates} \tag{5.37b}$$

The first term in Eq. (5.36) is the electron's kinetic energy operator, and e^2/r is its potential in the central, electrostatic field of the nucleus. This Hamiltonian is invariant to any rotation, reflection, or inversion about the origin: it is spherically symmetric.

The use of a trial solution of the form

$$\psi(\bar{r}) = F(r)Y(\theta, \phi) \tag{5.38}$$

reveals that the Schroedinger equation for our particular Hamiltonian can be separated into two distinct parts, involving, respectively, the variables r, and θ and ϕ. The (θ, ϕ) equation is

$$\left[-\hbar^2\left\{\frac{1}{\sin^2\theta}\frac{\partial^2}{\partial\phi^2} + \frac{1}{\sin\theta}\frac{\partial}{\partial\theta}\left(\sin\theta\frac{\partial}{\partial\theta}\right)\right\}\right]Y(\theta,\phi) = \hbar^2\lambda Y(\theta,\phi) \tag{5.39}$$

†For simplicity we shall disregard relatively small terms such as spin-orbit coupling.

REVIEW OF SOME QUANTUM MECHANICS

Solution of this eigenequation in accordance with proper boundary conditions yields $Y(\theta, \phi)$, the angular dependent part of the electron orbital.

The bracketed term of Eq. (5.39) is the quantum mechanical operator corresponding to the square of the classical orbital angular momentum $\bar{L} = \bar{r} \times \bar{p}$. This is no coincidence; we shall see that as a consequence of the spherical symmetry of our Hamiltonian $[\hat{L}^2, \mathscr{H}] = 0$, and functions can therefore be found which are simultaneously eigenvectors of both operators.

The possible solutions to Eq. (5.39) are the well-known spherical harmonics, $Y_l^m(\theta, \phi)$.

Table 5.3

$$Y_0^0 = (1/4\pi)^{1/2}$$
$$Y_1^0 = (3/4\pi)^{1/2} \cos\theta = (3/4\pi)^{1/2} z/r$$
$$Y_1^{\pm 1} = \mp(3/8\pi)^{1/2} \exp(\pm i\phi) \sin\theta = \mp(3/8\pi)^{1/2}(x \pm iy)/r$$
$$Y_2^0 = (5/16\pi)^{1/2}(3\cos^2\theta - 1) = (5/16\pi)^{1/2}(2z^2 - x^2 - y^2)/r^2$$
$$Y_2^{\pm 1} = \mp(15/8\pi)^{1/2} \exp(\pm i\phi) \cos\theta \sin\theta = \mp(15/8\pi)^{1/2}(x \pm iy)z/r^2$$
$$Y_2^{\pm 2} = (15/32\pi)^{1/2} \exp(\pm 2i\phi) \sin^2\theta = (15/32\pi)^{1/2}(x \pm iy)^2/r^2$$

Physically, spherical harmonics of different l describe electrons with different orbital angular momenta $\bar{L}$. The m label, where $-l < m < l$, refers to the magnitude of the component of $\bar{L}$ that one would measure along an externally applied magnetic field.

The spherical harmonics form a complete set, and the angular dependent part of any hydrogenic wave function can be expanded as

$$|Y(\theta, \phi)\rangle = \sum_{l,m} |Y_l^m\rangle \langle Y_l^m | Y(\theta, \phi)\rangle \qquad (5.40\text{a})$$

They are, moreover, orthonormal, where the scalar product is defined in

$$\int_0^{2\pi} \int_0^{\pi} Y_l^m(\theta, \phi) Y_{l'}^{m'}(\theta, \phi) \sin\theta \, d\theta \, d\phi = \delta_{ll'} \delta_{mm'} \qquad (5.40\text{b})$$

These equations, which lead directly to the Clebsch–Gordan coefficients used in the addition of angular momenta, are fundamental to the study of bound atomic states.

The functions of Table 5.3 are natural for the description of a spherically symmetric system. When a time-dependence factor of $\exp(-i\omega t)$ is included, they describe 'running waves' circulating about the z-axis. And because of their clear-cut dependence upon l and m, they are still useful (though no longer degenerate) for a system situated in a magnetic field aligned along z.

In many applications, however, it is easier to visualize and work with real-valued linear combinations, Eq. (5.40a), of the spherical harmonics. The following 'standing wave' sets are orthogonal and normalized in the sense of Eq. (5.40b):

Table 5.4

$$|s\rangle = (1/4\pi)^{1/2} = Y_0^{\,0}(\bar{r})$$

$$|p_1\rangle = (3/4\pi)^{1/2} x/r$$
$$|p_2\rangle = (3/4\pi)^{1/2} y/r$$
$$|p_3\rangle = (3/4\pi)^{1/2} z/r = Y_1^{\,0}(\bar{r})$$

$$|d_1\rangle = (15/4\pi)^{1/2}(x^2 - y^2)/2r^2$$
$$|d_2\rangle = (15/4\pi)^{1/2} xy/r^2$$
$$|d_3\rangle = (15/4\pi)^{1/2} xz/r^2$$
$$|d_4\rangle = (15/4\pi)^{1/2} yz/r^2$$
$$|d_5\rangle = (15/4\pi)^{1/2}(3z^2 - r^2)/2\sqrt{3}\, r^2 = Y_2^{\,0}(\bar{r})$$

and some of the functions are illustrated in Fig. 5.5. If, however, one wishes to use the functions of Table 5.4 to solve an essentially two-dimensional problem, the z dependence may not simply be ignored; rather, one must find new normalization factors.

PROBLEM 5.12 Show that the complex- and real-valued forms of Y_l^m are of either even or odd parity

$$\hat{P}_p Y_l^m(\theta, \phi) = Y_l^m(\pi - \theta, -\phi) = \pm Y_l^m(\theta, \phi) \qquad (5.40c)$$

Why is this?

PROBLEM 5.13 Find the proper normalization factors for the following set of two-dimensional functions. Are they orthogonal in x–y space?

$$\begin{cases} (x^2 - y^2)/r^2 = \cos 2\phi \\ 2xy/r^2 = \sin 2\phi \\ x/r = \sin \phi \\ y/r = \cos \phi \\ (x^2 + y^2)/r^2 = 1 \end{cases} \qquad (5.41)$$

While the possible angular dependences of $\psi(\bar{r})$ of Eq. (5.38) reside within the $\{Y_l^m(\theta, \phi)\}$, to find the system's stationary energy levels one must solve the radial equation for $F(r)$. In the case of the single-electron atom with $\mathcal{H} = -(\hbar^2/2m)\nabla^2 + e^2/r$, one obtains exactly the same energies $\{E_n\}$ as are predicted by the simple

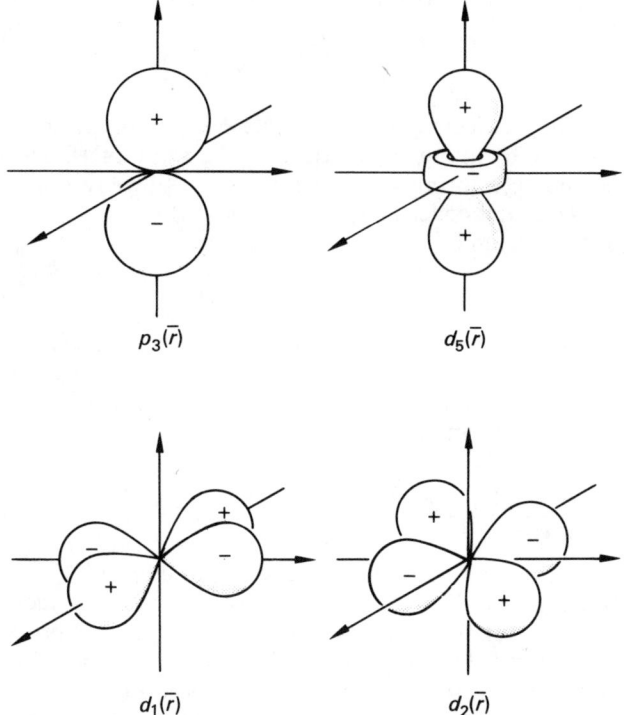

$p_3(\bar{r})$ $d_5(\bar{r})$

$d_1(\bar{r})$ $d_2(\bar{r})$

Figure 5.5.

Bohr-atom model. In particular, states of different l but the same principal quantum number n are degenerate.

One can talk about single-electron states even with a multi-electron atom. Then the total Hamiltonian is of the form

$$\mathcal{H} = \sum_i^N \mathcal{H}_i \tag{5.42a}$$

$$\mathcal{H}_i = -\frac{\hbar^2}{2m}\nabla_i^2 - \frac{Ze^2}{r_i} + \sum_{j \neq i}^N \frac{e^2}{|\bar{r}_j - \bar{r}_i|} \tag{5.42b}$$

$\mathcal{H}_i$ involves (except for the electron-electron interaction part) only the co-ordinates of the ith electron. It might seem that the last term in Eq. (5.42b) would destroy the useful spherical symmetry of the Hamiltonian. This is not necessarily the case, however, since the ith electron senses a roughly spherical cloud of charge due to the other electrons. We can then replace the mutual repulsion potential with a much simpler approximate radial potential $V'(r)$, and solve the less accurate but strictly one-electron

Schroedinger equation of effective Hamiltonian

$$\mathcal{H}' = -\frac{\hbar^2}{2m}\nabla^2 + \frac{Ze^2}{r} + V'(r) \tag{5.43a}$$

for a set of single-electron states $\{\psi_i(\bar{r})\}$. With these, we construct an atom or ion by filling levels, allowing for the Pauli Exclusion Principle and accounting for electron spin, until we run out of electrons. To a first order of approximation, the state of the entire system is then given by the product function

$$\psi(\bar{r}_1, \bar{r}_2, \ldots \bar{r}_N) \equiv \psi_1(\bar{r}_1)\psi_2(\bar{r}_2) \ldots \psi_N(\bar{r}_N) \tag{5.43b}$$

The accuracy of our solution can be improved considerably by a Hartree self-consistent treatment. Here Eq. (5.43b) is used to find the expectation value of the inter-electronic electrostatic potential felt by the ith electron,

$$V(r_i) = \left\langle \sum_{j \neq i}^{N} e^2/|\bar{r}_i - \bar{r}_j| \right\rangle$$

This expectation value in turn is roughly of the form $V''(r)$, again centrally symmetric. This revised effective potential replaces $V'(r)$ in Eq. (5.43a), the new equation is solved, and the procedure is repeated until little change in effective potential is noted from generation to generation of the iteration procedure,

Further improvement may arise from a Hartree–Foch treatment, antisymmetrizing Eq. (5.43b) in accord with the Pauli Exclusion Principle and directly acknowledging the fermion nature of an electron.

Self-consistent field calculations, such as the Hartree and Hartree–Foch types, have been immensely useful in the calculation of atomic properties, and lead to results in close agreement with experiment. But while the radial dependences of wave functions depend critically upon the details of the system we are examining and the method of calculation, the angular dependences are determined largely by the geometric symmetry properties of the system and its Hamiltonian, and are of the same general nature for all spherically symmetric objects. We shall return to this topic in Section 10.3.

†5.12 Construction and Diagonalization of a Hamiltonian—a Simple Example

We illustrate more of the ideas of this chapter by discussing briefly the 21 centimetre line of atomic hydrogen. This is a topic of interest to solid state chemists (hydrogen as a point defect in a solid) and to astrophysicists (in the study of inter-stellar gases), and it provides us with an especially transparent, non-trivial example.

It has long been known from the anomalous Zeeman effect of optical spectroscopy and from the Stern–Gerlach experiment that an electron displays behaviour similar to that of a classical spinning charged body. In particular, a free electron appears to possess an inherent magnetic dipole moment $\bar{\mu}$, and consequently an orientational potential energy $-\bar{\mu}\cdot\bar{H}$ in a magnetic field. It is an experimentally observed fact of life that this energy is quantized: $-\bar{\mu}\cdot\bar{H}$ can take on two, and only two, values.

In analogy to the classical case, and to be consistent with what was already known about the orbital angular momenta of electrons in atoms, it was argued that the magnetic moment is proportional to the electron's intrinsic 'spin' angular momentum $\bar{S}$

$$\bar{\mu} = -g_e \beta_e \bar{S} \tag{5.44}$$

where g_e is a convenient pure number of the order of 2, and $\beta_e = e\hbar/2mc$ is the Bohr magneton, a constant of proportionality linking $\bar{\mu}$ and $\bar{S}$. An unbound electron situated in an external magnetic field along $|3\rangle$, $\bar{H} = H|3\rangle$, can exist only in 'spin up' or 'spin down' orientations relative to H: if $\hat{S}_3$ is an operator associated with the component of $\hat{S}$ along $|3\rangle$, then

$$\hat{S}_3|\alpha_e\rangle = \tfrac{1}{2}|\alpha_e\rangle$$
$$\hat{S}_3|\beta_e\rangle = -\tfrac{1}{2}|\beta_e\rangle \tag{5.45}$$

define, respectively, the spin up and down states.

With $-\bar{\mu}\cdot\bar{H} = g_e\beta_e S_3 H$, the electron Hamiltonian $\mathscr{H}^e = g_e\beta_e H\hat{S}_3$ also has $|\alpha_e\rangle$ and $|\beta_e\rangle$ as eigenstates:

$$\mathscr{H}^e \left\{ \begin{array}{c} |\alpha_e\rangle \\ |\beta_e\rangle \end{array} \right\} = g_e\beta_e H \hat{S}_3 \left\{ \begin{array}{c} |\alpha_e\rangle \\ |\beta_e\rangle \end{array} \right\} = \pm\tfrac{1}{2} g_e\beta_e H \left\{ \begin{array}{c} |\alpha_e\rangle \\ |\beta_e\rangle \end{array} \right\} \tag{5.46}$$

The energy splitting of the two states is proportional to H, as we might expect.

Similar considerations apply to a free proton, for which the spin operator, spin-up and -down eigenstates, and eigenenergies are, respectively, $\hat{I}$, $|\alpha_n\rangle$ and $|\beta_n\rangle$, and $\mp\tfrac{1}{2}g_n\beta_n H$. The difference from Eq. (5.46) on the signs of the energies occurs because the electron and proton have opposite charge, and consequently dissimilar behaviour in a magnetic field.

The total Hamiltonian for a system consisting of non-interacting electron and proton is

$$\mathscr{H}^{en} = g_e\beta_e H\hat{S}_3 - g_n\beta_n H\hat{I}_3 \tag{5.47}$$

This suggests that we select as basis vectors for the Hilbert space of our system the product states

$$\{|\alpha_e\rangle|\alpha_n\rangle, |\beta_e\rangle|\beta_n\rangle, |\alpha_e\rangle|\beta_n\rangle, |\beta_e\rangle|\alpha_n\rangle\} \tag{5.48a}$$

An electron operator such as $\hat{S}_3$, then, would affect only the $|\alpha_e\rangle$ or $|\beta_e\rangle$ part of each state, and be oblivious to the nuclear spin alignment. The four states of Eq. (5.48a) are pure states with respect to the three commuting operators $\hat{S}_3$, $\hat{I}_3$, and $\mathcal{H}^{en}$, and eigenstates of all three. We shall say more about the general properties of 'product spaces' spanned by functions such as those of Eq. (5.48a) in Section 9.5.

We shall abbreviate the four state functions of Eq. (5.48a) by

$$\{|\alpha_e\alpha_n\rangle, |\beta_e\beta_n\rangle, |\alpha_e\beta_n\rangle, |\beta_e\alpha_n\rangle\} \tag{5.48b}$$

Bear in mind, however, that an electron operator does not affect the nuclear quantum numbers, and vice versa. (And do not confuse the Bohr magneton β_e and the nuclear magneton β_n with the two spin-down states!)

In a ground state hydrogen atom, the electron and proton are *not* independent particles, but rather influence one another through the hyperfine interaction:

$$\mathcal{H}_{\text{hfi}} = a\hat{S} \cdot \hat{I} = a(\hat{S}_1\hat{I}_1 + \hat{S}_2\hat{I}_2 + \hat{S}_3\hat{I}_3)$$

$$a = \frac{8\pi}{3} g_e\beta_e g_n\beta_n |\psi(0)|^2 \tag{5.49}$$

$\psi(\bar{r})$ is the wave function of the electron, with the nucleus at the origin, $\bar{r}=0$. It would take us far afield to derive this term, and it will suffice to say that it describes a quantum mechanical effect associated with interaction of the magnetic moments of electron and proton. In addition to Eq. (5.49), there is a classical magnetic dipole-dipole hyperfine interaction, but this vanishes in systems of spherical symmetry

The full 'spin Hamiltonian' $\mathcal{H}$ for the spin states of our hydrogen atom in a magnetic field is

$$\mathcal{H} = g_e\beta_e H\hat{S}_3 - g_n\beta_n \hat{H}\hat{I}_3 + a\hat{S}_3\hat{I}_3 + a(\hat{S}_1\hat{I}_1 + \hat{S}_2\hat{I}_2) \tag{5.50}$$

*PROBLEM 5.14 Show that Eq. (5.48) are eigenstates also of $a\hat{S}_3\hat{I}_3$.
PROBLEM 5.15 In constructing the spin Hamiltonian Eq. (5.50), why have we ignored the kinetic energies of electron and proton, and their Coulombic interaction?

Our basic problem is to find the eigenvalues and eigenstates of the full Hamiltonian, Eq. (5.50). We know, however, that the product functions Eq. (5.48) are already perfectly good eigensolu-

REVIEW OF SOME QUANTUM MECHANICS

tions to the simpler, though less accurate, incomplete Hamiltonian

$$\mathscr{H}^0 = g_e\beta_e H \hat{S}_3 - g_n\beta_n H \hat{I}_3 + a\hat{S}_3\hat{I}_3 \tag{5.51}$$

We are therefore inclined to construct a representation matrix of the true $\mathscr{H}$ using the basis set appropriate for the approximate $\mathscr{H}^0$, and hope that we can diagonalize it without too much difficulty.

Matrix elements of the first three terms in Eq. (5.50) are trivial; for example

$$\langle \alpha_e\beta_n | g_e\beta_e H \hat{S}_3 | \alpha_e\beta_n \rangle = g_e\beta_e H \langle \beta_n|\beta_n\rangle \langle \alpha_e|\hat{S}_3|\alpha_e\rangle = \tfrac{1}{2}g_e\beta_e H \tag{5.52}$$

The standard procedure for evaluating the matrix elements of $(\hat{S}_1\hat{I}_1 + \hat{S}_2\hat{I}_2)$ involves† the use of 'raising' operators $\hat{S}^+$ and $\hat{I}^+$, and 'lowering' operators $\hat{S}^-$ and $\hat{I}^-$:

$$\hat{S}^+ = \hat{S}_1 + i\hat{S}_2, \quad \hat{S}^- = \hat{S}_1 - i\hat{S}_2$$
$$\hat{I}^+ = \hat{I}_1 + i\hat{I}_2, \quad \hat{I}^- = \hat{I}_1 - i\hat{I}_2 \tag{5.53a}$$

which exhibit the properties

$$\hat{S}^+|\alpha_e\rangle = 0$$
$$\hat{S}^+|\beta_e\rangle = |\alpha_e\rangle$$
$$\hat{S}^-|\beta_e\rangle = 0$$
$$\hat{S}^-|\alpha_e\rangle = |\beta_e\rangle \tag{5.53b}$$

and similarly for $\hat{I}^+$ and $\hat{I}^-$

*PROBLEM 5.16 Prove the identity

$$\hat{S}_1\hat{I}_1 + \hat{S}_2\hat{I}_2 = \tfrac{1}{2}(\hat{S}^+\hat{I}^- + \hat{S}^-\hat{I}^+)$$

The representation of the complete Hamiltonian Eq. (5.50) in the $\{|\alpha_e\alpha_n\rangle, \ldots\}$ basis system is

(5.54)

	$|\alpha_e\alpha_n\rangle$	$|\beta_e\beta_n\rangle$	$|\alpha_e\beta_n\rangle$	$|\beta_e\alpha_n\rangle$
$\langle\alpha_e\alpha_n|$	$\tfrac{1}{2}(g_e\beta_e H - g_n\beta_n H) + \tfrac{1}{4}a$	0	0	0
$\langle\beta_e\beta_n|$	0	$-\tfrac{1}{2}(g_e\beta_e H - g_n\beta_n H) + \tfrac{1}{4}a$	0	0
$\langle\alpha_e\beta_n|$	0	0	$\tfrac{1}{2}(g_e\beta_e H + g_n\beta_n H) - \tfrac{1}{4}a$	$\tfrac{1}{2}a$
$\langle\beta_e\alpha_n|$	0	0	$\tfrac{1}{2}a$	$-\tfrac{1}{2}(g_e\beta_e H + g_n\beta_n H) - \tfrac{1}{4}a$

We find the energy eigenvalues and eigenstates of our system by diagonalizing this matrix.

For a hydrogen atom in high magnetic fields, $g_e\beta_e H$ will be

---

†Discussions of angular momentum operators and the associated ladder operators can be found in all quantum mechanics texts.

much larger than either $g_n\beta_n H$ or a. In this situation one can immediately find a crude approximate solution by simply throwing away the relatively small off-diagonal terms, $\frac{1}{2}a$. This gives the same result as is obtained by treating $a\hat{S}\hat{I}$ through first order perturbation theory.

*PROBLEM 5.17 Show that $|\alpha_e\alpha_n\rangle$ and $|\beta_e\beta_n\rangle$ are eigenvectors of the complete Hamiltonian $\mathscr{H}$. Verify Eq. (5.54).

An exact diagonalization of Eq. (5.54) entails a re-alignment of Hilbert space to a principal axis orientation. Since $|\alpha_e\alpha_n\rangle$ and $|\beta_e\beta_n\rangle$ already lie along principal axis directions of the full Hamiltonian, we need concern ourselves now only with diagonalizing the remaining two-dimensional subspace spanned by $|\alpha_e\beta_n\rangle$ and $|\beta_e\alpha_n\rangle$. By proper rotation of this subspace, we should be able to find two functions of the form

$$\lambda|\alpha_e\beta_n\rangle + \eta|\beta_e\alpha_n\rangle \tag{5.55}$$

which not only span the subspace but are also eigenvectors of $\mathscr{H}$. From the lower right-hand quadrant of Eq. (5.54), the representation of the energy eigenequation generated by only the $\{|\alpha_e\beta_n\rangle, |\beta_e\alpha_n\rangle\}$ basis set reads

$$\frac{1}{2}\begin{pmatrix} x-\dfrac{a}{2} & a \\ a & -x-\dfrac{a}{2} \end{pmatrix}\begin{pmatrix} \lambda \\ \eta \end{pmatrix} = E\begin{pmatrix} \lambda \\ \eta \end{pmatrix} \tag{5.56a}$$

where $x \equiv (g_e\beta_e H + g_n\beta_n H)$, and λ and η are defined in Eq. (5.55). Multiplying this out yields the two coupled equations

$$\begin{cases} \lambda x/2 - \lambda a/4 + \eta a/2 = \lambda E \\ \lambda a/2 - \eta x/2 - \eta a/4 = \eta E \end{cases} \tag{5.56b}$$

whose solutions are

$$E = \pm\{(x^2 + a^2)^{1/2}/2\} - a/4 \tag{5.57a}$$

The four exact eigenenergies of $\mathscr{H}$ are therefore

$$\left.\begin{aligned} &\tfrac{1}{2}g_e\beta_e H - \tfrac{1}{2}g_n\beta_n H + \dfrac{a}{4} \\ &-\tfrac{1}{2}g_e\beta_e H + \tfrac{1}{2}g_n\beta_n H + \dfrac{a}{4} \\ &\pm\{(g_e\beta_e H + g_n\beta_n H)^2/4 + a^2\}^{1/2} - \dfrac{a}{4} \end{aligned}\right\} \tag{5.57b}$$

illustrated in Fig. 5.6.

REVIEW OF SOME QUANTUM MECHANICS

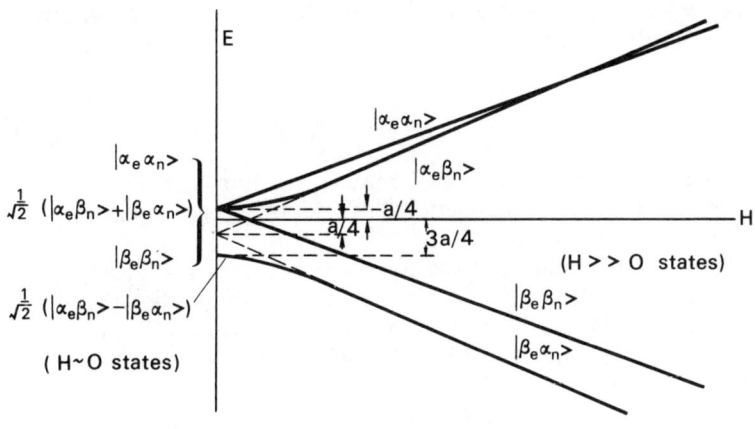

Figure 5.6.

*PROBLEM 5.18 Show that a similar treatment of the 4×4 matrix of Eq. (5.54) also gives these four eigenvalues.

PROBLEM 5.19 What are the four eigenvectors of $\hat{\mathscr{H}}$? What happens to them in the $x \gg a$ and the $x = 0$ limiting cases?

*PROBLEM 5.20 Show that Eq. (5.5a) can also be found by solving the 'secular determinant' for the Hamiltonian matrix of Eq. (5.56a) as in Eq. (5.20)

$$\begin{vmatrix} x/2 - a/4 - E & a/2 \\ a/2 & -x/2 - a/4 - E \end{vmatrix} = 0 \tag{5.58}$$

In the high field case, $x \gg a$, and Eq. (5.57a) reduces to $\pm x/2 - a/4 \pm a^2/4x$; the same form is obtained through taking $a\hat{S} \cdot \hat{I}$ to second order with ordinary perturbation theory. This result is the Breit–Rabi equation for spin-$\frac{1}{2}$ particles.

PROBLEM 5.21 Show that a singlet-triplet transition at $H = 0$ involves a 21 cm photon. Is Fig. 5.6 drawn realistically to scale? At what magnetic field strength will the two upper levels cross?

5.13 Time-Independent Perturbation Theory

If it is not possible to solve exactly the time-independent Schroedinger equation for the Hamiltonian $\hat{\mathscr{H}}$,

$$\hat{\mathscr{H}}|k\rangle = E_k|k\rangle \tag{5.59}$$

where we have condensed the notation a bit with $|\psi_k\rangle \equiv |k\rangle$, then one must resort to approximation schemes. Often it happens that the true Hamiltonian can be broken into two parts, $\hat{\mathscr{H}}^0$ and $\Delta \hat{V}$, with $\mathcal{H}^0$ nearly the same as $\mathcal{H}$, but such that the eigenequation of $\mathcal{H}^0$ is easily solved:

$$\mathcal{H}^0 \sim \mathcal{H} \gg \Delta\hat{V}$$
$$\mathcal{H}^0 |k^0\rangle = E_k{}^0 |k^0\rangle \quad (5.60)$$

$\Delta\hat{V}$, the small correction to $\mathcal{H}^0$, may represent a complication (produced by a weak magnetic field, crystal field, etc) to a simple system described accurately by $\mathcal{H}^0$.

Time-independent perturbation theory describes what happens to the eigensolutions of $\mathcal{H}^0$ when $\Delta\hat{V}$ is turned on adiabatically (very slowly). With the application of $\Delta\hat{V}$, the states $|k^0\rangle$ and energies $E_k{}^0$ are shifted slightly, becoming eigensolutions to $\mathcal{H}$, approximated as

$$|k\rangle = |k^0\rangle - \sum_{n \neq k} \frac{\langle n^0 | \Delta\hat{V} | k^0 \rangle}{E_n{}^0 - E_k{}^0} |n^0\rangle + \ldots \quad (5.61a)$$

$$E_k = E_k{}^0 + \langle k^0 | \Delta\hat{V} | k^0 \rangle - \sum_{n \neq k} \frac{1}{E_n{}^0 - E_k{}^0} |\langle n^0 | \Delta\hat{V} | k^0 \rangle|^2 + \ldots \quad (5.61b)$$

To find the shift in the kth level, one need only evaluate integrals such as $\langle k^0 | \Delta V | k^0 \rangle$ and $\langle n^0 | \Delta V | k^0 \rangle$.

But we run into troubles if two eigenstates of $\mathcal{H}^0$, such as $|k^0\rangle$ and $|m^0\rangle$, happen to be degenerate. (For simplicity, we shall consider only a doubly degenerate level.) Then the denominator in

$$\frac{1}{E_k{}^0 - E_m{}^0} |\langle m^0 | \Delta\hat{V} | k^0 \rangle|^2 \quad (5.62)$$

goes to zero, and the contribution of Eq. (5.62) to the relevant terms in Eq. (5.61) blows up.

The resolution of this dilemma is to try to make the numerator of any such offending term vanish also, leaving only the innocous indeterminate 0/0 form. Since problems with Eq. (5.62) arise because of the presence of finite *off-diagonal* terms in the 2×2 matrix of $\Delta\hat{V}$ generated by $|k^0\rangle$ and $|m^0\rangle$, our objective is achieved by diagonalizing this small matrix. The process of finding linear combinations of $|k^0\rangle$ and $|m^0\rangle$ for which a representation of $\Delta\hat{V}$ is diagonal is equivalent to rotating the $k^0 - m^0$ subspace to a principal axis alignment. The solution of the secular determinant equation

$$\begin{vmatrix} \langle m^0 | \Delta\hat{V} | m^0 \rangle - \Delta E & \langle m^0 | \Delta\hat{V} | k^0 \rangle \\ \langle k^0 | \Delta\hat{V} | m^0 \rangle & \langle k^0 | \Delta\hat{V} | k^0 \rangle - \Delta E \end{vmatrix} = 0 \quad (5.63)$$

yields two new states $|k'\rangle$ and $|m'\rangle$ good for replacing $|k^0\rangle$ and $|m^0\rangle$ in Eq. (5.61), and our difficulty clears up.

This raises an important related question: will $\Delta\hat{V}$ break the original $|k^0\rangle - |m^0\rangle$ degeneracy? Transforming to the $|k'\rangle - |m'\rangle$ basis system causes the second order term of Eq. (5.61b) to disappear, but the degeneracy can still remain if the new first order energy contributions are the same:

$$\langle m'|\Delta\hat{V}|m'\rangle = \langle k'|\Delta\hat{V}|k'\rangle \tag{5.64a}$$

In that case, $|k^0\rangle$ and $|m^0\rangle$ will simply be shifted an identical amount by $\Delta\hat{V}$. Eq. (5.63), however, indicates that the degeneracy will be *broken unless* it happens that

$$\langle m^0|\Delta\hat{V}|m^0\rangle = \langle k^0|\Delta\hat{V}|k^0\rangle$$
$$\langle m^0|\Delta\hat{V}|k^0\rangle = \langle k^0|\Delta V|m^0\rangle = 0 \tag{5.64b}$$

The connection between the symmetries of perturbations and states and the vanishing of integrals such as those of Eq. (5.64b) is the subject matter of Section 11.2.

PROBLEM 5.22 If $|a\rangle$ and $|b\rangle$ are orthonormal eigenvectors of $\mathcal{H}$, under what conditions will the matrix element $\langle\psi_1|\mathcal{H}|\psi_2\rangle$ vanish, where $|\psi_i\rangle = a_i|a\rangle + b_i|b\rangle$?

PROBLEM 5.23 Is anything wrong with the following argument? The two linear combinations of $|k^0\rangle$ and $|m^0\rangle$ that lead to a diagonal form for the matrix of $\Delta\hat{V}$ are eigenvectors of $\Delta\hat{V}$; they are also eigenvectors of $\mathcal{H}^0$. Therefore they are exact solutions of $\mathcal{H}$. (Examine Eq. (5.61) carefully.)

*****PROBLEM 5.24** Imagine three electrons in atomic *p*-states aligned, respectively, along the *x*, *y*, and *z* axes of a cubic crystal. $+1$ charges are placed at $z = \pm a$. Which of the *p*-state degeneracies will be broken, and why?

PROBLEM 5.25 Use perturbation theory to solve the Schroedinger equation for the example of the previous section in the $H \gg 0$ case, breaking $\mathcal{H}$ into

$$\mathcal{H}^0 = g_e\beta_e H + g_n\beta_n H + a\hat{I}_3\hat{S}_3$$
$$\Delta\hat{V} = a(\hat{I}_1\hat{S}_1 + \hat{I}_2\hat{S}_2) \tag{5.65}$$

PROBLEM 5.26 Show that exact eigenvectors to the zero-field limit, $H = 0$, of Eq. (5.50) are

$$\left.\begin{array}{l}|\alpha_e\alpha_n\rangle\\ \{|\alpha_e\beta_n\rangle + |\beta_e\alpha_n\rangle\}/2^{1/2}\\ |\beta_e\beta_n\rangle\end{array}\right\} \text{ spin triplet} \tag{5.66a}$$

$$\{|\alpha_e\beta_n\rangle - |\beta_e\alpha_n\rangle\}/2^{1/2} \quad \text{spin singlet} \tag{5.66b}$$

What are the eigenvalues? What will a small perturbation of the form $g_e\beta_e H$ do to these energies?

5.14 Summary

1. The physical nature of a quantum mechanical system can be expressed as a vector in Hilbert space. Performing a measurement in the laboratory is mirrored mathematically by operating on the state vector with the appropriate linear Hermitian operator. This operation may or may not alter the state. In either case, measurement of the attribute A yields only one of the eigenvalues a_n of operator $\hat{A}$, and leaves the system in the corresponding eigenstate. The probability of measuring A and finding a_n for a system known to be in the state $|\psi\rangle$ is $|\langle a_n|\psi\rangle|^2$; this leads to $\langle A\rangle = \langle\psi|\hat{A}|\psi\rangle$.

2. A pure state is the simultaneous eigenvector of a complete set of compatable, commuting Hermitian operators. Pure states often form convenient basis systems in which to expand arbitrary states. Moreover, since

$$\frac{d}{dt}\langle\hat{A}\rangle = \frac{-i}{\hbar}[\hat{A}, \mathcal{H}],$$

those Hermitian operators which commute with the Hamiltonian are associated with constants of the motion.

3. If $\hat{R}$ is a symmetry operator, $\mathcal{H}(\hat{R}\vec{r}) = \mathcal{H}(\vec{r})$. $[\hat{P}_R, \mathcal{H}] = 0$ for all unitary operators $\hat{P}_R$ corresponding to the symmetry group $\{\hat{R}\}$. These $\{\hat{P}_R\}$, however, do not all commute among themselves unless $\{\hat{R}\}$ happens to be abelian. If $\psi_n(\vec{r})$ is an eigenvector of $\mathcal{H}$ of eigenvalue E_n, then $\hat{P}_R\psi_n(\vec{r})$ is also an eigenvector of $\mathcal{H}$, with the same eigenvalue.

4. The fundamental problem of quantum mechanics is to learn the energy eigenstates of $\mathcal{H}$. To find these, one chooses a complete basis set spanning Hilbert space, constructs the matrix representation of $\mathcal{H}$, and diagonalizes this matrix by solving its secular equation. Symmetry arguments may help in this process.

5. A spherically symmetric Hamiltonian has eigensolutions whose angular dependences are given by the spherical harmonics. One can construct an essentially equivalent set of real-valued linear combinations of these.

6. A perturbation $\Delta \hat{V}$ will break the degeneracy of the eigenstates $|k^0\rangle$ and $|m^0\rangle$ of an unperturbed Hamiltonian $\mathcal{H}^0$ *unless* $\Delta V_{k^0k^0} = \Delta V_{m^0m^0}$ and $\Delta V_{k^0m^0} = \Delta V_{m^0k^0} = 0$. This condition is determined by symmetry considerations.

CHAPTER 6

Representations of an Operator in Function Space

Many vectors in function space, such as atomic orbitals in their familiar form, may be expressed as functions of the real-space position vector $\bar{r}$. If an observer changes his orientation or position in the laboratory (i.e. if the co-ordinate axes of real space are altered) or if some externally applied factors actually move the physical system, a function describing the system will change in appearance:

$$|f(\bar{r})\rangle \rightarrow |f(\hat{R}\bar{r})\rangle = |g(\bar{r})\rangle$$

The link between happenings in real space and transformations in function space is provided by $\hat{P}_R$, *defined* to operate in function space in such a way that if $\bar{r} \rightarrow \hat{R}\bar{r}$ in real space, then in function space

$$\hat{P}_R |f(\hat{R}\bar{r})\rangle = |f(\bar{r})\rangle$$

That is, for any $\hat{R}$ in real space, $\hat{P}_R$ acts in function space to annul the effect of $\hat{R}$. Thus there is a clear one-to-one connection between any given transformation in real space and a particular transformation in function space.

One can use a closure relation for the atomic orbitals $\{|\psi_k(\bar{r})\rangle\}$

$$\sum_k |\psi_k(\bar{r})\rangle \langle \psi_k(\bar{r})| = \hat{1}$$

to generate a representation of $\hat{P}_R$; the matrix elements of this representation are

$$\Gamma_{ij}^{(\psi)}(\hat{R}) = \langle \psi_i(\bar{r})|\hat{P}_R|\psi_j(\bar{r})\rangle$$

The superscript on Γ refers to the type of basis set used in generating the representation. Because of the orthogonality of atomic orbitals, the matrix $\bar{\bar{\Gamma}}^{(\psi)}(\hat{R})$ can be broken into the direct sum of independent sub-matrices $\bar{\bar{\Gamma}}^{(nl)}(\hat{R})$ which lie along a diagonal of the original matrix; each sub-block will correspond to a combination of principal and angular momentum quantum number, n and l, and will be of dimension $2l+1$.

With suitable alignment of function space relative to the symmetry of the physical system in real space, these sub-matrices may perhaps be further reduced into even smaller blocks along the diagonal. A five-dimensional d-state ($l=2$) function space, for example, can be broken into three distinct invariant subspaces such that under a rotation about $|3\rangle$ in real space, $d_1(\bar{r})$ and $d_2(\bar{r})$ mix only among themselves, $d_3(\bar{r})$ and $d_4(\bar{r})$ mix only among themselves, and $d_5(\bar{r})$ is not changed. Thus $\Gamma^{(d)}(\hat{R})$ is reduced into two 2×2 blocks and a one-dimensional block.

6.1 The Link between Real Space and Function Space: the State $\psi(\bar{r})$

Most of the functions with which we deal in quantum mechanics are functions of the position vector. Atomic states, for example, are most familiar as $\psi(\bar{r})$, and operators are generally seen in the form $\hat{F}=\hat{F}(\bar{r}, \bar{\nabla}, t)$.

As described in Sections 3.8 and 5.1, it is assumed that complete information about a quantum mechanical system resides in its state $|\psi\rangle$. If the state is represented in the continuous $|r\rangle$ basis system,

$$|\psi\rangle = \int |r\rangle \langle r|\psi\rangle \, dr \qquad (6.1)$$

then the complete information content of $|\psi\rangle$ is passed on to the elements of the representation, $\langle r|\psi\rangle \equiv \psi(\bar{r})$. Thus what we have been calling the 'state $\psi(\bar{r})$' is, in fact, the $\{|r\rangle\}$ representation of a more abstract state $|\psi\rangle$. We can think of the representation $\psi(\bar{r})$ as itself being a vector, and to emphasize the point, we write $\psi(\bar{r})$ in some situations as $|\psi(\bar{r})\rangle$. $|\psi\rangle$ and $|\psi(\bar{r})\rangle$ do not reside within the same vector space, but the link between the two is obviously close.†

The vector space inhabited by the $\{|\psi(\bar{r})\rangle\}$ will itself be spanned by some complete, orthonormal set $\{|\psi_i(\bar{r})\rangle\}$ such as, for example, atomic orbitals. Then a general state in this Hilbert space can be

†Recall that the m-tuple representing $|v\rangle$ in Euclidian space, $(v_1, v_2, \ldots v_m)$, generated by the basis system $\{|1\rangle, |2\rangle, \ldots m\rangle\}$, can also be considered a vector It exists in a space slightly different from that of $|v\rangle$ itself, namely representation-vector space.

decomposed as

$$|\psi(\bar{r})\rangle = \sum_i |\psi_i(\bar{r})\rangle \langle \psi_i(\bar{r})|\psi(\bar{r})\rangle \qquad (6.2a)$$

$$= \sum_i c_i |\psi_i(\bar{r})\rangle$$

or, more simply

$$\psi(\bar{r}) = \sum_i c_i \psi_i(\bar{r}) \qquad (6.2b)$$

6.2 The Link between Real Space and Function Space: the Operator $\hat{P}_R$

A question fundamental to our development of the theory of group representations is: What happens in function space to $\psi(\bar{r})$ when for some reason we decide either to rotate or displace the real space co-ordinate axis system, or to rotate or displace the physical system under examination? What is the effect upon $\psi(\bar{r})$ if the alignment between the physical system and an initially selected basis vector set is altered? If the co-ordinate axes of real space are chosen with an observer or measuring instrument as the origin, then a re-orientation of these axes would arise simply from a movement of the observer or instrument during the examination; we assume, of course, that the measuring device creates *no* magnetic fields, etc., which could directly produce an absolute orientation of the system. Alternatively, we could actually rotate the sample in a field-free region. By either approach, the physics of the problem does not change, nor should the results of any experiment.

But consider what happens to the description $\psi(\bar{r})$ of the system, which assigns to each point in space the numerical value $\psi(\bar{r})$. Here it may help to call upon the analogy of a rigid canopy or umbrella hung above a plane of two-dimensional vectors $\{\bar{r}\}$. The function describes the canopy of inherent 'shape' $|\psi\rangle$, by reporting its height $\psi(\bar{r})$ above each point $\bar{r}$ in our two-dimensional 'real' space. $\bar{r}$ is given in terms of its components relative to the basis system of 'real' space, and if the components (x_1, x_2) vary, the representation $\psi(x_1, x_2)$ of $|\psi\rangle$ will change also. The problem is to reconcile a changeable and somewhat arbitrary $\psi(\bar{r})$ with a physically invariant $|\psi\rangle$.

How does a transformation in real space, defined through

$$\hat{R}|i\rangle = \sum_j^3 |j\rangle R_{ji}, \qquad R_{ji} = \langle j|\hat{R}|i\rangle \qquad (6.3)$$

influence $\psi(\bar{r})$ in function space? Following Eq. (6.3), we might at first guess that some good could come of trying to construct the entries $\langle\psi_i(\bar{r})|\hat{R}|\psi_j(\bar{r})\rangle$ from the $\{|\psi_i(\bar{r})\rangle\}$ of function space. But the $\{|\psi_i(\bar{r})\rangle\}$ lie in an abstract, multi-dimensional function space, and $\hat{R}$ operates only on vectors in real-space. How, then, do we find an operator, say $\hat{P}_R$, which *works in function space*, but which nonetheless has a well-defined link with $\hat{R}$ itself?

A simple procedure is to generalize the relatively palatable relationship, Eq. (5.25), between the real-space inversion operator $\hat{R}_p$ and the function-space operator $\hat{P}_p$:

$$\hat{P}_p f(\hat{R}_p \bar{r}) = f(\bar{r}) \tag{5.25e}$$

This suggests the following approach to selecting $\hat{P}_R$: if $\hat{R}$ is a geometric operator in real space, then $\hat{P}_R$ is *defined* to be an associated operator which acts within the function space spanned by $\{\psi_i(\bar{r})\}$ such that

$$\boxed{\hat{P}_R \psi(\hat{R}\bar{r}) = \psi(\bar{r})} \tag{6.4a}$$

$\hat{P}_R$, like the general operator $\hat{A}$ of Section 3.2, is defined through the way it works, rather than by any explicit expression. Given an operator $\hat{R}$ which acts upon the position vectors $\{\bar{r}\}$ of real space, we build or choose the corresponding entity $\hat{P}_R$ whose effect upon all functions $\{|\psi(\bar{r})\rangle\}$ is to counteract or nullify the consequences in function space of using $\hat{R}$ in real space.†

*PROBLEM 6.1 Show that if $\hat{P}_R \psi(\hat{R}\bar{r}) = \psi(\bar{r})$, then

$$\hat{P}_R \psi(\bar{r}) = \psi(\hat{R}^{-1}\bar{r}) \tag{6.4b}$$

6.3 Linear Displacements

Because of the importance of $\hat{P}_R$ to what will follow, and because its significance may not be fully apparent from a definition alone, we shall now offer two examples of its use. First we shall examine the state function $\psi(x)$ of a free particle in a one-dimensional space. The term 'free-particle' implies that the space is homogeneous, and any potential influencing the system is a constant: $V(x) = V_c$. If $V(x)$ were not constant, the particle's dynamic condition would depend upon x; but our assumption here of $V(x) = V_c$ means that a linear displacement from x to $x + a$, say, has no physical sig-

---

†Some authors define $\hat{P}_R$ through

$$\hat{P}_R \psi(\bar{r}) = \psi(\hat{R}\bar{r}) \tag{6.4c}$$

The end result of any calculation is, of course, the same.

nificance, and is a symmetry transformation. This does not mean, however, that $\psi(x)$ is also position independent!

Consider a particle whose wave packet happens to be peaked at x_0, Fig. 6.1a. After the particle moves the distance a to the right, it is described by some different function $\psi'(x)$, Fig. 6.1b. Comparison of Figs 6.1a and 6.1b for $x = x_0$ will dispel any notion that $\psi'(x)$ might equal $\psi(x+a)$. During the displacement process, $\psi(x)$ evolves continuously into $\psi'(x)$, and this development† may be described by some appropriate function-space operator $\hat{P}_a$:

$$\psi(x) \to \psi'(x) = \hat{P}_a \psi(x) \tag{6.5a}$$

Under displacement of the particle to the right, its position changes from x_0 to $x_0 + a$. The same apparent change in x-coordinate would occur if we left the particle immobile, but relabelled the *origin* and shifted it by the amount $-a$—i.e. through the same distance, but to the left, Fig. 6.1c. The system should now be described by a state which has the shape of $\psi(x)$ but which peaks

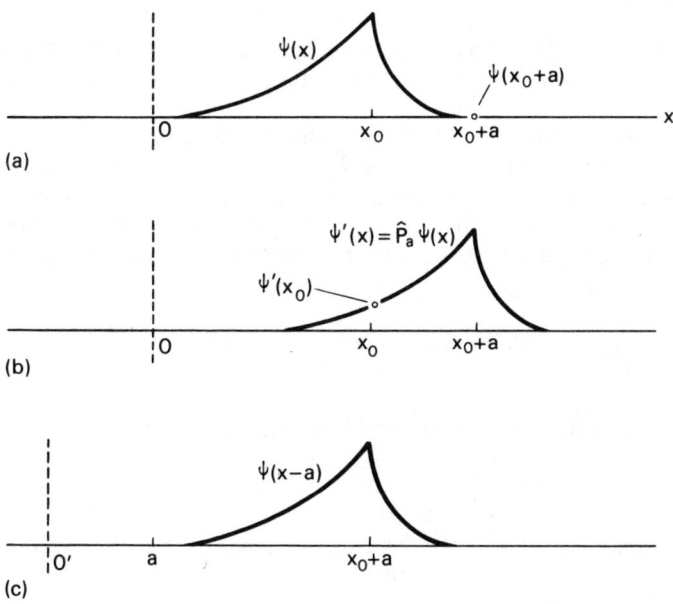

Figure 6.1.

†One should distinguish between the motion of the particle and the time development of its state function, which is more reminiscent of the movement of a ruck in a rug, or of a water wave.

at $x_0 + a$, and $\psi(x-a)$ fits the bill perfectly. Thus under the shift of origin, the representation of the particle state undergoes

$$\psi(x) \rightarrow \psi(x-a) \tag{6.5b}$$

Since moving the particle to the right by a is essentially equivalent to shifting the origin to the left, and should have the same effect upon $\psi(x)$, we can equate Eqs (6.5a) and (6.5b) to obtain

$$\psi(x) \rightarrow \psi'(x) = \hat{P}_a \psi(x) = \psi(x-a) \tag{6.6}$$

Now let us define a symmetry operator which displaces every position vector in real (i.e. x-) space by the amount a:

$$\hat{R}_a x = x + a \tag{6.7}$$

The set $\{\hat{R}_a\}$ forms a group, with unit operator $\hat{R}_0$ and inverses $\hat{R}_a^{-1} = \hat{R}_{-a}$. Combination of Eqs (6.6) and (6.7) now yields

$$\hat{P}_a \psi(\hat{R}_a x) = \psi(x) \tag{6.8}$$

in agreement with our fundamental definition Eq. (6.4a).

*PROBLEM 6.2 Do Eqs (6.5) depend upon the fact that $\hat{R}_a$ and/or $\hat{P}_a$ is a *symmetry* operator for the system?

We can produce an explicit form for this generator of finite displacements in function space, $\hat{P}_a$. Let us first consider what happens to $\psi(x)$ under an *infinitesimal* displacement Δa either of coordinate system to the left, or of the physical system in the other direction. $\psi(x)$ becomes the new function $\psi'(x) \equiv \psi(x - \Delta a)$, as with Eq. (6.6). But

$$\psi'(x) = \psi(x - \Delta a) = \psi(x) - \Delta a \frac{\partial \psi}{\partial x} + \ldots = \left(1 - \Delta a \frac{\partial}{\partial x} + \ldots\right) \psi(x) \tag{6.9a}$$

In the limit $\Delta a \rightarrow 0$, from comparison with Eq. (6.6),

$$\hat{P}_{\Delta a} \equiv \left(1 - \Delta a \frac{\partial}{\partial x}\right) \tag{6.9b}$$

is the *generator of infinitesimal translations*. Since the quantum mechanical momentum operator is

$$\hat{p} = -i\hbar \frac{\partial}{\partial x}, \tag{6.9c}$$

Eq. (6.9b) becomes

$$\boxed{\hat{P}_a = (1 - i\hat{p}\Delta a/\hbar)} \tag{6.10a}$$

A *finite* translation by the amount a may be obtained by taking n steps of size $\Delta a = a/n$ each:

$$\hat{P}_a = (\hat{P}_{\Delta a})^n = (\hat{1} - i\hat{p}a/n\hbar)^n \tag{6.10b}$$

Then, in the limit $n \to \infty$,

$$\boxed{\begin{aligned}\hat{P}_a &= \exp(-i\hat{p}a/\hbar) &&(6.11a)\\ \hat{P}_a\psi(x) &= \psi(x-a) &&(6.11b)\\ \hat{P}_a\psi(x+a) &= \hat{P}_a\psi(\hat{R}_a x) = \psi(x) &&(6.11c)\end{aligned}}$$

The operator $\hat{P}_a$ is meaningful in terms of its power series expansion:

PROBLEM 6.3 Define the operator $\exp(\hat{A})$ by

$$\exp(\hat{A}) = 1 + \hat{A} + \hat{A}^2/2! + \ldots \tag{6.12}$$

Prove that if $\hat{H}$ is hermitian, then $\hat{P} = \exp(i\hat{H})$ is unitary.

*__PROBLEM 6.4__ Show that $[\hat{P}_a, \hat{V}(x)] = 0$ implies the conservation of linear momentum.

If $\psi(x)$ should happen to be an eigenstate of $\hat{p}$,

$$\hat{p}\psi(x) = p\psi(x)$$

then the operator $\hat{p}$ in the exponent of $\hat{P}_a$ can be replaced by the eigenvalue p.

As our second example of the workings of $\hat{P}_R$, we consider a wave function of the form

$$\zeta(z) = \sin z \tag{6.13a}$$

If $z = z + 2\pi$, then we are discussing rotations, otherwise linear displacements. Suppose we displace or rotate the axes of z-space by the amount b. Our system is now described by

$$\begin{aligned}\zeta'(z) &= \sin(z+b) &&(6.13b)\\ &= \sin z \cos b + \cos z \sin b\end{aligned}$$

We associate the operator $\hat{R}_b$ with such transformations in z-space.

Sin z and cos z are orthogonal, in the sense that

$$\int_{-\pi/2}^{\pi/2} \sin z \cos z \, dz = 0 \tag{6.14a}$$

Any sinusoidal function of period 2π, moreover, can be written as a linear combination of these two functions. Therefore we choose to think of

$$X = \sin z \tag{6.14b}$$
$$Y = \cos z$$

as a complete, orthogonal basis set spanning a two-dimensional, abstract function space. Equations (6.13a) and (6.13b) become, respectively, Figs 6.2a and 6.2b, and see Problem 3.7,

$$\xi(z) = X \tag{6.14c}$$
$$\xi'(z) = X \cos b + Y \sin b$$

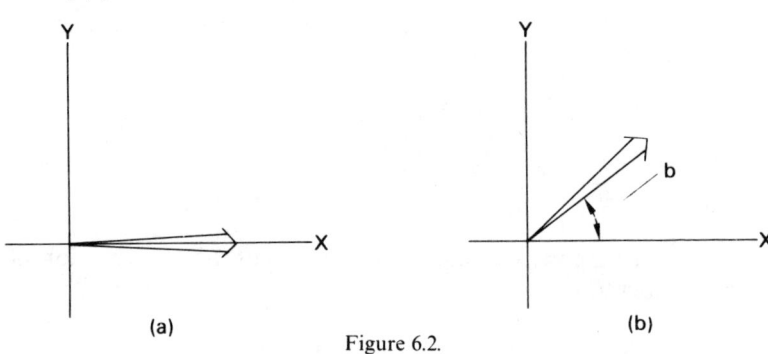

Figure 6.2.

These equations, however, describe a rotation by the angle b in XY space generated by the simple rotation operator $\hat{P}_b$, which acts in XY space, such that

$$\hat{P}_b \xi(\hat{R}_b z) = \xi(z) \tag{6.15}$$

We close this section with the example of another symmetry operator with an explicit form

$$\hat{P}_R = \exp(-i\hat{J} \cdot \bar{n}\phi/\hbar) \tag{6.16}$$

where $\bar{n}$ is a unit vector in real space. Equation (6.16) is the rotation operator appropriate to a space spanned by the eigenvectors of $\hat{J}$, the angular momentum operator (i.e. $\hat{P}_R$ operates in the space of spinors or spherical harmonics); it corresponds to, and mirrors in function space, a rotation in *real* space by ϕ about $\bar{n}$.

*PROBLEM 6.5 Derive Eq. (6.16) with $\hat{n} = |3\rangle$.

6.4 The Operators $\{\hat{P}_R\}$ form a Group Isomorphic to $\{\hat{R}\}$

To summarise: A system associated with the position vector $\bar{r}$ is described by $\psi(\bar{r})$. $\psi(\bar{r})$ exists in function space, but depends upon

REPRESENTATIONS OF AN OPERATOR IN FUNCTION SPACE

the real-space variable $\bar{r}$. If either sample or the co-ordinate system of real space is re-oriented, and $\bar{r} \to \hat{R}\bar{r} = \bar{r}'$, then we need a way of transferring this information to function space. Consequently we define $\hat{P}_R$ as that operator which exactly counteracts or nullifies the effects in function space of performing $\hat{R}$ in real space: $\hat{P}_R \psi(\hat{R}\bar{r}) \equiv \psi(\bar{r})$, Eq. (6.4a). By a slight reshuffling of labels, this is equivalent to $\hat{P}_R \psi(\bar{r}) = \psi(\hat{R}^{-1}\bar{r})$, Eq. (6.4b). Thus to every $\hat{R}$, there directly corresponds in a one-to-one fashion a unique $\hat{P}_R$. Although we have provided an explicit form for two particular $\hat{P}_R$'s, (Eqs (6.11) and (6.16)), we have defined the operator in general by how it behaves. Equation (6.4) presents one of the fundamental ideas of the theory of group representations, and we shall make much use of it.

In Chapter 1 we noted that if two rotations of a vector in real space occur in sequence, a third single transformation can be found which is equivalent to this sequence. If our construct $\hat{P}_R$ is, in fact, to offer function space a true mirror of events in real space, the same situation should apply to the $\hat{P}_R$'s. Thus if $\hat{P}_{R,1}$, $\hat{P}_{R,2}$, and $\hat{P}_{R,3}$ correspond, respectively, to $\hat{R}_1$, $\hat{R}_2$, and $\hat{R}_3$, then

$$\left.\begin{array}{l} \hat{P}_{R,3}\psi(\bar{r}) = \hat{P}_{R,2}\{\hat{P}_{R,1}\psi(\bar{r})\} \\ \text{if and only if} \\ \hat{R}_3\bar{r} = \hat{R}_2(\hat{R}_1\bar{r}) \end{array}\right\} \quad (6.17)$$

PROOF

$$\hat{P}_{R,1}\psi(\bar{r}) = \quad (6.18a)$$
$$\psi(\hat{R}_1^{-1}\bar{r}) \equiv g(\bar{r}) \quad (6.18b)$$

from Eq. (6.4b). Then

$$\hat{P}_{R,2}(\hat{P}_{R,1}\psi(\bar{r})) = \hat{P}_{R,2}g(\bar{r}) = g(\hat{R}_2^{-1}\bar{r}) = g(\bar{r}'') \quad (6.18c)$$

where

$$\bar{r}'' \equiv \hat{R}_2^{-1}\bar{r}$$

But from Eq. (6.18b)

$$g(\bar{r}'') = \psi(\hat{R}_1^{-1}\bar{r}'') = \psi\{\hat{R}_1^{-1}(\hat{R}_2^{-1}\bar{r})\} = \psi(\hat{R}_1^{-1}\hat{R}_2^{-1}\bar{r}) \quad (6.18d)$$
$$= \psi\{(\hat{R}_2\hat{R}_1)^{-1}\bar{r}\} = \psi(\hat{R}_3^{-1}\bar{r}) = \hat{P}_{R,3}\psi(\bar{r}) \quad \text{QED}$$

The validity of the step $(\hat{R}_1^{-1}\hat{R}_2^{-1}) = (\hat{R}_2\hat{R}_1)^{-1}$ is seen by left-multiplying both sides of this equation by $\hat{R}_1$, then $\hat{R}_2$.

Equation (6.17) verifies the desirable property that compound operations in real space are mirrored by compound operations in function space, and is just the group closure condition required if the set $\{\hat{P}_R\}$ is to form a group isomorphic to $\{\hat{R}\}$. We shall consider the nature of this isomorphism further in the next chapter.

6.5 Representations of $\hat{P}_R$

We have just formulated a method for keeping track of changes in function space due to transformations in real space. We can now turn to the problem of finding *representations* of the relevant operator $\hat{P}_R$ generated by the basis vectors $\{|\psi_i(\bar{r})\rangle\}$ of function space†; such representations are central to the application of group theory to quantum mechanics. In direct analogy to our path to

$$\hat{R}|i\rangle = \sum_j^3 |j\rangle R_{ji}, \qquad R_{ji} = \langle j|\hat{R}|i\rangle \qquad (6.3)$$

we consider the effects of $\hat{P}_R$ upon the *basis* states of Hilbert space. We can perform the fourier decomposition of $\hat{P}_R|\psi_i(\bar{r})\rangle$ as

$$\begin{aligned} \hat{P}_R|\psi_i(\bar{r})\rangle &= \sum_j |\psi_j(\bar{r})\rangle \langle \psi_j(\bar{r})|\hat{P}_R|\psi_i(\bar{r})\rangle \\ &= \sum_j |\psi_j(\bar{r})\rangle \Gamma_{ji}^{(\psi)}(\hat{R}) \end{aligned} \qquad (6.19a)$$

The superscript 'ψ' on

$$\Gamma_{ji}^{(\psi)}(\hat{R}) = \langle \psi_j(\bar{r})|\hat{P}_R|\psi_i(\bar{r})\rangle \qquad (6.19b)$$

is used to indicate that this representation $\Gamma^{(\psi)}(\hat{R})$ was generated via the $\{|\psi_i\rangle\}$ basis set; we can therefore call it the ψth representation of $\hat{P}_R$.‡ Γ was first introduced in Eqs (4.16), (5.29c), and (5.34b).

Along with the definition Eq. (6.4) of $\hat{P}_R$, Eq. (6.19) is fundamental to all that follows. Equation (6.19) is valid, of course, only if the $\{|\psi_k(\bar{r})\rangle\}$ are complete, and obey a closure relationship.

There are two reasons for wanting to find the matrices $\bar{\bar{\Gamma}}(\hat{R})$. First, knowledge of $\bar{\bar{\Gamma}}(\hat{R})$ allows us to determine, through the results of Problem 6.10 below, how a general state-representation vector $\psi(\bar{r})$ behaves with a change of basis system in real space. And second, the application of group theoretic methods to quantum mechanical problems depends upon unusual properties of those particular representations of $\hat{P}_R$ which happen to be irreducible. We shall consider these properties in detail in the next two chapters.

†We have been reserving $\psi_i(\bar{r})$ for the ith eigenvector of the Hamiltonian. This will generally still be the case in what follows, since nearly any of the complete sets we consider will constitute the eigenspectrum of some $\mathcal{H}$. But it is not necessarily part of the development, and $\{\psi_i(\bar{r})\}$ could represent some other complete basis set.

‡See footnote page 13.

†6.6 The Link between Real Space and Function Space: $\hat{R}$ and $\hat{P}_R$

Now that we have introduced $\hat{P}_R$ and its representations of the form $\bar{\bar{\Gamma}}(\hat{R})$, as an aside we shall return briefly to the question of the connection between $\hat{P}_R$ and $\hat{R}$. Although we shall provide a far from complete picture, this should shed a bit more light on the inter-relationship among the space containing† three-vectors $\bar{r}$, abstract state-vectors $|\psi\rangle$ and $|r\rangle$, and the state-representation functions $\psi(\bar{r})$. We begin by reviewing the properties of three-vectors, and work from there by analogy. Some of the ideas discussed here were first presented in Section 3.9.

Real space is spanned by three unit vectors $\{|i\rangle\}$, and an arbitrary three-vector $|x\rangle$ can be decomposed as

$$|x\rangle = \sum_i^3 |i\rangle\langle i|x\rangle = \sum_i^3 x_i |i\rangle \tag{6.20a}$$

where $\langle i|x\rangle = x_i$ is the component of $|x\rangle$ lying along the ith coordinate axis; the components x_i can be assembled to form the representative column vector $\begin{pmatrix} x_1 \\ x_2 \\ x_3 \end{pmatrix}$ of $|x\rangle$, and the representation-space associated with three-space is spanned by unit vectors $\begin{pmatrix} 1 \\ 0 \\ 0 \end{pmatrix}, \begin{pmatrix} 0 \\ 1 \\ 0 \end{pmatrix}$ and $\begin{pmatrix} 0 \\ 0 \\ 1 \end{pmatrix}$; these can, of course, be rotated into three other orthonormal column vectors, as a change of basis of representation space. (Problem 2.8.)

The Hilbert space containing the general states $|\psi\rangle$ can similarly be decomposed in terms of unit vectors $\{|r\rangle\}$. These unit vectors are not simply the $|x\rangle$ or $\bar{r}$ of three-space; rather, they inhabit, and span, abstract *Hilbert* space, and are quantum mechanically significant states. The arbitrary state $|\psi\rangle$ can be decomposed as

$$|\psi\rangle = \int |r\rangle\langle r|\psi\rangle \, d^3r = \int \psi_r |r\rangle \, d^3r \tag{6.20b}$$

where we have temporarily written $\langle r|\psi\rangle$ as ψ_r to emphasize the analogy with Eq. (6.20a); ψ_r is the component of $|\psi\rangle$ along the 'rth axis' of Hilbert space.

But although $|r\rangle$ is not itself a three-vector, there exists an intimate tie between the $|r\rangle$'s spanning Hilbert space and the

---

†In this section we shall use $|x\rangle$ and $\bar{r}$ as general vectors in three-space, and $\{|r\rangle\}$ as the continuum of unit vectors of abstract Hilbert space.

three-vectors $\bar{r}$ of real-space: every $|r\rangle$ in Hilbert space corresponds in a one-to-one fashion to a point in three-space; equivalently, every $\bar{r}$ in real-space labels a basis vector $|r\rangle$ in Hilbert space. The truth of this is seen from Eq. (3.38a), in which we defined a Hermitian operator $\hat{r}$ corresponding to a quantum mechanical measurement of position, such that

$$\hat{r}|r\rangle = \bar{r}|r\rangle \tag{3.38a}$$

The eigenvectors of $\hat{r}$ are our $\{|r\rangle\}$, and the corresponding eigenvalues are the ordinary three-space vectors $\bar{r}$. Therefore we can interpret the variable r both as a Hilbert space axis-label, in $|r\rangle$, and also in the real-space sense, in $\bar{r}$, and it is thus meaningful to write

$$\langle r|\psi\rangle = \psi_r \equiv \psi(\bar{r}) \tag{6.21}$$

Moreover, just as the components $\{x_1, x_2, \text{ and } x_3\}$ of $|x\rangle$ combined to produce a new kind of vector, the representation $\begin{pmatrix} x_1 \\ x_2 \\ x_3 \end{pmatrix}$, so also the 'components' $\{\psi(\bar{r}), \text{ all } \bar{r}\}$ of $|\psi\rangle$ can be considered to describe a vector in a new space, state-representation space. And as $\begin{pmatrix} 1 \\ 0 \\ 0 \end{pmatrix}$, etc. constitute the basis vectors of the representation space associated with real-space, likewise one can find basis vectors $\{\psi_i(\bar{r})\}$, such as s-, p-, d-states spanning the representation-space associated with abstract Hilbert space.†

On to $\hat{P}_R$. Previously on several occasions, we have examined the consequences of a change of basis upon the description of a system. We now shall consider the way in which a rotation of the coordinate axes of real space affects the vectors $\psi(\bar{r})$ of state-representation space. We shall approach this topic rather indirectly, however, and first define an operator $\hat{\mathscr{R}}$ which acts in *abstract* Hilbert space to produce the transformation of basis vectors

$$\hat{\mathscr{R}}|r\rangle = |r'\rangle \tag{6.22a}$$

$\mathscr{R}$ transforms an arbitrary vector, of course, as $\hat{\mathscr{R}}|\psi\rangle = |\psi'\rangle$.

The new basis states $\{|r'\rangle\}$ obey $\hat{r}|r'\rangle = r'|r'\rangle$. But from Chapter 4 there is a real-space operator $\hat{R}$ for which

$$\hat{R}\bar{r} = \bar{r}' \tag{6.22b}$$

Thus the transformations of eigenvectors and eigenvalues of $\hat{r}$,

---

†The subscript i indicates that our analogy here is not exact. What has happened?

Eqs (6.22a) and (6.22b), go hand in hand, and there exists an isomorphism between $\{\hat{R}\}$ and $\{\hat{\mathscr{R}}\}$:

$$\hat{R} \leftrightarrow \hat{\mathscr{R}} \tag{6.22c}$$

The interested reader is encouraged to examine this isomorphism in greater detail.

Finally we turn to state-representation Hilbert space. If we begin with a $\{|r\rangle\}$ representation $\psi(\vec{r})$ of some state $|\psi\rangle$, but perform a rotation of the physical system in question, or of the coordinate axes of real space, then $\{|r\rangle\}$ will be transformed into $\{|r'\rangle\}$ in accordance with Eq. (6.22a). With the new basis set,

$$|\psi\rangle = \int |r'\rangle \langle r'|\psi\rangle \mathrm{d}^3 r' = \int |r'\rangle \psi(\vec{r}') \, \mathrm{d}^3 r' \tag{6.23a}$$

and the representation of $|\psi\rangle$ changes from $\psi(\vec{r})$ to $\psi(\vec{r}')$. Since $\hat{R}\vec{r} = \vec{r}'$ provides a clear-cut relationship between $\vec{r}'$ and $\vec{r}$, we can express $\psi(\vec{r}')$ as

$$\begin{aligned} \psi(\vec{r}') &= \psi(x_1', x_2', x_3') = \psi[x_1'(x_1, x_2, x_3), x_2'(x_1, x_2, x_3), \\ & \qquad x_3'(x_1, x_2, x_3)] \\ &\equiv \psi'(x_1, x_2, x_3) = \psi'(\vec{r}) \end{aligned} \tag{6.23b}$$

which defines a completely new function $\psi'(\vec{r})$. In short,

$$\psi(\vec{r}') = \psi(\hat{R}\vec{r}) \equiv \psi'(\vec{r}) \tag{6.23c}$$

$\psi'(\vec{r})$ is a function of $\vec{r}$ which is totally different from our original $\psi(\vec{r})$. We can interpret this as saying that under an operation $\hat{R}$ on the vectors of real space, mirrored in the corresponding transformation $\mathscr{R}|r\rangle = |r'\rangle$ of abstract Hilbert space, the representation $\psi(\vec{r})$ is rotated, reflected, or translated into a totally new vector $\psi'(\vec{r})$ in state-representation space. It is appropriate now to introduce a third operator $\hat{P}_R$ which carries the information content of $\hat{R}$ or $\hat{\mathscr{R}}$, but which acts directly in state-representation space. We choose or define $\hat{P}_R$ to be that special operator which performs in function space so as to counteract the effects of $\hat{R}$ in real space; it is specifically designed to convert $\psi(\hat{R}\vec{r}) = \psi'(\vec{r})$ into a function $\hat{P}_R \psi(\hat{R}\vec{r}) \equiv \hat{P}_R \psi'(\vec{r})$ identical to $\psi(\vec{r})$, the original representation of $|\psi\rangle$ before the $\hat{R}$ rotation ever occurred. That is, $\hat{P}_R \psi(R\vec{r}) = \psi(\vec{r})$. $\hat{P}_R$ thus establishes a natural, and indeed, the simplest possible, one-to-one correspondence between activities in real space and events in function space.

We end up with a three-way isomorphism

$$\begin{array}{c} \hat{R} \\ \nearrow \searrow \\ \hat{\mathscr{R}} \leftrightarrow \hat{P}_R \end{array} \tag{6.24}$$

which provides a symmetrical chain of transformations linking the three spaces of interest.

We now return to the central issue of the chapter, the generation of representations $\bar{\bar{\Gamma}}$ of $\hat{P}_R$ through the use of basis vectors $\{\psi_i(\bar{r})\}$ of state-representation space.

6.7 Representation of $\hat{P}_R$ by d-States: $\Gamma^{(d)}(\hat{R})$

As an important example of the use of Eq. (6.19), in forming function space representations of the transformation operators $\hat{P}_R$, we consider the effects of an $\hat{R}_\phi$ rotation of the real-space co-ordinate system upon hydrogenic d-states, whose angular dependences are those of spherical harmonics of order two, Table 5.4. Since rotations do not alter functions of the type $F(|\bar{r}|)$, we disregard the multiplicative radial factor and normalization.

The d-states, expressed in their real-valued forms with $x_i = \langle i|x\rangle$, are

$$\psi_i(r) = \begin{cases} d_1(\bar{r}) = \tfrac{1}{2}(x_1{}^2 - x_2{}^2) \\ d_2(\bar{r}) = x_1 x_2 \\ d_3(\bar{r}) = x_1 x_3 \\ d_4(\bar{r}) = x_2 x_3 \\ d_5(\bar{r}) = (3x_3{}^2 - r^2)/2\sqrt{3} \end{cases} \quad (6.25)$$

Then from Eq. (6.4b)

$$\hat{P}_R \psi_i(\bar{r}) = \psi_i(\hat{R}^{-1}\bar{r}) = \psi_i(\bar{r}'') = \begin{cases} d_1(\bar{r}'') = \tfrac{1}{2}(x_1''{}^2 - x_2''{}^2) \\ d_2(\bar{r}'') = x_1'' x_2'' \\ \text{etc.} \end{cases} \quad (6.26)$$

where $\bar{r}'' = \hat{R}^{-1}\bar{r}$. From the matrix representation of $\hat{R}$, Eq. (4.7), and since $\hat{R}$ is unitary, Eq. (5.9),

$$\bar{\bar{R}}^{-1} = \bar{\bar{R}}^t = \begin{pmatrix} \cos\phi & \sin\phi & 0 \\ -\sin\phi & \cos\phi & 0 \\ 0 & 0 & 1 \end{pmatrix} \quad (6.27a)$$

$\bar{r}'' = \hat{R}^{-1}\bar{r}$ can thus be expressed in terms of its components as

$$\begin{pmatrix} x_1'' \\ x_2'' \\ x_3'' \end{pmatrix} = \bar{\bar{R}}^{-1}\begin{pmatrix} x_1 \\ x_2 \\ x_3 \end{pmatrix} = \begin{pmatrix} x_1 \cos\phi + x_2 \sin\phi \\ -x_1 \sin\phi + x_2 \cos\phi \\ x_3 \end{pmatrix} \quad (6.27b)$$

or

$$\begin{aligned} x_1'' &= x_1 \cos\phi + x_2 \sin\phi \\ x_2'' &= -x_1 \sin\phi + x_2 \cos\phi \\ x_3'' &= x_3 \end{aligned} \quad (6.27c)$$

Employing Eq. (6.27c) directly on $d_1(\bar{r}'')$ of Eq. (6.26), we find that

$$\begin{aligned}\hat{P}_R d_1(\bar{r}) &= \tfrac{1}{2}(x_1''^2 - x_2''^2) \\ &= \tfrac{1}{2}(x_1^2 \cos^2\phi + x_2^2 \sin^2\phi + 2x_1 x_2 \sin\phi \cos\phi) \\ &\quad -\tfrac{1}{2}(x_1^2 \sin^2\phi + x_2^2 \cos^2\phi - 2x_1 x_2 \sin\phi \cos\phi) \quad (6.28)\\ &= \tfrac{1}{2}(x_1^2 - x_2^2)(\cos^2\phi - \sin^2\phi) + x_1 x_2 (2\sin\phi \cos\phi) \\ &= d_1(\bar{r}) \cos 2\phi + d_2(\bar{r}) \sin 2\phi\end{aligned}$$

Thus the operator $\hat{P}_R$ transforms $d_1(\bar{r})$ into a totally *new function* of $\bar{r}$, $d_1(\bar{r}) \cos 2\phi + d_2(\bar{r}) \sin 2\phi$, in function space! Similarly,

$$\boxed{\begin{aligned}\hat{P}_R d_1(\bar{r}) &= d_1(\bar{r}) \cos 2\phi + d_2(\bar{r}) \sin 2\phi \\ \hat{P}_R d_2(\bar{r}) &= -d_1(\bar{r}) \sin 2\phi + d_2(\bar{r}) \cos 2\phi \\ \hat{P}_R d_3(\bar{r}) &= d_3(\bar{r}) \cos \phi + d_4(\bar{r}) \sin \phi \\ \hat{P}_R d_4(\bar{r}) &= -d_3(\bar{r}) \sin \phi + d_4(\bar{r}) \cos \phi \\ \hat{P}_R d_5(\bar{r}) &= d_5(\bar{r})\end{aligned}} \quad (6.29)$$

Just as in Section 5.2 we can now either (1) compare Eq. (6.29) term by term with Eq. (6.19) to find $\Gamma_{ji}^{(d)}(\hat{R}_\phi)$ or, (2) we can actually calculate

$$\begin{aligned}\Gamma_{11}^{(d)}(\hat{R}_\phi) &= \langle d_1|(\hat{P}_R|d_1\rangle) = \langle d_1|\hat{P}_R|d_1\rangle = \cos 2\phi \\ \Gamma_{21}^{(d)}(\hat{R}_\phi) &= \langle d_2|(\hat{P}_R|d_1\rangle) = \langle d_2|\hat{P}_R|d_1\rangle = \sin 2\phi\end{aligned} \quad (6.30)$$

etc.

where we use $\langle d_i(\bar{r})|d_j(\bar{r})\rangle = \langle d_i|d_j\rangle = \delta_{ij}$

Either way we find

$$\bar{\bar{\Gamma}}^{(d)}(\hat{R}_\phi) = \begin{pmatrix} \cos 2\phi & -\sin 2\phi & 0 & 0 & 0 \\ \sin 2\phi & \cos 2\phi & 0 & 0 & 0 \\ 0 & 0 & \cos \phi & -\sin \phi & 0 \\ 0 & 0 & \sin \phi & \cos \phi & 0 \\ 0 & 0 & 0 & 0 & 1 \end{pmatrix} \quad (6.31a)$$

We add the superscript 'd' to Γ to indicate that this particular representation of $\hat{P}_R$ (and indirectly, of $\hat{R}_\phi$) was generated using a d-orbital basis system. The procedure we have used is completely analogous to that used in leading up to Eq. (4.7), the representation matrix $\bar{\bar{R}}$ of the real-space operator $\hat{R}$.

*PROBLEM 6.6 The derivation of Eq. (6.31a) depended critically upon the use of Eq. (6.27), which follows from Eq. (4.7). Now use Eq. (4.8), instead, to prove that the representation generated by

d-states of the mirror operator $\hat{P}_{M,\phi}$ is

$$\Gamma^{(d)}(\hat{M}_\phi) = \begin{pmatrix} \cos 4\phi & \sin 4\phi & 0 & 0 & 0 \\ \sin 4\phi & -\cos 4\phi & 0 & 0 & 0 \\ 0 & 0 & \cos 2\phi & \sin 2\phi & 0 \\ 0 & 0 & \sin 2\phi & -\cos 2\phi & 0 \\ 0 & 0 & 0 & 0 & 1 \end{pmatrix} \quad (6.31b)$$

*PROBLEM 6.7 $\bar{\bar{R}}$ is a unitary matrix; is $\bar{\bar{\Gamma}}^{(d)}(\hat{R})$?

*PROBLEM 6.8 Try to obtain the matrix representation Eq. (6.31a) from Eq. (6.16); i.e. evaluate matrix elements of the sort $\langle \psi_i | \exp(-i\hat{J}_3\phi/\hbar) | \psi_j \rangle$. Does it help to use the complex-valued spherical harmonics of order 2 instead?

PROBLEM 6.9 If $\mathscr{H}$ is the Hamiltonian of a system, what does $[\mathscr{H}, \hat{J}_3] = 0$ tell us about $\mathscr{H}$ and $\hat{J}_3$? See Problem 6.4 and Eq. (6.16).

It is interesting that while the generator $\hat{P}_R = \exp(-i\hat{J}\cdot\bar{n}\phi/\hbar)$ of finite rotations, Eq. (6.16) acts only in the abstract function space spanned by the eigenvectors of $\hat{J}$, $\langle J \rangle$ the *expectation value* of $\hat{J}$ behaves like a conventional axial vector. Thus if we simultaneously examine both in real space and in function space an entity with angular momentum $\hat{J}$, Eq. (6.19) will tell us how the eigenvectors of $\hat{J}$ (as basis vectors of function space) change, but Eq. (6.3) will describe the change in the expectation value of $\hat{J}$. This is the sort of connection between real space and function space that is offered by our definition Eq. (6.4) of $\hat{P}_R$.

PROBLEM 6.10 Equation (6.19) tells us how to rotate the unit vectors $\{\psi_i(\bar{r})\}$ of state-representation space. How do we, in analogy to Eq. (4.12), transform the representation $\psi(\bar{r})$ of an *arbitrary* state $|\psi\rangle$, where

$$\psi(\bar{r}) = \sum_i c_i \psi_i(\bar{r})? \quad (6.2b)$$

PROBLEM 6.11 In a fashion analogous to the one yielding Eq. (6.31), use p-states to find $\bar{\bar{\Gamma}}^{(p)}(\hat{R}_\phi)$ and $\bar{\bar{\Gamma}}^{(p)}(\hat{M}_\phi)$ and compare with Eq. (4.7 and 4.8).

†PROBLEM 6.12 With respect to C_{4v} (the square-based pyramid), it is legitimate to describe the influence of $\hat{R}_{90}$ upon the point with coordinates (x, y, z), by means of

$$\bar{\bar{R}}_{90} \begin{pmatrix} x \\ y \\ z \end{pmatrix} = \begin{pmatrix} x' \\ y' \\ z' \end{pmatrix} = \begin{pmatrix} y \\ -x \\ z \end{pmatrix} = \begin{pmatrix} 0 & 1 & 0 \\ -1 & 0 & 0 \\ 0 & 0 & 1 \end{pmatrix} \begin{pmatrix} x \\ y \\ z \end{pmatrix}$$

Why does it follow immediately that

$$\Gamma^{(p)}(\hat{R}_{90}) = \begin{pmatrix} 0 & -1 & 0 \\ 1 & 0 & 0 \\ 0 & 0 & 1 \end{pmatrix}$$

[HINT: Reconsider Problem 6.11.]

*PROBLEM 6.13 Show that the representations Eq. (6.31a) and (6.31b) yield, with proper choices of ϕ, a group isomorphic to C_{3v}. Do the six matrices have the same block structure?

6.8 Block-Diagonal Form of $\Gamma^{(d)}(\hat{R}_\phi)$

Again, we note the significance of the block-diagonal form, this time in Eq. (6.31). $\Gamma_{31} = \Gamma_{41} = \Gamma_{51} = 0$ and the fundamental Eq. (6.19)

$$\hat{P}_R |d_i(\bar{r})\rangle = \sum_j |d_j(\bar{r})\rangle \Gamma_{ji}^{(d)}(\hat{R}_\phi) \tag{6.19a}$$

tell us that when we operate on $d_1(\bar{r})$ with $\hat{P}_{R,\phi}$—which is to say, when we rotate our co-ordinate system in real space about the $|3\rangle$ axis by any angle ϕ—the resultant transformed function in d-space will be a linear combination of d_1 and d_2 only, containing no components in the 'directions' of d_3, d_4, or d_5. Similarly, Γ_{13}, Γ_{14}, $\Gamma_{15} = 0$ says that $\hat{R}$ will leave $\hat{P}_R d_3(\bar{r})$, $\hat{P}_R d_4(\bar{r})$, and $\hat{P}_R d_5(\bar{r})$ free of any components along $d_1(\bar{r})$. The same arguments apply to $d_2(\bar{r})$, $d_3(\bar{r})$, and $d_4(\bar{r})$. Finally, no rotation about $|3\rangle$ mixes $d_5(\bar{r})$ with any other vector in d-function space.

We say that d-space has been separated into three subspaces, invariant with respect to $\hat{R}_\phi$. d_1 and d_2 mix under $\hat{P}_R$, d_3 and d_4 also mix, and d_5 remains unchanged. If we were to tilt the axis of rotation slightly, all the zeroes in Eq. (6.31) would vanish, and any rotation would admix contributions from all five basis functions. It is because we have exploited the special symmetry of the situation that Eq. (6.31) is block diagonalized: the five d-states are first and foremost functions of $\bar{r}$,

$$\begin{aligned} d_1(\bar{r}) &= \tfrac{1}{2}(x_1^2 - x_2^2) \\ d_2(\bar{r}) &= x_1 x_2 \\ d_3(\bar{r}) &= x_1 x_3 \\ d_4(\bar{r}) &= x_2 x_3 \\ d_5(\bar{r}) &= (3x_3^2 - r^2)/2\sqrt{3} \end{aligned} \tag{6.25}$$

and their form can be chosen (d-space can be aligned) to reflect the presence of an axis of rotation. Thus if we choose $|3\rangle$ such that x_3 is constant under $\hat{R}_\phi$, $d_5(\bar{r})$ is obviously invariant also. Likewise, d_3 and d_4 transform under $\hat{P}_{R,\phi}$ as do x_1 and x_2 under $\hat{R}_\phi$; that is, they are mixed exactly as are the $|1\rangle$ and $|2\rangle$ unit vectors of Fig. 4.1. This explains the presence in Eq. (6.31) of a sub-block of the form of $\overline{\overline{R}}$, Eq. (4.7). d_1 and d_2 seem to behave in a more complex fashion.

*PROBLEM 6.14 For a rotation of 120 degrees compare the transformational properties of (d_3, d_4), Eq. (6.29), with those of (x, y). Now do (d_1, d_2).

*PROBLEM 6.15 Construct a representation of $\{\hat{P}_R\}$ for C_{3v} using a p-state basis system, and compare with Eq. (4.18).

We could rephrase the above by saying that under $\hat{P}_{R}$, d_3 and d_4 display the symmetry behaviour of $|1\rangle$ and $|2\rangle$ under $\hat{R}_\phi$, and d_5 is, like $|3\rangle$, invariant. d_1 and d_2 clearly belong together, but their symmetry properties are not yet obvious. Problem 6.13 showed that if d-states are employed to generate a representation of all six $\hat{P}_R$ operators of C_{3v}, the six $\Gamma^{(d)}(\hat{R})$ matrices have the same block diagonal structure. Exactly as in Section (4.5), one can argue here also that this d-representation of the symmetry group breaks naturally into irreducible representations. d_5 generates an A_1 irreducible representation, d_3 and d_4 produce one of E symmetry, and d_1 and d_2 lead to something which appears to be new. (In fact, it is not!)

6.9 How can an Incomplete Basis Set Generate a Meaningful Representation?

One small point may have caused the reader a bit of consternation—namely that a set of five d-states is not complete. The d-state basis set alone does not obey a closure relation; only the *entire* set of *all* hydrogenic orbitals, $\{|1s\rangle, |2s\rangle, |2p\rangle\text{'s}, |3s\rangle, |3p\rangle\text{'s}$ $|3d\rangle\text{'s}, |4s\rangle, \ldots \}$, is complete (ignoring continuum states). Has it been meaningful, then, to produce a representation using d-states alone?

The answer is yes, since five d-states together themselves constitute a subspace invariant under any possible rotation or reflection symmetry operation of interest to us, and this vitally significant statement is demonstrated in

*PROBLEM 6.16 Show that

$$\Gamma_{ij}(\hat{R}) \equiv \langle \psi_i(\bar{r})|\hat{P}_{R,\phi}|d_j(\bar{r})\rangle \tag{6.32}$$

vanishes for any hydrogenic state $|\psi_i(\bar{r})\rangle$ which is not a d-state. Does this allow us to use Eq. (6.19) to establish the representations $\overline{\overline{\Gamma}}^{(d)}(\hat{R}_\phi)$ and $\overline{\overline{\Gamma}}^{(d)}(\hat{M}_\phi)$ with five d-states alone?

*PROBLEM 6.17 Does $\Gamma_{ij}^{(d)}(\hat{R}) = 0$ if $|d_i\rangle$ and $|d_j\rangle$ are hydrogenic d-states but of different principal quantum numbers, e.g. $|3d_1\rangle$ and $|4d_2\rangle$, and $\hat{R}$ is any symmetry operator of C_{3v}?

REPRESENTATIONS OF AN OPERATOR IN FUNCTION SPACE 111

If we use the complete set of atomic orbitals $\{|1s\rangle, |2s\rangle, |2p\rangle$'s, $|3s\rangle, |3p\rangle$'s, $|3d\rangle$'s, ... $\}$ to generate an infinite dimensional representation, $\Gamma_{\text{atomic}}(\hat{R})$, the very fact that these basis vectors have the familiar atomic orbital orthogonality properties† breaks function space into subspaces, and leads automatically to a block-diagonal form for $\Gamma_{\text{atomic}}(\hat{R})$:

$$\Gamma_{nln'l'}^{\text{atomic}}(\hat{R}) = \langle nl|\hat{P}_R|n'l'\rangle$$
$$= \langle nl|\hat{P}_R|nl\rangle \delta_{nn'}\delta_{ll'} \qquad (6.33)$$

An l-sub-block is, of course, $2l+1$ dimensional. A suitable alignment of each individual (nl)-subspace (such as a d-space when used in generating a representation of C_{3v}) may cause a further separation of function space into sub-spaces of different symmetry types. As shown in Section 5.10, the basis vectors for each of these subspaces are degenerate.

PROBLEM 6.18 How much of the content of this chapter depends upon the fact that $\hat{P}_R$ is a symmetry operator? Where have we assumed that $\{\hat{P}_R\}$ is a continuous group of rotations, containing infinitesimal rotations?

6.10 Summary

1. If a system is in the state $|\psi\rangle$, then motion of the system or of the observer will lead to a change in the *description* $\langle r|\psi\rangle = \psi(\bar{r})$ of the system to some new description $\psi(\hat{R}\bar{r}) = \psi'(\bar{r})$. To account for the change, we define $\hat{P}_R$, an operator in function space which nullifies the consequences of variations in real space.

$$\hat{P}_R \psi(\hat{R}\bar{r}) = \psi(\bar{r}) \qquad (5.4)$$

The operators $\{\hat{P}_R\}$ form a group isomorphic to $\{\hat{R}\}$.

2. One can form representations of $\hat{P}_R$ with the basis set $\{|\psi_i(\bar{r})\rangle\}$

$$\hat{P}_R|\psi_i(\bar{r})\rangle = \sum_j |\psi_j(\bar{r})\rangle \Gamma_{ji}^{(\psi)}(\hat{R}) \qquad (6.19)$$

†We may seem to be cheating somewhat in this regard; atomic orbitals have uniquely helpful orthogonality properties which lead to an exceedingly transparent analysis. But the ideas and approaches presented are quite broad, even though we reach them via example rather than by vigorous proof; and the treatment becomes somewhat self-consistent later when we show that the proprieties of atomic orbitals exploited here can in turn be explained by symmetry arguments—d-space, for example, is invariant to any $\hat{P}_R$ precisely because of the spherical symmetry of atoms!

Also, in the actual applications of group representation theory to quantum mechanical problems, it is true that we frequently do use atomic orbitals as basis vectors.

Such representations will be seen to be of importance later.

3. The representations of $\hat{P}_R$ for $\hat{R}_\phi$ and $\hat{M}_\phi$ by d-states are worked out in a fashion directly analagous to the way in which we found the matrix representation $\bar{\bar{R}}_\phi$ of $\hat{R}_\phi$ generated by the $\{|1\rangle, |2\rangle, |3\rangle\}$ of real space. Thus we can find a set of six representative matrices for, say, the group C_{3v} as generated by d-states.

4. The block-diagonal form says that for our system of axial symmetry, we have found an advantageous set of basis vectors—or alignment of co-ordinate axes—of function space, which breaks it into subspaces invariant under the $\hat{P}_R$ for $\hat{R}_\phi$ and $\hat{M}_\phi$. The basis vectors $d_1(\bar{r})$ and $d_2(\bar{r})$ span one such subspace, $d_3(\bar{r})$ and $d_4(\bar{r})$ a second, and $d_5(\bar{r})$ a third. $d_5(\bar{r})$ is invariant to the $\{\hat{P}_R\}$ of C_{3v}, as $|3\rangle$ was invariant to the $\{\hat{R}\}$ of C_{3v}; the pairs $\{d_1, d_2\}$ and $\{d_3, d_4\}$, on the other hand, display the symmetry behaviour of $\{|1\rangle, |2\rangle\}$ under the $\{\hat{P}_R\}$ of C_{3v}.

5. Two $\hat{P}$-type operators with explicit forms are the generators of finite symmetry displacements and rotations:

$$\hat{P}_{\bar{a}} = \exp(-i\hat{p} \cdot \bar{a}/\hbar)$$
$$P_{n,\phi} = \exp(-i\hat{J} \cdot \bar{n}\phi/\hbar)$$

CHAPTER 7

Representations of Groups of Operators

This chapter provides a reconsideration, in greater depth, of the ideas of groups, representations, and representations of groups.

A p-state ($l=1$) representation of the six operators of C_{3v} is presented, and shown to be isomorphic to the original group; it is found, however, that all the matrices generated by an s-state are of the form $\Gamma^{(s)}(\hat{R})=1$. These observations lead to the definitions of equivalence of representations and of homomorphism. It is argued that for quantum mechanical purposes, an isomorphic representation is *not* necessarily to be considered more important than a homomorphic representation; while an isomorphic representation obviously contains complete information about the original symmetry group, a collection of several homomorphic representations may have the same information content, but in much more useful a form.

A set of large matrices representing a group may be built up out of smaller matrix representations. Similarly, sets of large matrices can be simultaneously reduced optimally, into diagonal blocks of *irreducible* representations.

7.1 More about Groups

Let us review a bit of what we expect of rotations $\hat{R}_\phi$ about $|3\rangle$ in real space. First of all, $\hat{R}_\alpha$ followed by $\hat{R}_\beta$ will surely give us $\hat{R}_{(\alpha+\beta)}$. Second, $\hat{R}_0$, a rotation through 0 degrees, is as meaningful as any other. And finally, $\hat{R}_\phi$ followed by $\hat{R}_{-\phi}$ is equivalent to $\hat{R}_0$. Just as we defined vectors painlessly by generalizing the notion of a directed line segment in Chapter 2, so we can also define a group by generalizing the behaviour of the set of rotations $\{\hat{R}_\phi\}$.

DEFINITION 7.1 *Group.* A set of elements $\{X_i\}$ for which a binary operation × (called the 'composition' or 'product') is defined constitutes the group G if the following four properties obtain:

(i) $X_i, X_j \in G \Rightarrow X_i \times X_j \in G$ (7.1a)

(ii) $X_i \times (X_j \times X_k) = (X_i \times X_j) \times X_k$ (7.1b)

(iii) there exists a unique 'identity' element $I \in G$ such that for every $X_i \in G$,

$$X_i \times I = I \times X_i = X_i$$ (7.1c)

(iv) for every $X_i \in G$ there exists a unique 'inverse' $X_i^{-1} \in G$ such that

$$X_i \times X_i^{-1} = X_i^{-1} \times X_i = I$$ (7.1d)

The first of these is called the *group closure* property, and states that the product of any two elements in a group is an element also in the group. The second is the law of associativity and not of much interest to us here (although it is critical in the study of continuous groups). The *identity* element and the presence of *inverses* for all elements are crucial.

DEFINITION 7.2 The *order* of a group is the number of distinct, non-equivalent elements in the group.

DEFINITION 7.3 An *abelian* or *commutating* group is one in which

$$X_i, X_j \in G \Rightarrow X_i \times X_j = X_j \times X_i, \text{ or } [X_i, X_j] = 0.$$ (7.1e)

The symmetry groups of molecules, and of isolated localized defects in crystalline solids, are called *point groups*; it is easy to see that all rotation and reflection operations of such a symmetry group will leave at least a single point (the centre of symmetry) unmoved. The symmetry group of a crystal, on the other hand, which includes not only rotations and reflections but also (neglecting boundary effects) the finite spatial displacements of a periodic lattice, is called its *space group*.

PROBLEM 7.1 Is it possible for a two-dimensional lattice to display C_{3v} symmetry? C_{4v}? C_5?

A group can be described, and in fact is completely and uniquely specified, by its *multiplication table*. For example, typical is the group of order 3 whose elements are α, $\beta = \alpha^2$, $\varepsilon = \alpha^3$, where $\alpha^4 \equiv \alpha$. Since $\alpha \varepsilon = \alpha^4 = \alpha$, ε is the identity element, and $\alpha \varepsilon = \varepsilon \alpha = \alpha$, $\beta \varepsilon = \varepsilon \beta = \beta$,

REPRESENTATIONS OF GROUPS OF OPERATORS   115

and $\varepsilon^2 = \varepsilon$. Similarly $\beta^2 = \alpha$. All this information can be tabulated as

Table 7.1

	α	β	ε
α	β	ε	α
β	ε	α	β
ε	α	β	ε

The group is abelian and there is no distinction between $\alpha\beta$ and $\beta\alpha$.

In non-abelian groups, however, to construct a table one must adapt a convention regarding the ordering of terms in a product. Consider the group of six elements $\{A, B, C, D, E, F\}$ whose multiplication table is displayed below.

Table 7.2

	A	B	C	D	E	F
A	E	D	F	B	A	C
B	F	E	D	C	B	A
C	D	F	E	A	C	B
D	C	A	B	F	D	E
E	A	B	C	D	E	F
F	B	C	A	E	F	D

This group was modeled on a well-known physical system which displays group properties (we shall soon see which one); we adopt the convention that in the product $X_i \times X_j$, the term on the right is written along the top row of the table, and the left-hand term is on the left-edge column. Thus in Table 7.2, $D \times C = B$ is noted.

Of course, it doesn't matter how we letter the elements, or in which order we place them in the table; what is important is that the product of two elements be uniquely defined and that the internal consistency of the table be clear.

DEFINITION 7.4 Groups that are *isomorphic* have the same multiplication table (disregarding any possible relabeling or re-ordering of the elements).

If two groups are isomorphic, there is a unique one-to-one correspondence between the elements of the two groups, and this correspondence is preserved even under a product or composition of elements.

PROBLEM 7.2 Show that $\{\alpha, \beta, \gamma\}$ of Table 7.1 is isomorphic to $\{\hat{R}_0, \hat{R}_{120}, \hat{R}_{240}\}$, the group C_3.

PROBLEM 7.3 Convince yourself that the following set of elements satisfies Table 7.2 under normal matrix multiplication, and thus forms a group isomorphic to the group $\{A, B, C, D, E, F\}$. α does not necessarily correspond to A, etc.

$$\bar{\bar{\alpha}} = \begin{pmatrix} 1 & 0 \\ 0 & -1 \end{pmatrix} \quad \bar{\bar{\beta}} = \tfrac{1}{2}\begin{pmatrix} -1 & \sqrt{3} \\ \sqrt{3} & 1 \end{pmatrix} \quad \bar{\bar{\gamma}} = \tfrac{1}{2}\begin{pmatrix} -1 & -\sqrt{3} \\ -\sqrt{3} & 1 \end{pmatrix}$$

$$\bar{\bar{\delta}} = \tfrac{1}{2}\begin{pmatrix} -1 & \sqrt{3} \\ -\sqrt{3} & -1 \end{pmatrix} \quad \bar{\bar{\varepsilon}} = \begin{pmatrix} 1 & 0 \\ 0 & 1 \end{pmatrix} \quad \bar{\bar{\rho}} = \tfrac{1}{2}\begin{pmatrix} -1 & -\sqrt{3} \\ \sqrt{3} & -1 \end{pmatrix} \quad (7.2)$$

PROBLEM 7.4 Prove that the groups $\{A, B, C, D, E, F\}$ and $\{\hat{I}, \hat{R}_{120}, \hat{R}_{240}, \hat{M}_I, \hat{M}_{II}, \hat{M}_{III}\}$, whose multiplication tables are Table 7.2 and 7.3, respectively, are isomorphic. Note that '$\hat{I}$' of Table 7.3 does not necessarily correspond to 'A' of Table 7.2, etc.

Table 7.3

	$\hat{I}$	$\hat{R}_{120}$	$\hat{R}_{240}$	$\hat{M}_I$	$\hat{M}_{II}$	$\hat{M}_{III}$
$\hat{I}$	$\hat{I}$	$\hat{R}_{120}$	$\hat{R}_{240}$	$\hat{M}_I$	$\hat{M}_{II}$	$\hat{M}_{III}$
$\hat{R}_{120}$	$\hat{R}_{120}$	$\hat{R}_{240}$	$\hat{I}$	$\hat{M}_{III}$	$\hat{M}_I$	$\hat{M}_{II}$
$\hat{R}_{240}$	$\hat{R}_{240}$	$\hat{I}$	$\hat{R}_{120}$	$\hat{M}_{II}$	$\hat{M}_{III}$	$\hat{M}_I$
$\hat{M}_I$	$\hat{M}_I$	$\hat{M}_{II}$	$\hat{M}_{III}$	$\hat{I}$	$\hat{R}_{120}$	$\hat{R}_{240}$
$\hat{M}_{II}$	$\hat{M}_{II}$	$\hat{M}_{III}$	$\hat{M}_I$	$\hat{R}_{240}$	$\hat{I}$	$\hat{R}_{120}$
$\hat{M}_{III}$	$\hat{M}_{III}$	$\hat{M}_I$	$\hat{M}_{II}$	$\hat{R}_{120}$	$\hat{R}_{240}$	$\hat{I}$

There is nothing mystical about a group multiplication table—it is simply a record that must be generated term by term; but once on paper, it allows us to find any product of elements $X_i \times X_j$ of the group instantaneously.

PROBLEM 7.5 Is Table 7.4 isomorphic to Table 7.2? Is it abelian?

Table 7.4

	a	b	c	d	e	f
a	b	c	d	e	f	a
b	c	d	e	f	a	b
c	d	e	f	a	b	c
d	e	f	a	b	c	d
e	f	a	b	c	d	e
f	a	b	c	d	e	f

*__PROBLEM 7.6__ In Tables 7.1, 7.2, 7.3, and 7.4, any element of the group appears exactly once in each row and each column. Why?

REPRESENTATIONS OF GROUPS OF OPERATORS

*PROBLEM 7.7 If a subset of a group is itself a group, it is called a *subgroup*. A useful theorem, which we shall not prove, states that the order of any subgroup is an integral divisor of the order of the whole group. What are the subgroups of Tables 7.1 through 7.4, and what are their orders?

*PROBLEM 7.8 A smallest possible set of elements whose powers and mutual products generate a group is called a set of *generators* of the group. Are $\{\hat{R}_{120}, \hat{M}_I\}$ generators of C_{3v}? Are there others?

*PROBLEM 7.9 A *cyclic* group is one in which there is an element X_1 such that the sequence $X_2 \equiv X_1 X_1 = X_1^2, X_3 \equiv X_1^3, \ldots, X_{N+1} = X_1^{N+1} = X_1$ yields the entire group. The group is of order N, and abelian. Is a cyclic group of order N isomorphic to C_N? Is any element in the group a generator for the group? (Compare C_4 and C_5.) Do any of the above tables describe cyclic groups?

*PROBLEM 7.10 Figure 7.1a shows a molecule slightly different from one of the C_{3v} variety.

This new system is invariant to all the symmetry operations of C_{3v}, but also to a 'horizontal' reflection, M_h, across the plane of the triangle. The elements $\hat{M}_h$ and $\hat{I}$ constitute a group, albeit a rather small one, $\{\hat{I}, \hat{M}_h\}$. Both the elements of $\{I, \hat{M}_h\}$ commute with all the elements of C_{3v}. We can construct the *product group* D_{3h} of the system of Fig. 7.1 by listing term by term products of the elements of $\{\hat{I}, \hat{M}_h\}$ and C_{3v}.

$$\{\hat{I}, \hat{M}_h\} \times C_{3v} = \{\hat{I} \times \hat{I}, \hat{I} \times \hat{R}_{120}, \hat{I} \times \hat{R}_{240}, \hat{I} \times \hat{M}_I, \ldots, \hat{M}_h \times \hat{I}, \hat{M}_h \times \hat{R}_{120} = \hat{R}_{120} \times \hat{M}_h, \ldots, \hat{M}_h \times \hat{M}_{III} = M_{III} \times M_h\}$$
$$= \{\hat{I}, \hat{R}_{120}, \hat{R}_{240}, \hat{M}_I, \ldots \hat{M}_h, \hat{M}_h \times \hat{R}_{120}, \ldots, M_{III} \times \hat{M}_h\} \quad (7.3)$$

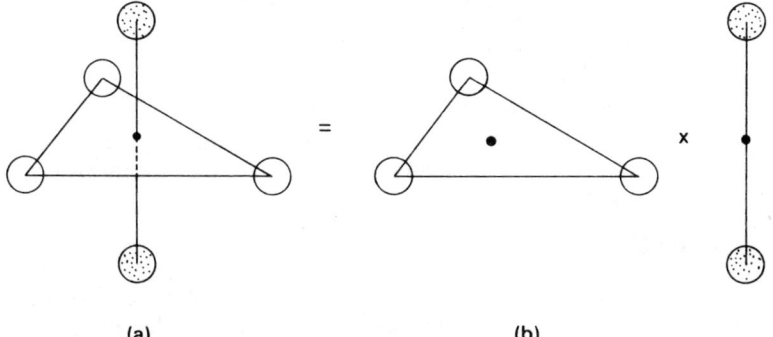

(a) (b)

Figure 7.1.

Prove that the elements of Eq. (7.3) form a group. Are any of its subgroups cyclic? Is C_{3v} a subgroup of D_{3h}? What are the generators of D_{3h}?

In the next section we shall examine two kinds of symmetry groups which are *not* composed of rotations, reflections, or translations.

7.2 Permutation and Colour Groups

The set of possible re-arrangements of n objects constitutes the *permutation group* for the objects, S_n.

A simple way to approach this group of transformations is to label the n objects as $X_1, X_2, \ldots, X_n$, and a set of n bins as 1 through n. The situation in which object X_1 is in bin 1, X_2 in bin 2, etc., is denoted

$$(X_1, X_2, \ldots X_n) \tag{7.4a}$$

and is illustrated in Fig. 7.2a.

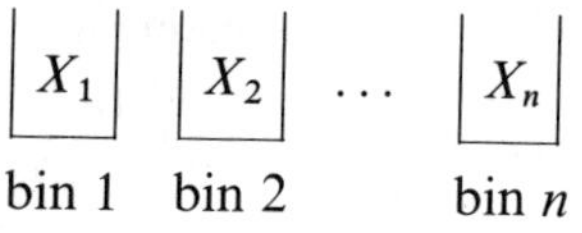

$$\lfloor X_1 \rfloor \quad \lfloor X_2 \rfloor \quad \ldots \quad \lfloor X_n \rfloor$$

bin 1   bin 2   bin n

Figure 7.2(a).

Another configuration,

$$(X_2, X_1, \ldots X_n) \tag{7.4b}$$

is shown in Fig. 7.2b.

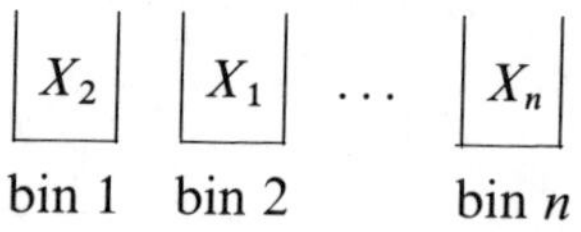

$$\lfloor X_2 \rfloor \quad \lfloor X_1 \rfloor \quad \ldots \quad \lfloor X_n \rfloor$$

bin 1   bin 2   bin n

Figure 7.2(b).

For simplicity we shall consider the specific example of a system of five objects. A typical element of the group of permutations of five objects, S_5, is the *operation*†

$$[\mathbf{54312}] \tag{7.5}$$

which is shorthand for 'take the object in bin number 5 and place it in bin number 1; the object in bin 4 goes to bin 2; the objects in

†Here, we use bold typeface to denote the operator.

REPRESENTATIONS OF GROUPS OF OPERATORS   119

bin 3, 1, and 2 go to bins 3, 4, and 5, respectively.' Thus if we start off as in Fig. 7.2a, the action of [**54312**] is manifest in

$$[\mathbf{54312}](X_1, X_2, X_3, X_4, X_5) = (X_5, X_4, X_3, X_1, X_2) \tag{7.6}$$

A second element in the group of permutations might be [**24315**]. Then†

$$[\mathbf{24315}][\mathbf{54312}](X_1, X_2, X_3, X_4, X_5)$$
$$= [\mathbf{24315}](X_5, X_4, X_3, X_1, X_2)$$
$$= \phantom{[\mathbf{24315}]}(X_4, X_1, X_3, X_5, X_2)$$
$$= [\mathbf{41352}](X_1, X_2, X_3, X_4, X_5) \tag{7.7a}$$

or

$$[\mathbf{24315}][\mathbf{54312}] = [\mathbf{41352}] \tag{7.7b}$$

This is how the product works for a permutation group.

*PROBLEM 7.11 Find the six elements of the group S_3 of permutations on three objects, and write out the group's multiplication table. Show why C_{3v} is isomorphic to this group.

PROBLEM 7.12 Show that the group of permutations on n objects is of order $n!$ Can it be cyclic?

The Hamiltonian of an atom is insensitive to the interchange of any of its n electrons. The set of all possible interchanges of electrons is isomorphic to S_n, and we shall consider the implications of such *exchange symmetry* in Section 9.6.

We close this section by mentioning another category of groups. The object of Fig. 7.3 has C_{3v} symmetry, but if we ignore shades of darkness, it becomes C_{6v}; C_{3v} is of course a subgroup of C_{6v}. Let

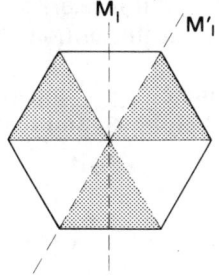

Figure 7.3.

†We drop, as in Chapter 1, the '×' sign for the product, when convenient.

us define an operator $\hat{R}_c$ which interchanges the colours white and black. Although neither $\hat{R}_c$ nor $\hat{R}_{60}$ is an element of C_{3v}, $\hat{R}_c\hat{R}_{60}$ is; the elements $\{\hat{I}, \hat{R}_{120}, \hat{R}_{240}, \hat{M}_I, \hat{M}_{II}, \hat{M}_{III}, \hat{R}_c\hat{R}_{60}, \hat{R}_c\hat{R}_{180}, \hat{R}_c\hat{R}_{300}, \hat{R}_c\hat{M}'_I, \hat{R}_c\hat{M}'_{II}, \hat{R}_c\hat{M}'_{III}\}$ constitute the *colour* or Shubnikov group of C_{3v}. Colour groups have found use in the study of magnetic solids; dark and light regions might correspond, for example, to spin-up and spin-down magnetic moments. In more complex systems, more than two colours might be employed.

PROBLEM 7.13 Is the colour group of C_{3v} isomorphic to C_{6v}?
PROBLEM 7.14 Analyse the colour group structure for Fig. 7.4.

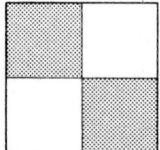

Figure 7.4.

7.3 Classes

C_{3v}, the symmetry group of the ammonia molecule (Table 7.3), is composed of elements which seem to fall naturally into three classes. $\hat{M}_I$, $\hat{M}_{II}$, and $\hat{M}_{III}$ are all reflections across mirror planes (or mirror lines), and are, for all intents and purposes, physically similar. With a co-ordinate system (or the molecule itself) rotated through 120 degrees, $\hat{M}_I$ becomes $\hat{M}_{II}$; the difference is only one of point of view. Similarly, $\hat{R}_{120}$ and $\hat{R}_{240} = \hat{R}_{120}^{-1}$ are surely alike. $\hat{I}$ is like both (or neither) in that it refers to the absence of a rotation or reflection. We can put this notion of class onto firmer footing as follows.

Consider operators $\hat{\theta}$ and $\hat{S}$ (e.g. symmetry operators for some object) in three-space. Let $\bar{r}_1, \bar{r}_2, \bar{r}'_1,$ and $\bar{r}'_2$ be vectors in three-space such that

$$\hat{\theta}\bar{r}_1 = \bar{r}_2 \tag{7.8a}$$

If $\bar{r}_1$ and $\bar{r}_2$ are related to $\bar{r}'_1$ and $\bar{r}'_2$ by the same simple rotation $\hat{S}$:

$$\hat{S}\bar{r}_1 = \bar{r}'_1, \quad \hat{S}\bar{r}_2 = \bar{r}'_2 \tag{7.8b}$$

then $\bar{r}'_i$ is just the new name and description after some rotation $\hat{S}$ of a vector initially called $\bar{r}_i$.

REPRESENTATIONS OF GROUPS OF OPERATORS

Since $\bar{r}_1$ and $\bar{r}_2$ are related via the operator $\hat{\theta}$, Eq. (7.8a), $\bar{r}'_1$ and $\bar{r}'_2$ should be related also by some (as of yet undefined) operator $\hat{\theta}'$:

$$\hat{\theta}'\bar{r}'_1 = \bar{r}'_2 \tag{7.9a}$$

$$\hat{\theta}'(\hat{S}\bar{r}_1) = (\hat{S}\bar{r}_2) \tag{7.9b}$$

$$\bar{r}_2 = (\hat{S}^{-1}\hat{\theta}'\hat{S})\bar{r}_1 \tag{7.9c}$$

but

$$\bar{r}_2 = \hat{\theta}\bar{r}_1, \tag{7.9d}$$

so

$$\hat{\theta} = \hat{S}^{-1}\hat{\theta}'\hat{S}; \quad \hat{\theta}' = \hat{S}\hat{\theta}\hat{S}^{-1} \tag{7.10}$$

Thus $\hat{\theta}$ and $\hat{\theta}'$ represent the same operation, but viewed from different frames of reference; they are related by a similarity transformation and are said to be *conjugate* to one another. Note that this is close to the argument used in Eq. (4.14).

Now let us restrict our concern to operators, both of the $\hat{\theta}$, $\hat{\theta}'$ type and of the $\hat{S}$, $\hat{S}^{-1}$ type, which are elements of a symmetry group G. It is easy to show that if B and C are both conjugate to A, then B and C are conjugate to one another.

*PROBLEM 7.15 Prove this last statement. Does it make any difference that $\hat{S}$ and $\hat{\theta}$ are unitary?

For any element X_i in G, we can construct a list of the elements in G conjugate to X_i simply by evaluating all products of the form IX_iI^{-1}, $X_1X_iX_1^{-1}$, $X_2X_iX_2^{-1}$ .... This set of operators conjugate to X_i (and hence to one another) are said to form a *class*. The above procedure thus allows us to break a group into distinct, disjoint classes such that the operators of a class are related to one another by similarity transformations. The definition of class is valid for an arbitrary (not necessarily symmetry) group; there, however, the meaning is less transparent.

PROBLEM 7.16. Can a reflection and a rotation belong to the same class? What about two rotations about different axes? Two rotations through different angles? Must the elements in a set of generators of a group belong to different classes?

Generally it is not necessary to go through this use of the multiplication table, but rather it is possible to break a group into classes by inspection. But the class structure may be more complex than it appears at first glance, so exercise due caution. Note that the identity element alone always forms its own class, and is the only class which is also a subgroup.

Class structure is determined not only by the natures of the individual symmetry operators taken one at a time, but also by the overall group structure. Thus the three classes of the group C_3 of the object in Fig. 7.5a consist, respectively, of $\hat{I}$, $\hat{R}_{120}$, and $\hat{R}_{240}$; for C_{3v}, Fig. 7.5b, however, $\hat{I}$ is again in a class by itself, $\hat{M}_\text{I}$, $\hat{M}_\text{II}$, and $\hat{M}_\text{III}$ form another class, and $\hat{R}_{120}$ and $\hat{R}_{240}$ together constitute a third (confirm this!).

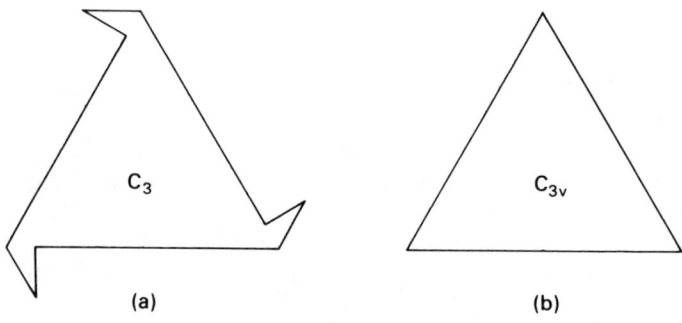

Figure 7.5.

PROBLEM 7.17 Prove that classes other than $\{\hat{I}\}$ cannot be subgroups, and that for an abelian group, every element forms a separate class. Why are the classes of a group disjoint—i.e. have no elements in common?

PROBLEM 7.18 What are the classes of S_3? How many objects remain in their original bins for each class?

PROBLEM 7.19 What are the classes of the group D_{3h} of Problem 7.10?

*PROBLEM 7.20 What are the symmetry operations and classes of O, the group of the cube?

7.4 Representation of a Group: $\Gamma^{(p)}$ of C_{3v}

The purpose of this section is to review what we have said about representations of the symmetry operators $\hat{P}_R$ of function space, and to connect with this what we know about groups.

Equation (6.31) is a five-by-five matrix, $\bar{\bar{\Gamma}}^{(d)}(R_\phi)$, which describes through

$$\hat{P}_R d_i(\bar{r}) = \sum_j d_j(\bar{r}) \Gamma^{(d)}_{ji}(\hat{R}) \qquad (6.19')$$

REPRESENTATIONS OF GROUPS OF OPERATORS

what $\hat{P}_R$ does to a set of five d-state functions. These particular d-states were specifically chosen to deal with a body possessing an axis of rotation and vertical mirror planes in three-space. When the co-ordinate system of an object is rotated about the $|3\rangle$ axis, for example, $d_5(\bar{r}) = (3z^2 - r^3)/2\sqrt{3}r^2$ is invariant. The other four basis vectors also reflect this axial symmetry. The matrix representation, Eq. (6.31), followed directly from

$$\hat{P}_R d_1(\bar{r}) = d_1(\bar{r}) \cos 2\phi + d_2(\bar{r}) \sin 2\phi, \quad \text{etc.} \quad (6.29)$$

which in turn came from a simple application of the fundamental link

$$\hat{P}_R \psi(\bar{r}) = \psi(\hat{R}^{-1}\bar{r}) \quad (6.4)$$

between real space and function space. We obtained $\bar{\bar{\Gamma}}^{(d)}(\hat{M}_\phi)$ in a similar fashion.

We can follow the same procedure using three p-states $\{|p_1\rangle, |p_2\rangle, |p_3\rangle\}$ as a basis system instead of five d-states; the representation of the operators $\hat{P}_{R,\phi}$ and $\hat{P}_{M,\phi}$ are easily found to be

$$\bar{\bar{\Gamma}}^{(p)}(\hat{R}_\phi) = \begin{pmatrix} \cos \phi & -\sin \phi & 0 \\ \sin \phi & \cos \phi & 0 \\ 0 & 0 & 1 \end{pmatrix} \quad (7.11a)$$

$$\bar{\bar{\Gamma}}^{(p)}(\hat{M}_\phi) \begin{pmatrix} \cos 2\phi & \sin 2\phi & 0 \\ \sin 2\phi & -\cos 2\phi & 0 \\ 0 & 0 & 1 \end{pmatrix} \quad (7.11b)$$

Comparison with Eqs (4.7) and (4.8) shows that these are identical to the representation based upon the three-unit vectors $\{|i\rangle\}$ of real space. This result is not surprising, since the three p-states are of the form

$$p_i(\bar{r}) = \begin{array}{c} x_1 \, F'(|r|) \\ x_2 \, F'(|r|) \\ x_3 \, F'(|r|) \end{array} \quad (7.11c)$$

where the radial factor $F'(|\bar{r}|)$ is spherically symmetric. Equation (7.11c) does *not* say that the $\{|p_i\rangle\}$ point in the $|1\rangle$, $|2\rangle$, and $|3\rangle$ directions; such a statement is meaningful only in real space, not in function space. Rather, we say that the *symmetry properties* of the $\{|p_i\rangle\}$ are identical to those of $\{|i\rangle\}$ of real space under the operations of, respectively, $\hat{P}_R$ and $\hat{R}$.

As we have seen in previous chapters, one can form the representation not only of a single operator such as $\hat{P}_{R,\phi}$ or $\hat{P}_{M,\phi}$, but also of an entire group of operators. From Eq. (7.11a) and (7.11b), for example, it is a trivial matter to write down a representation of

C_{3v} generated by the three real-valued p-states:

$$\overline{\overline{\Gamma}}^{(p)}(\hat{I})=\begin{pmatrix}1 & 0 & 0\\ 0 & 1 & 0\\ 0 & 0 & 1\end{pmatrix} \qquad \overline{\overline{\Gamma}}^{(p)}(\hat{R}_{120})=\tfrac{1}{2}\begin{pmatrix}-1 & -\sqrt{3} & 0\\ \sqrt{3} & -1 & 0\\ 0 & 0 & 2\end{pmatrix}$$

$$\overline{\overline{\Gamma}}^{(p)}(\hat{R}_{240})=\tfrac{1}{2}\begin{pmatrix}-1 & \sqrt{3} & 0\\ -\sqrt{3} & -1 & 0\\ 0 & 0 & 2\end{pmatrix} \qquad \overline{\overline{\Gamma}}^{(p)}(\hat{M}_{\mathrm{I}})=\begin{pmatrix}-1 & 0 & 0\\ 0 & 1 & 0\\ 0 & 0 & 1\end{pmatrix}$$

$$\overline{\overline{\Gamma}}^{(p)}(\hat{M}_{\mathrm{II}})=\tfrac{1}{2}\begin{pmatrix}1 & \sqrt{3} & 0\\ \sqrt{3} & -1 & 0\\ 0 & 0 & 2\end{pmatrix} \qquad \overline{\overline{\Gamma}}^{(p)}(\hat{M}_{\mathrm{III}})=\tfrac{1}{2}\begin{pmatrix}1 & -\sqrt{3} & 0\\ -\sqrt{3} & -1 & 0\\ 0 & 0 & 2\end{pmatrix}$$
(7.12)

The matrices of Eq. (7.12) form a group with a multiplication table identical to that of Table 7.3; the two groups are isomorphic to one another, and Eq. (7.12) forms the isomorphic, *true*, or *faithful* representation of C_{3v} generated by the $|p_i\rangle$ basis system.

7.5 Homomorphic Representations: $\Gamma^{(s)}$ of C_{3v}

We have just made use of

$$\hat{P}_R|\psi_i(\bar{r})\rangle = \sum_j |\psi_j(\bar{r})\rangle \Gamma_{ji}^{(\psi)}(\hat{R}) \qquad (6.19)$$

and a basis set of atomic p-states to construct a representation of C_{3v}. This p-state representation itself constitutes a group exactly isomorphic to the original symmetry group. Occurrences in function space mirror precisely what happens in real space; the symmetry group element $\hat{R}_{120}$, for example, is in a one-to-one correspondence with the matrix $\overline{\overline{\Gamma}}^{(p)}(\hat{R}_{120})$. The group properties and multiplication tables of the sets $\{\hat{I}, \hat{R}_{120} \dots\}$ and $\{\overline{\overline{\Gamma}}^{(p)}(\hat{I}), \overline{\overline{\Gamma}}^{(p)}(\hat{R}_{120}) \dots\}$ are identical.

Equation (7.12) is not, of course, the only representation of the group of C_{3v}. The five d-states, for example, generate another isomorphic representation of C_{3v}. In fact, bearing in mind the very important results of Problems 6.16 and 6.17, we might expect to find interesting representations generated by *all* the various atomic angular functions. This raises an intriguing question: What is the significance of the matrix generated by an atomic s-state alone,

$$\boxed{\overline{\overline{\Gamma}}_{ji}^{(s)}(\hat{R}) \equiv \langle s|\hat{P}_R|s\rangle = 1} \qquad (7.13a)$$

where $\hat{R}$ and $\hat{P}_R$ can refer to any symmetry operation in, say, C_{3v}?

*PROBLEM 7.21   Prove Eq. (7.13a). What is the multiplication table for the representation of C_{3v} generated by $|s\rangle$?

First of all, let us note that given some group $\{\hat{R}\}$ such as C_{3v}, the $\{\bar{\bar{\Gamma}}^{(s)}(\hat{R})\}$ also form a group, each element of which is the one-by-one unit matrix; but while the $\{\bar{\bar{\Gamma}}^{(s)}(\hat{R})\}$ do form a group, this new group is clearly *not* isomorphic to the original group $\{\hat{R}\}$. There is rather, a one-to-many correspondence, worthless though it may seem, between the elements of $\bar{\bar{\Gamma}}^{(s)}(\hat{R})\}$ and those of $\{\hat{R}\}$:

$$\bar{\bar{\Gamma}}^{(s)}(\hat{R}) = 1 \text{ for every } \hat{R} \in \{\hat{R}\} \tag{7.13b}$$

A group such as $\{\bar{\bar{\Gamma}}^{(s)}(\hat{R})\}$, which has a one-to-many correspondence to another group such as $\{\hat{R}\}$, is said to be *homomorphic* to it. The multiplication table of a homomorphic group is different from, but determined by, that of the symmetry group to which it is homomorphic. For our especially simple example of $\bar{\bar{\Gamma}}^{(s)}(\hat{R}) = 1$ for all $\hat{R} \in C_{3v}$

$$\hat{R}_i \hat{R}_j = \hat{R}_k \Rightarrow \bar{\bar{\Gamma}}^{(s)}(\hat{R}_i) \bar{\bar{\Gamma}}^{(s)}(\hat{R}_j) = \bar{\bar{\Gamma}}^{(s)}(\hat{R}_k) \tag{7.14}$$

The converse would also be true for all $\hat{R}$ only if the groups were isomorphic, rather than homomorphic.

The set of matrices $\{\bar{\bar{\Gamma}}^{(s)}(\hat{R})\}$ generated by an atomic *s*-state, then, forms a group homomorphic to a symmetry group such as C_{3v}. The fact that $\bar{\bar{\Gamma}}^{(s)}(\hat{R}) = 1$ for every $\hat{R}$ of *any* symmetry group might tempt us to discard such a triviality in favour of a representation which is isomorphic to the symmetry group. But *no representation may be ignored.* For application to quantum mechanics, any possible representation of a symmetry group might be of importance, and all must be examined carefully. It is a fundamental mistake to think of an isomorphic representation, such as the set Eq. (7.12), as being especially important; it is not! We shall see that $\{\bar{\bar{\Gamma}}^{(s)}(\hat{R}) = 1\}$ is *every bit as essential* to our development as are $\{\bar{\bar{\Gamma}}^{(p)}(\hat{R})\}$, $\{\bar{\bar{\Gamma}}^{(d)}(\hat{R})\}$, and other representations.

This perhaps surprising statement must be qualified now. It is obvious that one hundred per cent of what may be said about a group can be learned from an isomorphic representation. After all, an isomorphic representation is itself a group with a multiplication table identical (although the names of the elements, and the ordering of the rows and columns, may be different) to that of the original group; the information residing within each of the two tables is the same. But the information content of an isomorphic representation is not necessarily in a form most useful to us in our study of quantum mechanical systems. Certain other representations, each perhaps simpler in form than an isomorphic representation and each *individually* containing less information than does an isomorphic representation, may *together* offer the same total information as an isomorphic representation (or the symmetry group itself), and yet be in a much more tractable and useful form. If we may draw a rather crude analogy: the author of a technical text could write a chapter as a single, highly condensed sentence; in all probability, his ideas could be understood unambiguously, but only with a fair amount of effort. A fully punctuated chapter in which the basic ideas are broken into separate, normal paragraphs can be extracted from the original; the new version will probably contain much redundant information, but will be considerably easier to understand and use, each sentence being a small, self-contained, and perhaps interesting unit.

Our point is that an isomorphic representation contains all the information on its group, but a set of derived simpler (i.e. the irreducible) representations may be more immediately useful.

7.6 Equivalent Representations

The form of a group representation depends, of course, upon the choice of basis system used in generating the representation. We have seen that the five $|d\rangle$-states generate a five-dimensional representation of C_{3v} which happens to be isomorphic to the symmetry group itself. (Recall that all terms such as $\langle\psi_i|\hat{P}_R|d_j\rangle$ vanish unless $|\psi_i\rangle$ is also a d-state, and of the same principal quantum number as $|d_j\rangle$.) The three p-states also generate an isomorphic representation; and $|s\rangle$ alone generates a one-dimensional representation which is not isomorphic to the group, but rather homomorphic. Surely there are an infinite number of possible representations of our group. In fact, there are even an infinite number of possible three-dimensional representations coming from the basis set $\{|p_1\rangle, |p_2\rangle, |p_3\rangle\}$ alone! Simply rotating $|p_1\rangle$ and $|p_2\rangle$ slightly about $|p_3\rangle$ in function space, such that they become $|p_1'\rangle$ and $|p_2'\rangle$, leads to a new and different set of $\bar{\bar{\Gamma}}^{(p')}(\hat{R})$ matrices. The fundamental question arises: are these innumerable representations all distinct and non-equivalent, or do they fall into a few basic categories?

The information content of a representation generated by $\{|p_1\rangle, |p_2\rangle, |p_3\rangle\}$ should not change if we rotate $|p_1\rangle$ and $|p_2\rangle$ about $|p_3\rangle$ in function space. The matrices may look differently before and after the reorientation of the $|p\rangle$-coordinate axes, but all we have to do is rotate $|p_1\rangle$ and $|p_2\rangle$ the other way, and we have all our original matrices representing the group back again. We say, therefore, that two matrix representations of a group are *equivalent* if and only if the sets of matrices differ from one another by a similarity transformation, by a simple transformation of function space. The situation is similar to that of classes of group elements. In the discussion of Eq. (7.10), we defined two operators (group elements) as being in the same class if they correspond to the same physical operation, but as seen from different viewpoints (e.g. from co-ordinate systems rotated relative to one another). Here, we say that two group representations are equivalent if, for all the operators in the group, each set of matrices can be transformed into the other via a rotation and/or reflection of the vector space used in generating the representations. That is, the two representations $\bar{\bar{\Gamma}}^{(v)}$ and $\bar{\bar{\Gamma}}^{(\mu)}$ are equivalent if and only if there exists a single matrix $\bar{\bar{A}}$ such that

$$\bar{\bar{A}}^{-1}\bar{\bar{\Gamma}}^{(v)}(\hat{R})\bar{\bar{A}} = \bar{\bar{\Gamma}}^{(\mu)}(\hat{R}), \quad \text{all } \hat{R} \qquad (7.15)$$

where the same matrix $\bar{\bar{A}}$ is used for all the $\hat{R}$ in $\{\hat{R}\}$.

Each of the two sets of matrices $\{\bar{\bar{\Gamma}}^{(v)}(\hat{R})\}$ and $\{\bar{\bar{\Gamma}}^{(\mu)}(\hat{R})\}$ forms a group, and since they are equivalent, the two groups are clearly isomorphic. These two representation groups in turn may be (but are not necessarily) isomorphic to the original symmetry group represented.

PROBLEM 7.22 Show that if the complete, orthonormal basis sets $\{|i\rangle\}$ and $\{|i'\rangle\}$ are related by $\hat{B}|i\rangle = |i'\rangle$, then their respective representations of any group $\{\hat{P}_R\}$ are equivalent.

*PROBLEM 7.23 Can p-states and d-states yield equivalent representations of C_{3v}?

*PROBLEM 7.24 Are the representations of C_{3v} generated by real- and complex-valued p-states, $\{|p_1\rangle, |p_2\rangle, |p_3\rangle\}$ and $\{\sqrt{\frac{1}{2}}(|p_1\rangle \pm i|p_2\rangle), |p_3\rangle\}$, equivalent?

7.7 Irreducible Representations

Consider Eqs (7.16), (7.17), and (7.18), which are the familiar (by now) representations of the group of the equilateral triangle generated by real-valued atomic $|s\rangle$-states $|p\rangle$-states, and $|d\rangle$-states, respectively:

$$\bar{\bar{\Gamma}}^{(s)}(\hat{R}) = 1 \text{ for all } \hat{R} \text{ in } C_{3v} \qquad (7.16)$$

$$\bar{\bar{\Gamma}}^{(p)}(\hat{R}_\phi) = \begin{pmatrix} \cos\phi & -\sin\phi & 0 \\ \sin\phi & \cos\phi & 0 \\ 0 & 0 & 1 \end{pmatrix} \quad \phi = 0°, 120°, 240° \qquad (7.17)$$

$$\overline{\overline{\Gamma}}^{(p)}(\hat{M}_I) = \begin{pmatrix} -1 & 0 & 0 \\ 0 & 1 & 0 \\ 0 & 0 & 1 \end{pmatrix}, \text{ etc.}$$

$$\overline{\overline{\Gamma}}^{(d)}(\hat{R}) = \begin{pmatrix} \cos 2\phi & -\sin 2\phi & 0 & 0 & 0 \\ \sin 2\phi & \cos 2\phi & 0 & 0 & 0 \\ 0 & 0 & \cos \phi & -\sin \phi & 0 \\ 0 & 0 & \sin \phi & \cos \phi & 0 \\ 0 & 0 & 0 & 0 & 1 \end{pmatrix} \begin{array}{l} \phi = 0°, \\ 120°, 240° \end{array}$$

$$\overline{\overline{\Gamma}}^{(d)}(\hat{M}_I) = \begin{pmatrix} 1 & 0 & 0 & 0 & 0 \\ 0 & -1 & 0 & 0 & 0 \\ 0 & 0 & -1 & 0 & 0 \\ 0 & 0 & 0 & 1 & 0 \\ 0 & 0 & 0 & 0 & 1 \end{pmatrix}, \text{ etc.} \quad (7.18)$$

A glance at these three representations reveals an extraordinary interrelationship. $\overline{\overline{\Gamma}}^{(d)}(\hat{R})$ appears to *contain* $\overline{\overline{\Gamma}}^{(p)}(\hat{R})$ and $\overline{\overline{\Gamma}}^{(s)}(\hat{R})$ for every $\hat{R}$ in the group C_{3v}. For example, with $\phi = 120°$,

$$\overline{\overline{\Gamma}}^{(d)}(\hat{R}_{120°}) = \begin{pmatrix} -1 & \sqrt{3} & 0 & 0 & 0 \\ -\sqrt{3} & -1 & 0 & 0 & 0 \\ 0 & 0 & -1 & -\sqrt{3} & 0 \\ 0 & 0 & \sqrt{3} & -1 & 0 \\ 0 & 0 & 0 & 0 & \boxed{2} \end{pmatrix} \quad (7.19)$$

$$\overline{\overline{\Gamma}}^{(p)}(\hat{R}_{120°}) = \tfrac{1}{2} \begin{pmatrix} -1 & -\sqrt{3} & 0 \\ \sqrt{3} & -1 & 0 \\ 0 & 0 & \boxed{2} \end{pmatrix} \quad (7.20)$$

$$\overline{\overline{\Gamma}}^{(s)}(\hat{R}_{120°}) = (1) \quad (7.21)$$

And the same holds for all $\hat{R}$ in $\{\hat{R}\}$.

Thus the $|d\rangle$-state representations are matrices which are composed, for every $\hat{R}$, of $\overline{\overline{\Gamma}}^{(p)}(\hat{R})$, plus additional (though perhaps redundant) information in the upper left-hand block. Similarly, the $\overline{\overline{\Gamma}}^{(p)}(\hat{R})$ representation itself contains the $\overline{\overline{\Gamma}}^{(s)}(\hat{R}) = 1$ matrix for every $\hat{R}$. We can see then that complex, but valid, representations can be built up out of smaller representations.

As noted in Eq. (4.20), if the constituent, smaller representations lie along the diagonal of the composite, larger matrices, then such a block-diagonal structure is preserved under group (matrix) multiplication. For example $\overline{\overline{\Gamma}}^{(d)}(\hat{R}_1)\overline{\overline{\Gamma}}^{(d)}(\hat{R}_2) = \overline{\overline{\Gamma}}^{(d)}(\hat{R}_1\hat{R}_2)$, and this in turn implies that the vector subspaces of d-space which generate the various blocks in $\bar{\bar{\Gamma}}^{(d)}$ are not mixed under the symmetry operators $\{\hat{R}\}$, an observation discussed in Sections 4.3 and 6.8. The blocks are thus essentially independent of one another within a given large matrix, if all matrices have the same block-diagonal form.

Naturally it is easy to conceal the block structure of a block-diagonalized $(n \times n)$ matrix by rotating function space about an axis aligned arbitrarily in n-space. One can argue conversely that given a group of $(n \times n)$ matrices, there may (or may not) be some similarity transformation, or rotation of function space, which brings about a situation in which all the matrices of a group representation have the same block-diagonal form. That is, we can 'break down' the set of large matrices $\bar{\bar{\Gamma}}^{(\psi)}(\hat{R})$ generated by the n basis vectors $\{|\psi(\bar{r})\rangle\}$, into a form where they all have the same block structure, simply by performing a unitary transformation on the $\{|\psi(\bar{r})\rangle\}$. And there must be one (or several equivalent) optimal, *most fully* reduced form, beyond which one can go no farther by means of *any* similarity transformation in this block-diagonalization procedure. Thus it is not always true that the optimal situation realizable is one in which all the Γ's are simultaneously fully diagonal.

What was said about class structure in Section 7.3 applies here too: optimal block-diagonal structure—i.e. the optimal alignment of function space yielding the most complete separation of Hilbert space into invariant subspaces—depends both upon a group's individual symmetry operations each taken alone, and upon the nature of the group structure as a whole. Thus while in the transparent case of the abelian group C_3 one only must consider a single unique axis of rotation, in C_{3v} on the other hand there are a set of competing symmetry properties all of which influence and constrain the possible types of invariant subspaces. We are left with a delicate balance, manifest in the set of possible invariant subspaces, which mirrors the nature of the entire group while recognizing the contributions of the individual operators.

*PROBLEM 7.25 Can any matrix representation of an abelian group be totally diagonalized, rather than simply block-diagonalized?

Suppose we have a maximally diagonalized representation of a group $\{\hat{R}\}$, generated by some basis set $\{|\psi(\bar{r})\rangle\}$. That is, $\bar{\bar{\Gamma}}^{(\psi)}(\hat{R})$ for each $\hat{R}$ is of the form

$$\bar{\bar{\Gamma}}^{(\psi)}(\hat{R}) = \begin{pmatrix} \bar{\bar{\Gamma}}^{(1)}(\hat{R}) & & 0 \\ & \bar{\bar{\Gamma}}^{(2)}(\hat{R}) & \\ 0 & & \bar{\bar{\Gamma}}^{(m)}(\hat{R}) \end{pmatrix} \quad (7.22)$$

Then the sets of blocks $\bar{\bar{\Gamma}}^{(1)}(\hat{R})$ for all $\hat{R}$, $\bar{\bar{\Gamma}}^{(2)}(\hat{R})$ for all $\hat{R}$, etc., are called the *irreducible representations* of which the larger representation $\bar{\bar{\Gamma}}^{(\psi)}(\hat{R})$ is composed. The reduction process involved in achieving block-diagonal form may by symbolized as

$$\Gamma^{(\psi)} \to \Gamma^{(1)} + \Gamma^{(2)} + \ldots + \Gamma^{(m)} \tag{7.23}$$

We have already seen an example of such a separation in Eq. (4.23). If it should happen that some of these irreducible representations appear more than once in a reduction, Eq. (7.23) is written

$$\boxed{\Gamma^{(\psi)} = \sum_v a_v \Gamma^{(v)}} \tag{7.24}$$

Each of the irreducible representations is a valid representation of the original symmetry group; some irreducible representations are isomorphic to the symmetry group, and others are homomorphic. We shall see that both kinds are crucial for application to quantum mechanical problems, a subject which we shall soon attack. But first we must examine some of the remarkable properties of irreducible representations.

The moral of this chapter is very simple and vitally important: a listing of the (generally) few possible inequivalent irreducible representations of a group contains all the information obtainable about the group. And this is the most convenient, meaningful form in which that information can be presented.

PROBLEM 7.26 Why does it happen that $\Gamma^{(d)}$ contains $\Gamma^{(p)}$ for C_{3v}?

7.8 Summary

1. Two groups with the same multiplication tables (disregarding changes of ordering or labelling) are isomorphic to one another. The properties of a group are completely specified once its group multiplication table is presented. The class structure of a group, which lumps together those operations which are conjugate to one another, is dependent upon both the natures of the individual symmetry operations involved, and also upon the group structure as a whole.
2. Just as one can construct the representation of a single operator, one can form the matrix representation of an entire group of operators. We used p-states to generate a representation of C_{3v}, and found that the six matrices are identical to what one would find if working with $\{\hat{R}\}$ and $\{|1\rangle, |2\rangle, |3\rangle\}$.
3. A representation, such as $\{\bar{\bar{\Gamma}}^{(s)}(\hat{R}) = 1\}$ of C_{3v}, for which there is a one-to-many relationship with $\{\hat{R}\}$, is said to be homomorphic to $\{\hat{R}\}$; thus the matrices $\{\bar{\bar{\Gamma}}^{(s)}(\hat{R})\}$ display group structure (a very

REPRESENTATIONS OF GROUPS OF OPERATORS

simple one!), but this group is homomorphic, not isomorphic, to the original C_{3v}.

4. Two representations of a group of order N are equivalent if one of the sets of N matrices can be made identical to the other by a similarity transformation brought about by a unique transformation matrix $\bar{\bar{A}}$; that is $\bar{\bar{A}}^{-1}\bar{\bar{\Gamma}}^{(v)}(\hat{R})\bar{\bar{A}} = \bar{\bar{\Gamma}}^{(\mu)}(\hat{R})$, with the same $\bar{\bar{A}}$ for all $\hat{R}$ in $\{\hat{R}\}$.

5. Any representation can be built up out of the most fully reduced irreducible representations. Consequently, a listing of the (for finite-order groups) few possible non-equivalent irreducible representations of a group contains complete information concerning the group; this information, moreover, is in a form most convenient for application.

CHAPTER 8

Irreducible Group Representations

Chapter 8 begins with a brief review of the arguments which led to the concept of an irreducible representation of a group of operators. Two important theorems are then presented, stating

1. the number of non-equivalent irreducible representations of a group equals the number of classes c of the group, and

2. $\sum_{v}^{c} n_v^2 = N$, where n_v is the dimensionality of the vth distinct irreducible representation and N is the order of the group.

These two theorems reveal that because $\{|1\rangle, |2\rangle, |3\rangle\}$ span polar-vector space but not axial-vector space, we have overlooked one possible one-dimensional irreducible representation of C_{3v}. The Regular Representation and the Celebrated Theorem are presented as providing a method of generating *all* the possible irreducible representations of a group without even having to choose a basis set.

The Great Orthogonality Theorem and the character (trace)

$$\chi^{(\mu)}(\hat{R}) = \sum_{k} \Gamma_{kk}^{(\mu)}(\hat{R})$$

are introduced as useful computational tools. Reduction of a typical three-dimensional representation of C_{3v} offers a simple exercise in their application.

8.1 Irreducible Group Representations: A Review

A strong link between transformations in real space and occurrences in function space exists because the vector spaces of greatest interest to us are spanned by functions of the real-space three-vector $\bar{r}$. Consequently, we are able to define operators $\hat{P}_R$ which establish a one-to-one correspondence between transformations of an object (or co-ordinate axis system) in real space and transformations in function space.

Both in real space and in function space, there are optimal sets of co-ordinate axes (best choices for the orientations of basis vectors) in which the representation of a given single symmetry operator, $\hat{R}$ or $\hat{P}_R$, respectively, is of block-diagonal form. The alignment of this particular basis system depends upon the nature of the physical object under consideration and also upon the properties of the function space involved. Thus if we are thinking of an object with an axis of rotation, or if we are concerned with the consequences of rotation about some axis, it pays us to orient the axis of rotation along a co-ordinate axis of real space, and to choose basis vectors in function space also reflecting this special direction. An optimal choice of basis states $\{|\psi_i(\bar{r})\rangle\}$ yields the representation $\Gamma^{(\psi)}$ of $\hat{P}_R$ in block-diagonal form. Then

$$\hat{P}_R \psi_i(\bar{r}) = \sum_j \psi_j(\bar{r}) \, \Gamma^{(\psi)}_{ji}(\hat{R}) \tag{6.19}$$

tells us that function space is broken into subspaces, each invariant with respect to $\hat{P}_R$; vectors in the different invariant subspaces are not intermixed as a consequence of the associated $\hat{R}$ of real space.

To a whole group of symmetry operators $\{\hat{R}\}$ in real space, moreover, there corresponds an isomorphic group $\{\hat{P}_R\}$ operating in function space. Given a basis set in function space, it may be possible to align these basis vectors in such a way that function space is broken into subspaces invariant to *all* the operators in the group $\{\hat{P}_R\}$. Thus any operation of the symmetry group $\{\hat{R}\}$ of an object in real space will be mirrored as a mixing of vectors in function space, but only within mutually disjoint invariant subspaces. Any operation $\hat{R}$ of the symmetry group will cause $|\psi_i(\bar{r})\rangle$ to be transformed into a linear combination, by Eq. (6.19), of basis vectors inhabiting the same invariant subspace. The basis vectors spanning the various optimal invariant subspaces generate the *irreducible representations* of the symmetry group; those belonging to any particular invariant subspace are said to be of the symmetry type associated with the subspace.

The reduction of a representation of the symmetry group of an object corresponds to transforming to an optimal basis system in function space which most fully reflects the three-dimensional symmetry of the object, and by means of which there is a beneficial separation of the space into subspaces invariant under the symmetry operations of the group in question. The equivalence of block-diagonalization to a rotation in function space is directly analagous to the point made in Section 5.7: diagonalizing a Hamiltonian, or solving the eigenequation $\mathcal{H}|\psi_n\rangle = E_n|\psi_n\rangle$, is equivalent to rotating Hilbert space until we find the principal axis system for the Hermitian form

$$\langle \phi | \mathcal{H} | \phi \rangle = \sum_{ij} \mathcal{H}_{ij} \phi_i^* \phi_j = \langle E_\phi \rangle \to \sum_k \mathcal{H}_{kk} \phi_k'^2 = \langle E_\phi \rangle \qquad (5.23)$$

Some irreducible representations may be isomorphic to the original symmetry group in real space, others are homomorphic only. *All are equally important* to the study of quantum mechanics. Although an isomorphic representation will have, hidden within itself, complete information about the symmetry group, this information is not necessarily in immediately useful form. On the other hand, since *any* representation either is itself irreducible or can be constructed out of irreducible representations, a listing of all the possible non-equivalent irreducible representations provides *complete* information about the group. (For some symmetry groups, any representations isomorphic to the symmetry group would be reducible; therefore in a list of irreducible representations, none would be isomorphic to the symmetry group!)
The crucial question, then, is: how many distinct, non-equivalent irreducible representations must one list in order to present completely all the information residing within the multiplication table of a group?

8.2 The Number of Non-Equivalent Irreducible Representations Equals the Number of Distinct Classes

In Eq. (7.16) through (7.21), we saw that for C_{3v}, $\Gamma^{(d)}$ can be reduced. For example,

$$\bar{\bar{\Gamma}}^{(d)}(\hat{R}_{120}) = \tfrac{1}{2} \begin{pmatrix} -1 & \sqrt{3} & 0 & 0 & 0 \\ -\sqrt{3} & -1 & 0 & 0 & 0 \\ 0 & 0 & -1 & -\sqrt{3} & 0 \\ 0 & 0 & \sqrt{3} & -1 & 0 \\ 0 & 0 & 0 & 0 & 2 \end{pmatrix} \qquad (8.1a)$$

$$\bar{\bar{\Gamma}}^{(p)}(\hat{R}_{120}) = \tfrac{1}{2} \begin{pmatrix} -1 & -\sqrt{3} & 0 \\ \sqrt{3} & -1 & 0 \\ 0 & 0 & 2 \end{pmatrix} \qquad (8.1b)$$

IRREDUCIBLE GROUP REPRESENTATIONS

$$\bar{\bar{\Gamma}}^{(s)}(\hat{R}_{120}) = (1) \tag{8.1c}$$

and the representations of the other five elements of the group of the triangle have the same respective block forms. For each of the six $\hat{R}$'s in the group, $\bar{\bar{\Gamma}}^{(d)}(\hat{R})$ actually contains $\bar{\bar{\Gamma}}^{(p)}(\hat{R})$, and both $\bar{\bar{\Gamma}}^{(d)}(\hat{R})$ and $\bar{\bar{\Gamma}}^{(p)}(\hat{R})$ contain $\bar{\bar{\Gamma}}^{(s)}(\hat{R})$. Each of the $\Gamma^{(d)}$ matrices can be built up out of three smaller representations; the matrix $\Gamma^{(d)}(\hat{R}_{120})$, for example, has along its diagonal

$$\tfrac{1}{2}\begin{pmatrix} -1 & \sqrt{3} \\ -\sqrt{3} & -1 \end{pmatrix} \tag{8.2a}$$

$$\tfrac{1}{2}\begin{pmatrix} -1 & -\sqrt{3} \\ \sqrt{3} & -1 \end{pmatrix} \tag{8.2b}$$

and

$$(1) \tag{8.2c}$$

We have already seen that Eq. (8.2b) and its five sister matrices constitute $\{\bar{\bar{\Gamma}}^{(p)}(\hat{R})\}$, and likewise Eq. (8.2c) is the same as $\bar{\bar{\Gamma}}^{(s)}(\hat{R})$. It is not difficult to show that the six matrices corresponding to Eq. (8.2a) form a representation equivalent under a similarity transformation to Eq. (8.2b) and the other members of the representation generated by $\{|p_1\rangle, |p_2\rangle\}$.

PROBLEM 8.1 Demonstrate this.

Thus we end up with only two distinct irreducible representations of C_{3v} lurking within $\Gamma^{(d)}$. Can C_{3v} have any others, and if so, what might they be like?

We explore this question by calling upon four important theorems concerning irreducible representations. Their proofs[†] are given in any of the standard group theory texts, and you can return to them later if you wish. But the theorems themselves are extremely useful in application.

THEOREM 8.1 The number, c, of distinct, non-equivalent irreducible representations of a group is equal to the number of classes in the group.

THEOREM 8.2 Let N be the order of the group, n_v the dimensionality of the vth irreducible representation, and c the number of

[†]Problem 8.6 will suggest a plausability argument for Theorem 8.2.

non-equivalent irreducible representations (and therefore the number of classes) of the group. Then

$$\sum_{v}^{c} n_v^2 = N \tag{8.3}$$

The symmetry group C_{3v}, for example, is of order six and consists of three classes. We have already found one one-dimensional irreducible representation for C_{3v}, and one of dimension two. Theorems 8.1 and 8.2 tell us, then, that there remains one and only one irreducible representation to be found, and that it is one-dimensional. It is *not* isomorphic to $\{\overline{\overline{\Gamma}}^{(s)}(R)\}$.

*PROBLEM 8.2 Show that the A_2 representation

$$\{\overline{\overline{\Gamma}}^{A_2}(\hat{R})\} = \begin{cases} +1 \text{ for } \hat{I}, \hat{R}_{120}, \hat{R}_{240} \\ -1 \text{ for } \hat{M}_\text{I}, \hat{M}_\text{II}, \hat{M}_\text{III} \end{cases} \tag{8.4}$$

is a homomorphic representation of the group of the triangle. The *axial* vector $|4\rangle$ is a suitable vector for producing this representation, where $|4\rangle$ (like $|3\rangle$) is invariant under rotation but (unlike $|3\rangle$) changes sign under reflection, Fig. 8.1, across a mirror which contains the axis of rotation.

†PROBLEM 8.3 It does indeed seem remarkable that the number of non-equivalent irreducible representations of a group should simply equal the number of classes in the group. The formal proofs of this equality are rather elaborate. One is tempted, however, to

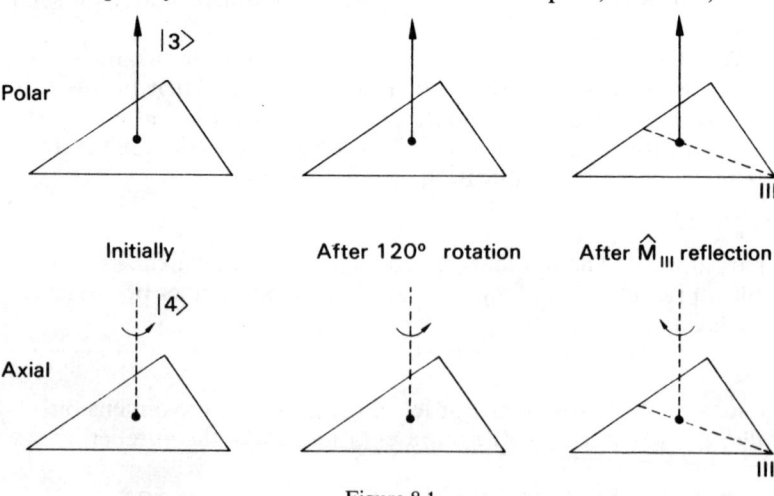

Figure 8.1.

IRREDUCIBLE GROUP REPRESENTATIONS

try to associate a particular type of invariant subspace with each individual class in some fashion. Can it be done?

Consider the image of a polar vector (such as $\bar{r}$ or $\bar{p}$) in a mirror, Fig. 8.2. If the polar vector $\bar{r}$ is normal to the plane of the mirror, the image $\bar{r}'$ is of opposite sign. If the polar vector $\bar{p}$ lies in a plane parallel to that of the mirror, the image $\bar{p}'$ is of the same sign.

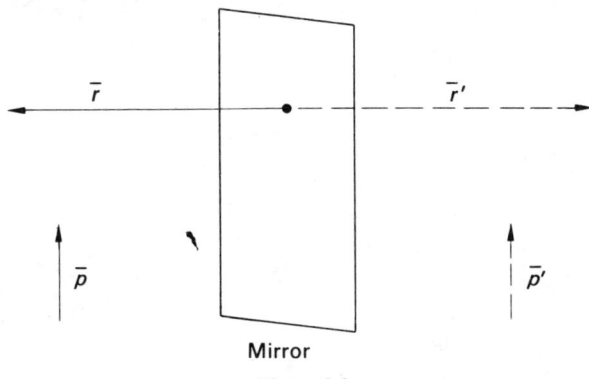

Figure 8.2.

The case of an axial vector such as $\bar{L} \equiv \bar{r} \times \bar{p}$ is different. As can be seen from Fig. 8.3a, if $\bar{r}$ and $\bar{p}$ lie parallel to a mirror, then $\bar{L} \equiv \bar{r} \times \bar{p}$ is normal to it, the alignment of $\bar{L}$ being given by the right-hand rule for the cross product. The images $\bar{r}'$ and $\bar{p}'$ are of the same sign as $\bar{r}$ and $\bar{p}$, and consequently $\bar{L} \equiv \bar{r}' \times \bar{p}'$ is of the same sign as $\bar{L}$. Similarly, the image of an axial vector lying parallel to a mirror (Fig. 8.3b) reverses sign.

Thus polar and axial vectors behave similarly under rotations but completely differently under reflection; clearly a basis set of three polar vectors alone, such as $\{|1\rangle, |2\rangle, |3\rangle\}$, is *not* complete and sufficient if we wish to consider a vector space containing both polar and axial vectors. This fact came out in the failure of $\{|1\rangle, |2\rangle, |3\rangle\}$ to generate A_2, the third irreducible representations of C_{3v}.

8.3 Labelling Irreducible Representations and Symmetry Types

A comment on notation. Basis vectors of the same given type may lead to irreducible representations for a number of different symmetry groups. Thus $\{|1\rangle, |2\rangle\}$ generate irreducible representations for both C_{3v} and C_{5v}. The basis pair $\{|1\rangle, |2\rangle\}$ behave similarly in both groups under rotations and reflections; i.e. they generate

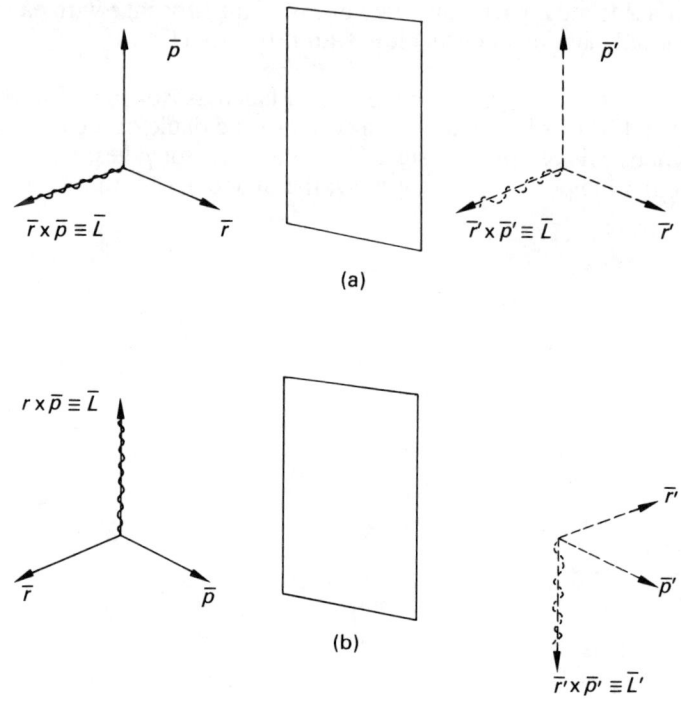

Figure 8.3.

invariant subspaces of the same general symmetry types for both groups. So also, $|3\rangle$ produces totally invariant (A_1) irreducible representations, and $|3\rangle$ is unaffected by all symmetry operations, of C_{3v} and C_{5v}.

Therefore it is often meaningful to speak of the symmetry types of functions, and of the associated irreducible representations, for whole families of groups. This lets us label certain symmetry behaviour, and irreducible representations, regardless of the specific group at hand.

There are several systems of notation. One makes use of gamma symbols with subscripts: $\Gamma_1, \Gamma_2, \ldots$, etc. correspond to various irreducible representations and types of symmetries. Another common system employs the letters A and B for one-dimensional representations, E for two-dimensional, and T for three-dimensional. The capital letters are generally accompanied by subscripted numbers or lower-case letters. A_1 is the totally invariant irreducible representation, and a typical basis function generating it might be an $|s\rangle$ state, since $\hat{P}_R|s\rangle = |s\rangle$ for any operation $\hat{P}_R$ of any symmetry group. If our physical system possesses an axis of rotation $|3\rangle$ containing a mirror plane, then $|3\rangle$ too generates A_1. In this case $|4\rangle$ would give what is labelled as A_2; conversely one says that a function has A_2 symmetry if it behaves under $\{\hat{P}_R\}$ as does $|4\rangle$ under $\{\hat{R}\}$. An E irreducible representation is generated *generally* by functions which behave like $\{|1\rangle, |2\rangle\}$ under rotations about $|3\rangle$. The T are more complex, and arise in dealing with objects with cubic or tetrahedral symmetry.

*PROBLEM 8.4 Show that any symmetry group has A_1 as one of its irreducible representations.

*PROBLEM 8.5 Prove that every irreducible representation of an abelian group is one-dimensional such that the fundamental Eq. (6.19) reduces to an eigenequation. See Problems 7.17 and 7.25 and Section 5.10. Can a cyclic group have an isomorphic irreducible representation?

Our development up to this point has been built upon the idea of representations generated by well-chosen basis sets. It might be disappointing, after all this effort, to learn that the forms of the entities of fundamental interest, namely the irreducible representations, are quite unconcerned with the *specific* basis set used for their generation.

But in actuality, this is exactly what we should *most* want—the irreducible representations, or equivalently, the transformation properties of the invariant subspaces, ought to depend *not* upon whether we use a $\{d_3, d_4\}$, a $\{p_1, p_2\}$, or a $\{|1\rangle, |2\rangle\}$ basis set; under the operations of C_{3v}, for example, all three pairs behave similarly! Thus it is not really the specific basis set which is of interest, but rather the general transformational properties shown by such sets under the various group operations $\{\hat{P}_R\}$; we should be looking for general symmetry-types of functions, not individual functions. What is perhaps most amazing is that there should be so *few* distinct types of invariant subspaces, or symmetries of basis functions.

†8.4 Regular Representation

The A_2 representation of C_{3v} has raised an awkward difficulty. We assumed earlier that the set $\{|1\rangle, |2\rangle, |3\rangle\}$ would form a complete basis for spanning real space; and yet we find it to be not suitable after all—it fails to produce the third, one-dimensional representation, A_2. $\{|1\rangle, |2\rangle, |3\rangle\}$ is complete only if it is called upon to span a space of polar vectors alone.

The apparent presence of a closure relation, then, does not guarantee that we have selected a large enough basis set with which to generate all the possible irreducible representations of a group. We may learn, too late, that our basis set was not sufficiently 'complete' after all. Fortunately, from Theorems 8.1 and 8.2, we can usually tell at least how many irreducible representations there should be for a group, and their dimensions; but how do we find any that are not obvious?

Trial and error may, or may not, work. There is, however, a straightforward way of generating every possible irreducible representation of a group in explicit form, and it requires no choices of basis vectors at all. For this we must construct the so-called *Regular Representation*, as follows: build a multiplication table by listing all the members of the group in a row, and the inverses of the elements vertically; then fill in the table.

Table 8.1

	$\hat{I}$	$\hat{R}_{120}$	$\hat{R}_{240}$	$\hat{M}_{I}$	$\hat{M}_{II}$	$\hat{M}_{III}$
$\hat{I}$	$\hat{I}$	$\hat{R}_{120}$	$\hat{R}_{240}$	$\hat{M}_{I}$	$\hat{M}_{II}$	$\hat{M}_{III}$
$(\hat{R}_{120})^{-1} = \hat{R}_{240}$	$\hat{R}_{240}$	$\hat{I}$	$\hat{R}_{120}$	$\hat{M}_{II}$	$\hat{M}_{III}$	$\hat{M}_{I}$
$(\hat{R}_{240})^{-1} = \hat{R}_{120}$	$\hat{R}_{120}$	$\hat{R}_{240}$	$\hat{I}$	$\hat{M}_{III}$	$\hat{M}_{I}$	$\hat{M}_{II}$
$\hat{M}_{I}^{-1} = \hat{M}_{I}$	$\hat{M}_{I}$	$\hat{M}_{II}$	$\hat{M}_{III}$	$\hat{I}$	$\hat{R}_{120}$	$\hat{R}_{240}$
$\hat{M}_{II}^{-1} = \hat{M}_{II}$	$\hat{M}_{II}$	$\hat{M}_{III}$	$\hat{M}_{I}$	$\hat{R}_{240}$	$\hat{I}$	$\hat{R}_{120}$
$\hat{M}_{III}^{-1} = \hat{M}_{III}$	$\hat{M}_{III}$	$\hat{M}_{I}$	$\hat{M}_{II}$	$\hat{R}_{120}$	$\hat{R}_{240}$	$\hat{I}$

The Regular Representation consists of matrices $\overline{\overline{\Gamma}}^{(reg)}(\hat{R}_i)$ obtained by replacing $\hat{R}_i$ in this sort of table by the number 1, and all other entries by 0. Thus for C_{3v} and Table 8.1

$$\Gamma^{(reg)}(\hat{R}_{120}) = \begin{pmatrix} 0 & 1 & 0 & 0 & 0 & 0 \\ 0 & 0 & 1 & 0 & 0 & 0 \\ 1 & 0 & 0 & 0 & 0 & 0 \\ 0 & 0 & 0 & 0 & 0 & 1 \\ 0 & 0 & 0 & 1 & 0 & 0 \\ 0 & 0 & 0 & 0 & 1 & 0 \end{pmatrix} \tag{8.5}$$

and so forth. Then the '*Celebrated Theorem*' states:

THEOREM 8.3: The Regular Representation $\Gamma^{(reg)}(\hat{R})$ of the group $\{\hat{R}\}$ contains each irreducible representation of $\{\hat{R}\}$ a number of times equal to the dimensionality of the irreducible representation.

And so it is a trivial matter to generate a reducible representation which contains all the irreducible representations at least once. The only problem consists of putting the $\Gamma^{(reg)}(\hat{R}_{120})$, etc. into the same irreducible block-diagonal form. This procedure and nearly every other applied group theoretic calculation involves the use of the *Great Orthogonality Theorem* and *characters*.

8.5 The Great Orthogonality Theorem

One of the most important computational tools of group representation theory is the *Great Orthogonality Theorem*, which we state without proof. Be it sufficient to say that the theorem has its roots in the mutual orthogonality of the basis vectors which span the distinct invariant subspaces (and generate inequivalent irreducible representations) associated with any group.

THEOREM : The Great Orthogonality Theorem:

$$\boxed{\sum_{R}^{N} \Gamma_{ij}^{(\mu)}(\hat{R})^* \, \Gamma_{i'j'}^{(\nu)}(\hat{R}) = \frac{N}{n_\nu} \delta_{\mu\nu} \delta_{ii'} \delta_{jj'}}$$ (8.6)

where N is the order of the group and n_ν the dimensionality of the νth irreducible representation.

The theorem says that we can take the ijth element of two irreducible representations, and consider what happens when the products $\Gamma_{ij}^{(\mu)}(\hat{R})^* \Gamma_{ij}^{(\nu)}(\hat{R})$ are summed over all elements of the group. This sum of products survives if and only if the μth and νth representations are equivalent. Even with equivalent irreducible representations, however, the sum is non-vanishing only if we use the same ijth element of each of the two matrices; $\sum_{R} \Gamma_{32}^{(\mu)}(\hat{R})^* \Gamma_{32}^{(\mu)}(\hat{R})$ will survive for the μth representation, for example, but

$\sum_{R} \Gamma_{32}^{(\mu)}(\hat{R})^* \Gamma_{21}^{(\mu)}(\hat{R})$ will not.

Equation (8.6) draws one immediately to an analogy: the various non-equivalent irreducible representations are in a sense, 'orthogonal'. Consider a set of ordinary but orthonormal vectors $\{|\mu\rangle, |\nu\rangle, \ldots\}$; in the $\{|R\rangle\}$ basis system,

$$\langle \mu | \nu \rangle = \sum_{R}^{N} \langle \mu | R \rangle \langle R | \nu \rangle = \sum_{R}^{N} \mu_R^* \nu_R = \delta_{\mu\nu}$$ (8.7a)

Here R is just an index assuming integer values. This shows that each vector $|\mu\rangle$, with the N components μ_R, inhabits an N-dimensional space. If the linearly independent vectors $\{|\mu\rangle, |\nu\rangle, \ldots\}$ are themselves to span this space, they too must be N in number. Comparison of Eqs (8.6) and (8.7a) suggests that we consider the functions

$$(n_\nu/N)^{1/2} \Gamma_{ij}^{(\nu)}(\hat{R})$$ (8.7b)

as being a set of orthonormal vectors spanning an abstract N-dimensional vector space. Just as 'R' labelled the co-ordinate axes of the $|\nu\rangle$ of Eq. (8.7a), so Eq. (8.7b) describes the $\hat{R}$th component, or the component in the $\hat{R}$th direction, of the $ij\nu$th vector in our space. This rather artificially constructed space is useful in that it leads into the proofs of several theorems, among them being that of Theorem 8.2.

PROBLEM 8.6 By an argument like that following Eq. (8.7a) prove Theorem 8.2. Think of Eq. (8.7b) as the ijth vector in the vth subspace of our space.

Many texts on symmetry in quantum mechanics begin by developing a rather formidable foundation in abstract group and group representation theory (Schur's lemma, etc.), mostly to prove Theorems 8.1 through 8.4. We feel that at this stage, the spade work required is a bit extensive, and the proofs too ponderous, to add much to one's real understanding of the theorems themselves. Little is lost here, and much confusion is avoided, by simply postulating the theorems, and beginning from there. Proofs are given by Joshi,[22] Tinkham,[39] etc., for those so inclined at this stage.

8.6 Characters

Another important device for performing group theoretic calculations, one which is used in practice almost exclusively, is the character.

It is well known that the trace (sum of the diagonal elements) of a unitary matrix is invariant to similarity transformations. Since equivalent representations are those which can be obtained from one another by means of such transformations of function space—equivalent matrices have the same traces—it would seem that the trace of a representation should be a useful label or characteristic invariant of it. Such is the case, and a trace is known as a *character*; nearly all tabulated information concerning symmetry groups is in the form of character tables (rather than as a listing of typical matrices of the various irreducible representations). There is a certain small loss of easily accessible information in presenting only the characters, but nearly all calculations can be done in terms of them alone.

The character corresponding to the μth representation of the symmetry operator $\hat{R}$ is written as $\chi^{(\mu)}(\hat{R})$:

$$\chi^{(\mu)}(\hat{R}) \equiv \sum_i \Gamma_{ii}^{(\mu)}(\hat{R}) \tag{8.8}$$

*PROBLEM 8.7 Show that, given a representation, all the operators in a class have the same character.

*PROBLEM 8.8 Construct a character table—list the character of each operator for each irreducible representation—for C_{3v}.

IRREDUCIBLE GROUP REPRESENTATIONS   143

The Great Orthogonality Theorem is expressed in terms of characters as

$$\sum_{R}^{N} \chi^{(\mu)}(\hat{R})^* \chi^{(\nu)}(\hat{R}) = N\delta_{\mu\nu} \tag{8.9}$$

PROBLEM 8.9 Confirm the validity of the character table worked out in Problem 8.8 by means of Eq. (8.9).

PROBLEM 8.10 Express Eq. (8.9) as a sum over classes rather than as a sum over the group elements $\{\hat{R}\}$.

8.7 Which Irreducible Representations Lie Hidden within a Given Reducible Representation?

We saw in Eq. (7.24) that the reduction of a reducible group representation $\Gamma^{(\text{red})}$ into its constituent irreducible representations may be presented as the direct sum

$$\Gamma^{(\text{red})} = \sum_{\nu}^{c} a_\nu \Gamma^{(\nu)} \tag{8.10}$$

c is the number of distinct classes in the group in question, and hence, by Theorem 8.1, of non-equivalent irreducible representations. In Eq. (4.23), for example, we found that a representation of C_{3v} generated by the three unit vectors of real space is reduced as

$$\Gamma^{(\text{red})} = A_1 + E \tag{4.23}$$

Similarly, from the discussion at the beginning of Section 8.2, for C_{3v}

$$\Gamma^{(d)} = A_1 + 2E$$
$$\Gamma^{(p)} = A_1 + E$$

We shall now employ the character form of the Great Orthogonality Theorem, Eq. (8.9), to provide a simple method for determining, given a reducible representation $\Gamma^{(\text{red})}$, the a_ν of Eq. (8.10) —this procedure will tell how many times a particular irreducible representation would occur in a group of large matrices if they were all brought to the same block-diagonal form.

Equation (8.10) describes the reduction of a set of N matrices representing a group of order N. Each of the matrices, consequently, individually undergoes the reduction

$$\Gamma^{(\text{red})}(\hat{R}_i) = \sum_{\nu}^{c} a_\nu \Gamma^{(\nu)}(\hat{R}_i) \tag{8.11a}$$

From the definition of character, Eq. (8.8), it follows that

$$\chi^{(\text{red})}(\hat{R}_i) = \sum_v^c a_v \chi^{(v)}(R_i) \qquad i = 1, 2, \ldots N \qquad (8.11b)$$

There is one such equation for each $\hat{R}_i$, though the equations corresponding to symmetry operators of the same class will be identical (Why?). Consider one particular irreducible representation, $\Gamma^{(\mu)}$, of the group, with characters $\{\chi^{(\mu)}(\hat{R}_i)\}$. Looking at the $\hat{R}_i$ of Eq. (8.11b) one at a time, we can use $\chi^{(\mu)}(\hat{R}_i)^*$ to form the product

$$\chi^{(\mu)}(\hat{R}_i)^* \chi^{(\text{red})}(\hat{R}_i) = \sum_v^c a_v \chi^{(\mu)}(\hat{R}_i)^* \chi^{(v)}(\hat{R}_i)$$

Summing over all $\hat{R}_i$:

$$\sum_R^N \chi^{(\mu)}(\hat{R})^* \chi^{(\text{red})}(\hat{R}) = \sum_v^c a_v \sum_R^N \chi^{(\mu)}(\hat{R})^* \chi^{(v)}(\hat{R})$$

But by the Great Orthogonality Theorem, Eq. (8.9), the right-hand side of this is

$$N \sum_v^c a_v \delta_{\mu v} = N a_\mu$$

Our final and very important result is

$$\boxed{a_\mu = \frac{1}{N} \sum_R^N \chi^{(\mu)}(\hat{R})^* \chi^{(\text{red})}(\hat{R})} \qquad (8.11c)$$

$\{\chi^{(\text{red})}(\hat{R})\}$ comes from the matrices we are given initially; and tabulated values of $\{\chi^{(\mu)}(\hat{R})\}$ are immediately accessible for virtually all symmetry groups. The problem of determining how many times an irreducible representation would be found in a given group representation after block-diagonalization, turns out to be no problem at all. And in the next three chapters, we shall see how very useful such information can be in attacking quantum mechanical problems. But first, we shall illustrate the mechanism of Eq. (8.11) with a simple example from C_{3v}.

8.8 Reduction of a Representation: A C_{3v} Example

In Problems 8.7 and 8.8, we constructed the character table for C_{3v}

Table 8.2

C_{3v}	$\hat{I}$	$\hat{R}_{120}, \hat{R}_{240}$	$\hat{M}_\text{I}, \hat{M}_\text{II}, \hat{M}_\text{III}$
A_1	1	1	1
A_2	1	1	-1
E	2	-1	0

IRREDUCIBLE GROUP REPRESENTATIONS 145

Thus $\chi^{(A_2)}(\hat{M}_{\text{II}}) = -1$, etc.

Suppose we are given the following set of matrices which supposedly forms a representation 'Γ' of C_{3v}; we have written down the extracted characters also.

$$I = \begin{pmatrix} 1 & 0 & 0 \\ 0 & 1 & 0 \\ 0 & 0 & 1 \end{pmatrix}; \quad \chi^{(\Gamma)}(\hat{I}) = 3 \qquad \bar{\bar{M}}_{\text{I}} = \begin{pmatrix} 1 & 0 & 0 \\ 0 & 0 & 1 \\ 0 & 1 & 0 \end{pmatrix}; \quad \chi^{(\Gamma)}(\hat{M}_{\text{I}}) = 1$$

$$\bar{\bar{R}}_{120} = \begin{pmatrix} 0 & 1 & 0 \\ 0 & 0 & 1 \\ 1 & 0 & 0 \end{pmatrix}; \chi^{(\Gamma)}(\hat{R}_{120}) = 0 \qquad \bar{\bar{M}}_{\text{II}} = \begin{pmatrix} 0 & 0 & 1 \\ 0 & 1 & 0 \\ 1 & 0 & 0 \end{pmatrix}; \quad \chi^{(\Gamma)}(\hat{M}_{\text{II}}) = 1$$

$$\bar{\bar{R}}_{240} = \begin{pmatrix} 0 & 0 & 1 \\ 1 & 0 & 0 \\ 0 & 1 & 0 \end{pmatrix}; \chi^{(\Gamma)}(\hat{R}_{240}) = 0 \qquad \bar{\bar{M}}_{\text{III}} = \begin{pmatrix} 0 & 1 & 0 \\ 1 & 0 & 0 \\ 0 & 0 & 1 \end{pmatrix}; \chi^{(\Gamma)}(\hat{M}_{\text{III}}) = 1$$

(8.12)

A quick check by forming a few matrix products, just to be on the safe side, reveals that Γ does, indeed, form an isomorphic representation of C_{3v}. And as expected, matrices of operators in the same class have the same character. $\{\bar{\bar{\Gamma}}(\hat{R})\}$ is, moreover, clearly reducible, since the largest irreducible representation of C_{3v} is two-dimensional.

If we knew the basis set by which these matrices were generated, we could determine the irreducible representations involved by inspection of their symmetry properties. Since these have not been given, we use Eq. (8.11):

$$a_\mu = \frac{1}{N} \sum_R^N \chi^{(\mu)}(\hat{R})^* \chi^{(\Gamma)}(\hat{R}) \qquad (8.11c)$$

For C_{3v}, the reduction will be of the form

$$\Gamma = a_{A_1} A_1 + a_{A_2} A_2 + a_E E \qquad (8.13)$$

From the character table Eq. (8.12) and Eq. (8.11c)

$$\left. \begin{aligned} a_{A_1} &= \frac{1}{N} \{\chi^{(A_1)}(\hat{I})^* \chi^{(\Gamma)}(\hat{I}) + \cdots + \chi^{(A_1)}(\hat{M}_{\text{III}})^* \chi^{(\Gamma)}(\hat{M}_{\text{III}})\} \\ &= \tfrac{1}{6}(1 \cdot 3 + 1 \cdot 0 + 1 \cdot 0 + 1 \cdot 1 + 1 \cdot 1 + 1 \cdot 1) = 1 \\ a_{A_2} &= \tfrac{1}{6}(1 \cdot 3 + 1 \cdot 0 + 1 \cdot 0 - 1 \cdot 1 - 1 \cdot 1 - 1 \cdot 1) = 0 \\ a_E &= \tfrac{1}{6}(2 \cdot 3 - 1 \cdot 0 - 1 \cdot 0 + 0 \cdot 1 + 0 \cdot 1 + 0 \cdot 1) = 1 \end{aligned} \right\}$$

(8.14a)

and so

$$\Gamma = A_1 + E \qquad (8.14b)$$

The use of projection operators, as we shall see in the next chapter, even allows us to put the matrices of Γ into block-diagonal form.

PROBLEM 8.11 Verify the Celebrated Theorem for the group C_{3v}.
PROBLEM 8.12 Do $\{|1\rangle, |2\rangle, |3\rangle, |4\rangle\}$ (see Fig. 8.1) form a complete, orthonormal basis set spanning axial-and-polar vector space?
PROBLEM 8.13 What are the irreducible representations and character tables of C_n and C_{nv}, prime integer n?
PROBLEM 8.14 The set of all rotations $\{\hat{R}_\phi\}$ about an axis forms a *Lie* or *continuous* group. What are the classes and characters of this group? How many irreducible representations does it have?
PROBLEM 8.15 A certain set of functions generates the representation Γ'', whose character table is given below, of C_{3v}. Show that $\Gamma'' = A_2 + E$.

Table 8.2

C_{3v}	$\hat{I}$	$\hat{R}_{120}$	$\hat{M}$
Γ''	3	0	-1

PROBLEM 8.16 Defining the *Dirac Character* $\hat{\Omega}_C$ of a class C as the sum† of symmetry operators (thus $\hat{\Omega}_C$ is by no means a scalar 'character' in the normal sense) in that class,

$$\hat{\Omega}_C = \sum_{R' \varepsilon C} \hat{R}' \qquad (8.15a)$$

show that all the Dirac Characters in a group commute. Find the C_{3v} multiplication table for $\{\hat{\Omega}_C\}$ and verify

$$\hat{\Omega}_C \hat{\Omega}_{C'} = \sum^{c} h_{C''}(C, C') \hat{\Omega}_{C''} \qquad (8.15b)$$

with scalar $h_{C''}(C, C') = h_{C''}(C', C)$.

8.9 Summary

1. The number of non-equivalent irreducible representations of a group of order N is equal to the number of distinct classes, c, in the group, Theorem 8.1. Also

$$\sum_\nu^c n_\nu^2 = N$$

where n_ν is the dimensionality of the νth irreducible representation, Theorem 8.2. These two theorems are usually sufficient to

---

†With the introduction of a second binary operation (the sum), the group $\{\hat{R}\}$ becomes an algebra.

IRREDUCIBLE GROUP REPRESENTATIONS

tell the number and the dimensionalities of the possible irreducible representations of a group.

2. If one cannot find suitable basis vectors to generate all the irreducible representations, then the Celebrated Theorem says that the regular representation will yield all of them. It is seen that the basis set $\{|1\rangle, |2\rangle, |3\rangle\}$ is *not*, after all, complete, if our vector space includes both polar and axial vectors; this is why $\{|1\rangle, |2\rangle, |3\rangle\}$ led to only two of the three irreducible representations of C_{3v}.

3. One can label irreducible representations and the symmetry types of functions regardless of the particular group at hand; standard conventions use either Γ or A, B, E, and T with subscripts.

4. The Great Orthogonality Theorem, Theorem 8.4, and characters are essential machinery in group theoretic manipulations. Through them, it is seen that if a reducible representation $\Gamma^{(\text{red})}$ is composed of irreducible representations $\{\Gamma^{(v)}\}$ such that $\Gamma^{(\text{red})} = \sum_v^c a_v \Gamma^{(v)}$, then

$$a_v = \frac{1}{N} \sum_R^N \chi^{(v)}(\hat{R})^* \chi^{(\text{red})}(\hat{R}) \tag{8.11c}$$

reminiscent of the Fourier decomposition of a vector.

CHAPTER 9

Quantum Mechanical States: How Symmetry in Real Space Predetermines Symmetry in Hilbert Space

We have finally reached the fundamental issue of our study: how the geometric symmetry group $\{\hat{R}\}$ of a quantum mechanical system determines the possible allowed symmetries of the functions in Hilbert space describing the system. The Wigner Theorem demands that every energy eigenstate displays the same transformational behaviour under $\{\hat{P}_R\}$ as is shown by a generator of one of the irreducible representations of $\{\hat{R}\}$. Or in other words, a sufficiently large set of orthonormal degenerate energy eigenvectors can serve as a basis set for generating one of the irreducible representations of the symmetry group. Thus once we have determined the system's group structure, we automatically know, from symmetry arguments alone, the possible degeneracies of the eigenlevels, and also much (sometimes everything) about the angular dependences of the corresponding eigenvectors. This is useful (1) in classifying states on the basis of their transformational properties, (2) in choosing zero-th order functions for setting up a matrix representation of a Hamiltonian to be diagonalized, (3) in discovering, without evaluating the relevant perturbation matrix elements, whether or not a perturbation will break a degeneracy or cause an electronic transition, and (4) much else.

The Wigner Theorem applies not only to groups of rotations and reflections, but also to non-geometric symmetries such as those involving time-reversal and the interchange of indistinguishable particles.

Projection operators and product spaces offer mathematical techniques for employing the above ideas in calculations. For example, we find molecular orbitals and vibrational modes of correct symmetry for a planar C_{3v} molecule. A re-consideration of the vibrational mode problem from a classical viewpoint shows

QUANTUM MECHANICAL STATES 149

that group representation theory can be of use in non-quantum mechanical situations as well.

9.1 The Wigner Theorem

In this chapter we continue the study, begun in Chapter 6, of the symmetry properties of functions. Here we shall restrict our attention (except in Section 9.10) to the function spaces of interest to quantum mechanics, and we shall draw upon what we know of irreducible representations.

We have covered a lot of ground so far, for the purpose of now proving what will surely seem to be a nearly trivial theorem. The significance of this theorem, however, cannot be over-stated, as it allows us to determine a tremendous amount about a quantum mechanical system with a relatively small input of information. All we need to know, in fact, is what the system looks like.

THEOREM 9.1, The Wigner Theorem. Let $\{\hat{R}\}$ be the symmetry group, in three-space, of a quantum mechanical system, and let $\mathcal{H}$ be the exact Hamiltonian of the system. Then:

1. $$\boxed{[\mathcal{H}, \hat{P}_R] = 0 \text{ for all } \hat{R} \in \{\hat{R}\}} \tag{9.1}$$

The group $\{\hat{R}\}$ such that $[\mathcal{H}, \hat{P}_R] = 0$ is called the *group of the Schroedinger equation*.

2. The eigenvectors $\{\psi_i^{(v)}(\bar{r})\}$ of $\mathcal{H}$ which correspond to the eigenvalue E_v constitute a subspace invariant with respect to the symmetry operators $\{\hat{P}_R\}$, and generate an irreducible representation of $\{\hat{R}\}$.

We have already seen Part 1 of this theorem in Section 5.8 as Eq. (5.27) and (5.32), but shall follow here a different approach to the proof†.

---

†We could, alternatively, extend the argument of Section 5.8 as follows.
If $\hat{R}$ is a real-space symmetry operator which, as with Eq. (5.26) or (5.28), leaves the Hamiltonian invariant,

$$\mathcal{H}(\hat{R}\bar{r}) = \mathcal{H}(\bar{r})$$

then for any $\psi(\bar{r})$

$$\hat{P}_R\{\mathcal{H}(\bar{r})\psi(\bar{r})\} = \mathcal{H}(\hat{R}^{-1}\bar{r})\psi(\hat{R}^{-1}\bar{r}) = \mathcal{H}(\bar{r})\psi(\hat{R}^{-1}\bar{r}) = \mathcal{H}(\bar{r})\hat{P}_R\psi(\bar{r})$$

Since $\psi(\bar{r})$ is arbitrary, then

$$\hat{P}_R \mathcal{H}(\bar{r}) = \mathcal{H}(\bar{r})\hat{P}_R \quad \text{or} \quad [\hat{P}_R, \mathcal{H}(\bar{r})] = 0$$

For this one must initially define a symmetry operator $\hat{R}$ by means of

$$\mathcal{H}(\hat{R}\bar{r}) = \mathcal{H}(\bar{r}) \tag{5.27b}$$

Why is this allowed for $\mathcal{H}(\bar{r})$ whereas generally $\psi(\hat{R}\bar{r}) \neq \psi(\bar{r})$?

If $|\phi(\bar{r})\rangle$ describes the state of our system, and $\hat{P}_R$ is a symmetry operator, then $\hat{P}_R|\phi(\bar{r})\rangle$ describes a situation physically indistinguishable from the original; although the mathematical representation of the state may be different, any physically measurable attribute must be the same.

Let us consider, in particular, the expectation values of $\mathscr{H}$, first using the arbitrary state $|\phi(\bar{r})\rangle$, and then with the state $\hat{P}_R|\phi(\bar{r})\rangle$ which results if the system undergoes the symmetry transformation $\hat{R}$. The energy of the system should not change, and

$$\langle\phi(\bar{r})|\mathscr{H}|\phi(\bar{r})\rangle = \langle\phi(\bar{r})|\hat{P}_R^+)\mathscr{H}(\hat{P}_R|\phi(\bar{r})\rangle) = \langle\phi(\bar{r})|\hat{P}_R^+\mathscr{H}\hat{P}_R|\phi(\bar{r})\rangle \tag{9.2a}$$

or

$$\langle\phi(\bar{r})|[\mathscr{H}, \hat{P}_R]|\phi(\bar{r})\rangle = 0 \tag{9.2b}$$

by the unitarity of $\hat{P}_R$.

Since this is true for any $|\phi(\bar{r})\rangle$ (be careful with this part of the argument) then

$$[\mathscr{H}, \hat{P}_R] = 0 \tag{9.1}$$

PROBLEM 9.1 Can this proof be generalized such that $\mathscr{H}$ may be replaced by any Hermitian operator corresponding to a physical observable, like $\hat{r}, \hat{p}, \hat{L}$?
Consider $\hat{r}$, the position operator, for a system with inversion symmetry.

We prove Part 2 of the theorem by extending the argument of Section 5.10. Consider the orthonormal set $\{\psi_i^{(\nu)}(\bar{r})\}$ of degenerate eigenvectors of $\mathscr{H}$ corresponding to the particular energy level E_ν; assume $\{\psi_i^{(\nu)}(\bar{r})\}$ to be 'level-complete', in the sense that any state of energy E_ν can be represented as a linear combination of the $\{\psi_i^{(\nu)}(\bar{r})\}$. For any such $\psi_j^{(\nu)}(\bar{r})$,

$$\mathscr{H}\psi_j^{(\nu)}(\bar{r}) = E_\nu \psi_j^{(\nu)}(\bar{r}) \tag{9.3a}$$

Then

$$\hat{P}_R \mathscr{H} \psi_j^{(\nu)}(\bar{r}) = \hat{P}_R E_\nu \psi_j^{(\nu)}(\bar{r}) \tag{9.3b}$$

But since $\hat{P}_R$ and $\mathscr{H}$ commute, as in Eq. (5.33)

$$\mathscr{H}(\hat{P}_R \psi_j^{(\nu)}(\bar{r})) = E_\nu(\hat{P}_R \psi_j^{(\nu)}(\bar{r})) \tag{9.3c}$$

Thus if $\psi_j^{(\nu)}(\bar{r})$ is an eigenvector of $\mathscr{H}$ with eigenvalue E_ν, then $\hat{P}_R \psi_j^{(\nu)}(\bar{r})$ is also an eigenvector of $\mathscr{H}$, associated with the *same* eigenvalue; $\hat{P}_R \psi_j^{(\nu)}(\bar{r})$ must therefore be some linear combination of the $\{\psi_i^{(\nu)}(\bar{r})\}$. (Why?) The elements of the group $\{\hat{P}_R\}$ will then

QUANTUM MECHANICAL STATES

operate on any function of the subspace spanned by $\{\psi_i^{(\nu)}(\bar{r})\}$ to produce another function still in this subspace; *the degenerate set $\{\psi_i^{(\nu)}(\bar{r})\}$ transform only among themselves under the operations of the group of the Schroedinger equation.* The $\psi_i^{(\nu)}(\bar{r})$ thus *inhabit a subspace invariant to the operations $\{\hat{P}_R\}$*, and can be used to generate a representation of the group $\hat{P}_R$:

$$\hat{P}_R \psi_i^{(\nu)}(\bar{r}) = \sum_j \psi_j^{(\nu)}(\bar{r}) \Gamma_{ji}^{(\nu)}(\hat{R})$$
$$\Gamma_{ji}^{(\nu)}(\hat{R}) \equiv \langle \psi_j^{(\nu)}(\bar{r}) | \hat{P}_R | \psi_i^{(\nu)}(\bar{r}) \rangle$$

(9.4)

The dimension of the representation matrices and of the invariant subspace is equal to the number of linearly independent (and preferably orthogonal) degenerate eigenvectors.

*PROBLEM 9.2 Why is the representation defined in Eq. (9.4) irreducible?

Theorem 9.1 is, for quantum mechanical purposes, the principal result of the theory of group representations: *the level-complete set of degenerate 'partner' eigenfunctions of energy E_ν, $\{\psi_i^{(\nu)}(\bar{r})\}$, forms a basis system for an irreducible representation $\Gamma^{(\nu)}$ of the group of the Schroedinger equation*; any vector in this subspace is said to 'belong' to its νth irreducible representation.

This leads us directly to an appreciation of why group representation theory is so invaluable in quantum mechanics. Once we have determined the symmetry of the object in real space, we know the group of the Schroedinger equation. We can then immediately calculate or look up the possible irreducible representations corresponding to this particular type of symmetry group. (This is not a matter of quantum mechanics. Tables of the irreducible representations of all the ordinary symmetry groups in real space are readily available; or we could use the methods of the last chapter to construct them.) But any level-complete set of degenerate eigenvectors generates one of the irreducible representations of the group; thus they must span a function subspace with the general symmetry properties and dimensionality of one of the irreducible representations of the symmetry group, as indicated by Eq. (9.4). Knowledge of the group structure and appropriate irreducible representations immediately tells us the possible degeneracies (which equals the number of basis vectors involved) and symmetry types of the system's energy eigenspectrum!

Consider for example, an ammonia molecule. We know that there are three possible irreducible representations of the symmetry group of C_{3v}. A_1 is one-dimensional, and the possible basis vectors which could generate such a representation behave under $\{\hat{P}_R\}$ as $|3\rangle$ behaves under the operations of C_{3v} in real space; i.e. they are invariant to any $\hat{P}_R$ of C_{3v}. Similarly, a second representation, A_2, is also one-dimensional, and is generated by a basis vector of A_2 symmetry. Finally, there is a two-dimensional irreducible representation E generated by two unit vectors which mix under $\{\hat{P}_R\}$ as $\{|1\rangle, |2\rangle\}$ mix under C_{3v}. Thus we know that for ammonia, there are three basic categories of eigenvectors; two of these categories contain non-degenerate states, and the third consists of doubly degenerate states. An ammonia molecule, then, can have no triply (or more) degenerate electronic orbital states! And the states that do exist must satisfy well-defined symmetry requirements as demonstrated in the extremely revealing

*PROBLEM 9.3 Show that if $\{\psi_1, \psi_2\} = \{d_3, d_4\}$, or $\{p_1, p_2\}$ are to be degenerate states of a C_{3v} system, then for $\hat{R}_{120}$

$$\hat{P}_R|\psi_1\rangle = -\tfrac{1}{2}|\psi_1\rangle + \tfrac{1}{2}\sqrt{3}\,|\psi_2\rangle$$

just as

$$\hat{R}_{120}|1\rangle = -\tfrac{1}{2}|1\rangle + \tfrac{1}{2}\sqrt{3}\,|2\rangle.$$

[HINT: see Eq. (6.29).] What about $\hat{R}_{119}$?

Consider, however, a spin system whose Hamiltonian, and consequently eigenvalues, are magnetic field, H, dependent, Fig. 9.1.

At a certain value of H, the γ and δ levels cross, and we have (at least over an infinitesimally narrow band of H) a degeneracy, noted at the point C. The degeneracy of $|\gamma\rangle$ and $|\delta\rangle$, unlike the

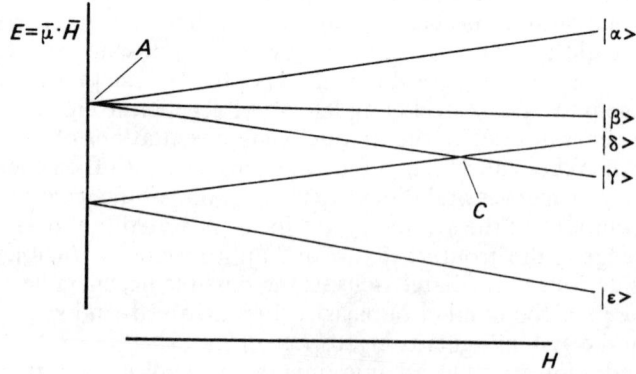

Figure 9.1.

QUANTUM MECHANICAL STATES   153

'essential' degeneracy of $|\alpha\rangle$, $|\beta\rangle$, and $|\gamma\rangle$ at A, is not attributable to any form of symmetry, and is known as an 'accidental' degeneracy.

*PROBLEM 9.4 Let Eq. (9.4) and Fig. 9.1 describe a system of well-defined symmetry. Will any sort of symmetry operation cause an admixture of states $|\gamma\rangle$ and $|\delta\rangle$? Why do $|\alpha\rangle$, $|\beta\rangle$ and $|\gamma\rangle$ become degenerate at the point A?

A well-known degeneracy is that between hydrogenic states of the same principal quantum number n. The nine states $|s\rangle$, $\{|p_i\rangle\}$, and $\{|d_i\rangle\}$ for any $n \geqslant 3$ have the same energy, and this is not interpretable by means of the rotational-reflectional symmetry of an atom; yet the degeneracy seems too systematic to be 'accidental' in the above sense. The solution to this dilemma was in terms of a rotation symmetry in four-dimensional momentum space; the degeneracy is not accidental after all, and can be lifted by adding a small, but still spherically symmetric, non-Coulombic perturbing potential to the system. The moral of this story is that 'accidental degeneracies' sometimes are, but often are not, accidental, and may indicate the presence of some previously overlooked symmetry.

*PROBLEM 9.5 Demonstrate Eq. (9.1) using the Schroedinger equation and the following definition: $\hat{R}$ is a symmetry operator if for any eigenstate $\psi_i(\bar{r})$ of $\mathscr{H}$, $\psi_i(\bar{r})$ and $\hat{P}_R\psi_i(\bar{r})$ are degenerate.

9.2 Symmetries of State Functions

We have argued that symmetry considerations place severe constraints upon the possible angular behaviour of a wave function. But can knowledge of a system's group structure give us *total* information on the angular dependences of its possible states? Only sometimes.

We shall illustrate in Section 10.2 the way in which full rotational symmetry about a single axis completely determines the angular dependences of the wave functions of a two-dimensional system. There we shall see that all of these states vary as $\exp(im\phi)$, integer m, where ϕ is the ordinary real-space azimuthal angle, and each m corresponds to a different irreducible representation. And one can also show that for a spinless quantum system with a spherically symmetric potential, the *only* allowed wave functions are linear combinations of the spherical harmonics. Thus Hilbert space for such a system is broken into subspaces spanned, respectively, by $\{Y_0^0\}$, $\{Y_1^1, Y_1^0, Y_1^{-1}\}$, $\{Y_2^2, \ldots\}$, ... times appropriate radial functions, i.e. the s, p, d, etc., states.

In the above two-dimensional and three-dimensional cases, it is crucial for a complete determination of the angular dependences of the states that the symmetry group in question be continuous, containing infinitesimally small rotations. In many cases of practical interest one is concerned with groups of finite order such as C_{3v}, and here group theory offers us less. The vitally important lesson to be learned from Problem 9.3 is that although two basis sets will yield equivalent representations of a finite-order group if they behave similarly *under the few particular symmetry operations of the group*, they may transform totally differently under an arbitrary rotation or reflection. In the case of a C_{3v} molecule, the symmetry of the system is sufficiently non-restrictive that numerous pairs of functions have E-type transformation properties under $\{\hat{I}, \hat{R}_{120}, \ldots\}$, but dissimiliar behaviour under other (non-symmetry) transformations. This is not necessarily the case in a situation of more stringent symmetry. A system with axial, say, symmetry imposes more extreme demands upon its wave functions: since $\hat{R}_\phi$ is a symmetry operator for all ϕ, so $\hat{P}_{R,\phi}\psi(\bar{r})$ is well-defined for all ϕ, and this is equivalent to having complete knowledge of $\psi(\phi)$, aside from a constant multiplicative phase factor. We shall consider further the case of axial symmetry and continuous groups in general in the next chapter.

Returning to finite order groups, suppose we wish to determine the wave functions of a C_{3v} molecule. First of all, we know that any likely candidate or trial function must display a precise behaviour under the few operations of the group, as in Problem 9.3. But an *exact* solution must be valid over *all* angles, and have well-defined transformational properties for *any* rotation of molecule or coordinates; unfortunately, only a complete diagonalization of $\mathcal{H}$ can yield this. Group theory can help in the process by narrowing the field to a certain range of possibilities which have the right behaviour *at least* under C_{3v}.

It may be disappointing that symmetry and group arguments cannot in general lead to the complete angular dependence of wave functions. Still, they can be immensely useful. In evaluating the matrix elements of importance to time-independent and time-dependent perturbation calculations, e.g. group theory is often powerful enough to tell whether the relevant matrix elements vanish or not. Thus symmetry considerations alone indicate when a perturbation will break a degeneracy or cause transitions between levels.

*PROBLEM 9.6 Suppose we solve the Schroedinger equation for a C_{3v} system *exactly* and find, among other things, a pair of E-

QUANTUM MECHANICAL STATES

states, ψ_1^E and ψ_2^E. Problem 9.3 tells us how they will transform under particular operations of C_{3v}; but can we employ

$$\hat{P}_R \psi_1^E(\vec{r}) = \psi_1^E(\hat{R}^{-1}\vec{r}) = c_{11}\psi_1^E(\vec{r}) + c_{21}\psi_2^E(\vec{r})$$

as in Eqs (6.26) and (6.19) to find the effect on ψ_1^E of a rotation through, say, $20°$, which is not a symmetry operation? Can one obtain the c_{ij} from one of the 2×2 blocks in Eq. (6.31a)?

*PROBLEM 9.7 Let $\{|\psi_i\rangle\}$ be an orthonormal basis set spanning Hilbert space for a C_{3v} molecule of Hamiltonian $\mathscr{H}$. Describe the matrix representation $\bar{\bar{\mathscr{H}}}$ of $\mathscr{H}$: 1. if $\{|\psi_i\rangle\}$ are all of mixed symmetry; 2. if $\{|\psi_i\rangle\}$ is broken into three categories belonging, respectively, to A_1, A_2, and E; and 3. if these three classes are further separated into subsets of degenerate states. Is $\bar{\bar{\mathscr{H}}}$ now of necessity fully diagonal? Compare with Sections 5.9 and 5.10.

The inversion operator (in contra-distinction to reflection and/or rotation operators) will commute with all other symmetry operators; and consequently the wave function of a system with inversion symmetry can always be made to be odd or even (why?). The wave functions will display other symmetry properties as well if the physical system does, but these are essentially independent of their oddness-evenness characteristics. It is conventional to label the parity of a representation with the subscript g ('gerade') or u ('ungerade') if it is generated by even or odd basis functions, respectively.

PROBLEM 9.8 Do the (d_1, d_2) and (d_3, d_4) basis vectors of C_{3v} all have the same parity? Why is parity not denoted as a subscript on the E representation of C_{3v}? Show that the parity operator commutes with any rotation or reflection.

9.3 Good Quantum Numbers

In this chapter we have been stressing the connection between the irreducible representations of a molecular system's symmetry group and the allowed angular dependences of its quantum mechanical state functions. In summary, a set of degenerate orthonormal states $\{|\psi_i^{(\nu)}(\vec{r})\rangle\}$ completely spanning the subspace associated with the level of energy $E^{(\nu)}$ transforms as

$$\hat{P}_R|\psi_i^{(\nu)}(\vec{r})\rangle = |\psi_i'^{(\nu)}(\vec{r})\rangle = \sum_j |\psi_j^{(\nu)}(\vec{r})\rangle\langle\psi_j^{(\nu)}(\vec{r})|\hat{P}_R|\psi_i^{(\nu)}(\vec{r})\rangle$$

$$= \sum_j \Gamma_{ji}^{(\nu)}(\hat{R})|\psi_j^{(\nu)}(\vec{r})\rangle, \quad \text{all } i, \hat{R}. \quad (9.4)$$

where the set of matrices $\{\bar{\bar{\Gamma}}^{(v)}(\hat{R})\}$ forms an irreducible group representation of $\{\hat{R}\}$. When the coordinate system of real space is affected by the *symmetry* operator $\hat{R}^{-1}$, or when the molecular object itself is moved such that any point $\bar{r}$ on it becomes $\hat{R}\bar{r}$, then the states of the vth level change in accordance with Eq. (9.4).

In this sense, a specification of the irreducible representation to which a state belongs is like the assignment of a 'good' quantum number to it. And in the case of an object with a radially symmetric Hamiltonian such as a hydrogen atom, the analogy becomes an equivalence. $s, p, d, f, \ldots$ are the names given to the distinct odd-dimensional irreducible representations of the full rotation group.

We can, moreover, use the irreducible group representations even in assigning quantum numbers to the different states within a degenerate level. Problem 4.5 showed that in a space spanned by $\{|i\rangle\}$, the basis vectors themselves are represented as $\begin{pmatrix} 1 \\ 0 \\ 0 \\ \vdots \end{pmatrix}, \begin{pmatrix} 0 \\ 1 \\ 0 \\ \vdots \end{pmatrix}, \ldots;$

since

$$|1\rangle = \sum_i |i\rangle\langle i|1\rangle = 1|1\rangle + 0|2\rangle + \cdots,$$

in the $\{|i\rangle\}$ system, the representation of $|1\rangle$ is $\begin{pmatrix} 1 \\ 0 \\ \vdots \end{pmatrix}$, etc. So also with a two-dimensional (for simplicity) function space, Eq. (9.4) leads, in matrix notation, to

$$\left. \begin{array}{l} \begin{pmatrix} \Gamma_{11} & \Gamma_{12} \\ \Gamma_{21} & \Gamma_{22} \end{pmatrix} \begin{pmatrix} 1 \\ 0 \end{pmatrix} = \begin{pmatrix} \Gamma_{11} \\ \Gamma_{21} \end{pmatrix} \\ \begin{pmatrix} \Gamma_{11} & \Gamma_{12} \\ \Gamma_{21} & \Gamma_{22} \end{pmatrix} \begin{pmatrix} 0 \\ 1 \end{pmatrix} = \begin{pmatrix} \Gamma_{11} \\ \Gamma_{22} \end{pmatrix} \end{array} \right\} \quad (9.5a)$$

This transcribes back to vector notation as

$$\begin{array}{l} \hat{P}_R|1\rangle = \Gamma_{11}|1\rangle + \Gamma_{21}|2\rangle \\ \hat{P}_R|2\rangle = \Gamma_{12}|1\rangle + \Gamma_{22}|2\rangle \end{array} \quad (9.5b)$$

Equation (9.5a) demonstrates that under the influence of the symmetry operator $\hat{P}_{R'}$, $\begin{pmatrix} 1 \\ 0 \end{pmatrix}$ becomes $\begin{pmatrix} \Gamma_{11} \\ \Gamma_{21} \end{pmatrix}$; $\begin{pmatrix} \Gamma_{11} \\ \Gamma_{21} \end{pmatrix}$, however, is identical to the first column of the irreducible representative

QUANTUM MECHANICAL STATES

matrix $\bar{\bar{\Gamma}}^{(\nu)}(\hat{R})$ itself. Thus one can associate the different degenerate states of a level, such as the one represented by $\binom{1}{0}$, not only with a particular irreducible representation of the group, but even with specific columns of the representative matrices.

PROBLEM 9.9 Show that Eq. (9.5a and b) follow from Eq. (9.4). Show that two functions belonging to non-equivalent irreducible representations and/or to different columns of the same irreducible representation are orthogonal. [Hint: consider the effect of a rotation upon the scalar product of basis functions $|\psi_i^{(\mu)}\rangle$ and $|\psi_j^{(\nu)}\rangle$:
$$\langle \psi_i^{(\mu)} | \psi_j^{(\nu)} \rangle = \langle \hat{P}_R \psi_i^{(\mu)} | \hat{P}_R \psi_j^{(\nu)} \rangle = \cdots]$$

*PROBLEM 9.10 If a C_{3v} molecule is floating in field-free space, its energy does not change under *any* rotation. Is any rotation a symmetry operation?

9.4 Projection Operators

As we showed in Section 5.6, to find the energy levels of a quantum system, one chooses a likely set of basis functions, generates a matrix representation of the Hamiltonian, and then diagonalizes this matrix. The diagonalization process amounts to transforming the original basis vectors into a set of new functions which have the desired property of being energy eigenvectors.

Here we shall show how one manipulates a given set of functions so as to create a new set which displays a similar but different desired property; we shall require that the new functions transform like the generators of an irreducible representation of a symmetry group. Thus the technique of projection operators will allow us to find functions which satisfy the Wigner Theorem.

One place where such symmetrized functions finds use is, in fact, as zero-th order states for diagonalizing a Hamiltonian.

THEOREM 9.2 Let $\Gamma^{(1)}$, $\Gamma^{(2)} \cdots \Gamma^{(c)}$ be all the distinct irreducible representations of a group $\{\hat{R}\}$. Then any function $|f(\bar{r})\rangle$ in the space in which the $\{\hat{P}_R\}$ operate can be decomposed into the sum

$$\boxed{|f(\bar{r})\rangle = \sum_{\mu=1}^{c} c_\mu |f^\mu(\bar{r})\rangle} \qquad (9.6)$$

where $|f^\mu(\bar{r})\rangle$ has the symmetry of the basis vectors of the μth irreducible representation.

Rather than offer a formal proof to this theorem, we shall explicitly show how to project out each of the terms in Eq. (9.6).

Let $\{|\psi_i^{(v)}(\bar{r})\rangle\}$ be an orthonormal basis set spanning the invariant subspace of the vth irreducible representation $\Gamma^{(v)}(\hat{R})$. Then our fundamental equation concerning the transformational properties of such basis vectors reads:

$$\hat{P}_R|\psi_k^{(v)}(\bar{r})\rangle = \sum_l |\psi_l^{(v)}(\bar{r})\rangle \Gamma_{lk}^{(v)}(\hat{R}) \tag{9.4}$$

Multiply this by $\Gamma_{ji}^{(\mu)}(\hat{R})^*$, where μ may or may not be equivalent to v, and sum over the group elements $\{\hat{R}\}$:

$$\sum_R^N \Gamma_{ji}^{(\mu)}(\hat{R})^* \hat{P}_R|\psi_k^{(v)}\rangle = \sum_l |\psi_l^{(v)}\rangle \sum_R^N \Gamma_{ji}^{(\mu)}(\hat{R})^* \Gamma_{lk}^{(v)}(\hat{R})$$

We can simplify the right-hand side by means of the Great Orthogonality Theorem, Eq. (8.6), and are left with

$$\sum_R^N \Gamma_{ji}^{(\mu)}(\hat{R})^* \hat{P}_R|\psi_k^{(v)}\rangle = \frac{N}{n_v} |\psi_j^{(v)}\rangle \delta_{v\mu} \delta_{ik} \tag{9.7}$$

Equation (9.7) is a set of equations, with one equation corresponding to every combination of i, j, k, and μ. Consider now a specific subset of these equations, belonging to the diagonal elements of $\Gamma_{ji}^{(\mu)}(\hat{R})^*$, in which $j=i$. Then

$$\sum_R^N \Gamma_{ii}^{(\mu)}(\hat{R})^* \hat{P}_R|\psi_k^{(v)}\rangle = \frac{N}{n_v} |\psi_i^{(v)}\rangle \delta_{v\mu} \delta_{ik}$$

Summing all such equations over i

$$\left\{\sum_i \sum_R^N \Gamma_{ii}^{(\mu)}(\hat{R})^* \hat{P}_R\right\} |\psi_k^{(v)}\rangle = \frac{N}{n_v} |\psi_k^{(v)}\rangle \delta_{v\mu} \tag{9.8}$$

defines the projection operator $\mathscr{P}^{(\mu)}$:

$$\boxed{\mathscr{P}^{(\mu)}|\psi_k^{(v)}\rangle = |\psi_k^{(v)}\rangle \delta_{\mu v}} \tag{9.9a}$$

This has led us to the explicit form of the projection operator

$$\mathscr{P}^{(\mu)} \equiv \sum_i \sum_R^N \frac{n_v}{N} \Gamma_{ii}^{(\mu)}(\hat{R})^* \hat{P}_R \tag{9.9b}$$

$$\boxed{\mathscr{P}^{(\mu)} = \frac{n_v}{N} \sum_R^N \chi^{(\mu)}(\hat{R})^* \hat{P}_R} \tag{9.9c}$$

which has the remarkable property described in Eq. (9.9a): $\mathscr{P}^{(\mu)}$ will annihilate any basis vector $|\psi_k^{(v)}(\bar{r})\rangle$ which does not belong to the μth invariant subspace, or the μth irreducible representation, of

QUANTUM MECHANICAL STATES

$\{\hat{R}\}$. But *any* function in the space operated on by the $\hat{P}_R$ can be described as a linear combination of the basis vectors of the various irreducible group representations. Thus given an arbitrary function $|f(\bar{r})\rangle$, through

$$\boxed{\hat{\mathscr{P}}^{(\mu)}|f(\bar{r})\rangle = c_\mu |f^\mu(\bar{r})\rangle} \tag{9.10}$$

we have a means for projecting out that portion which has the symmetry of basis vectors generating the μth irreducible representation. We have thus shown how to perform the separation in Eq. (9.6).

PROBLEM 9.11 In calculating electric dipole transition rates, we need to evaluate matrix elements of the form $\langle \psi_i(\bar{r})|x_j|\psi_k(\bar{r})\rangle$ where $x_j = \langle j|x\rangle$, $j = 1, 2, 3$. Separate the functions x_j, as in Eq. (9.6), for a system whose symmetry group is C_{3v}.

As a variation on the above theme, we can extract from Eq. (9.7) an operator of the form

$$\boxed{\hat{\mathscr{P}}_{ji}^{(\mu)} = \sum_R \frac{n_\mu}{N} \Gamma_{ji}^{(\mu)}(\hat{R})^* \hat{P}_R} \tag{9.11a}$$

Equation (9.7) says that if $\hat{\mathscr{P}}_{ji}^{(\mu)}$ operates on the basis function $|\psi_i^{(\mu)}(\bar{r})\rangle$ belonging to the ith column of the μth irreducible representation of $\{\hat{R}\}$, then the resultant will be the basis function corresponding to the jth column of the same irreducible representation, $|\psi_j^{(\mu)}(\bar{r})\rangle$. $\hat{\mathscr{P}}_{ji}^{(\mu)}$ operating on any other basis vector will annihilate it.

$$\boxed{\hat{\mathscr{P}}_{ji}^{(\mu)}|\psi_k^{(\nu)}\rangle = |\psi_j^{(\mu)}\rangle \delta_{ik}\delta_{\mu\nu}} \tag{9.11b}$$

This provides us with a useful prescription for generating all the degenerate partners of a basis vector, given only one.

PROBLEM 9.12 Separate the arbitrary function $f(x)$ into even and odd components. [HINT: What is the significance of evenness and oddness?]

PROBLEM 9.13 Given $|1\rangle = x$, find its partner $|2\rangle$ for an E representation of C_{3v}.

9.5 Direct Product Spaces

Equations (9.9) and (9.10) show how to project out components of various symmetries from an arbitrary function. This procedure is especially useful if we are given a complicated function of mixed symmetry. If, however, we are examining a function which is the product of several terms, and if we happen to know the symmetry type of each of the terms, then the simplest way to find the symmetry of the product function may be through the use of product spaces.

Consider a group of symmetry operators active in some function space appropriate to a quantum mechanical system at hand. The various subspaces invariant to the group are spanned by $\{\psi_i^{(\nu)}(\bar{r})\}$, $\{\psi_j^{(\mu)}(\bar{r})\}$, etc. We can, if we wish, employ product functions of the sort $\{\psi_i^{(\nu)}(\bar{r})\psi_j^{(\mu)}(\bar{r})\}$ to generate a representation of the group, and see what happens. A space containing such states is called a direct product space.

We shall proceed by example, examining a small molecule of known symmetry. We shall assume that this molecule has two electrons, that these electrons interact only weakly, and that we need not concern ourselves with the constraint of the Pauli Exclusion Principle, which demands that no two electrons may inhabit the exact same electronic orbital simultaneously. We can label the electrons number one and two, and their wave functions will be $\psi(\bar{r}_1)$ and $\phi(\bar{r}_2)$, respectively. It is possible, of course, that there may exist a partner state $\psi'(\bar{r}_1)$ degenerate with $\psi(\bar{r}_1)$, or $\phi'(\bar{r}_2)$ with $\phi(\bar{r}_2)$, if the symmetry allows. In fact, $\psi(\bar{r})$ and $\phi(\bar{r})$ might themselves belong to the same irreducible representation of the system's symmetry group.

The Hamiltonian for the two weakly interacting electrons may be written as $\mathcal{H} = \mathcal{H}_1(\bar{r}_1) + \mathcal{H}_2(\bar{r}_2) + \hat{V}(\bar{r}_1, \bar{r}_2)$. We shall neglect the interaction term, and note that $\mathcal{H}_1(\bar{r}_1)$ and $\mathcal{H}_2(r_2)$ have the same functional form:

$$\mathcal{H} = \mathcal{H}(\bar{r}_1) + \mathcal{H}(\bar{r}_2) \tag{9.12a}$$

$\mathcal{H}$, $\mathcal{H}(\bar{r}_1)$, and $\mathcal{H}(\bar{r}_2)$ should all be invariant under the symmetry operations of the group of the Schroedinger equation.

A product function $\psi(\bar{r}_1)\phi(\bar{r}_2)$ is an eigenfunction of $\mathcal{H}$ if $\psi(\bar{r}_1)$ and $\phi(\bar{r}_2)$ are individually eigenvectors of $\mathcal{H}(\bar{r}_1)$ and $\mathcal{H}(\bar{r}_2)$, respectively: for if

$$\mathcal{H}(\bar{r}_1)\psi(\bar{r}_1) = E_1 \psi(\bar{r}_1)$$
$$\mathcal{H}(\bar{r}_2)\phi(\bar{r}_2) = E_2 \phi(\bar{r}_2) \tag{9.12b}$$

then

$$\mathcal{H}(\psi(\bar{r}_1)\phi(\bar{r}_2)) = \phi(\bar{r}_2)\mathcal{H}(\bar{r}_1)\psi(\bar{r}_1) + \psi(\bar{r}_1)\mathcal{H}(\bar{r}_2)\phi(\bar{r}_2)$$
$$= (E_1 + E_2)(\psi(\bar{r}_1)\phi(\bar{r}_2)) \tag{9.12c}$$

We know that $\psi(\bar{r}_1)$ (and its partner $\psi'(\bar{r}_1)$ if $\psi(\bar{r}_1)$ is, say, doubly degenerate) generates an irreducible representation of the symmetry group, as does $\phi(\bar{r}_2)$. From Eq. (9.12c), however, $\psi(\bar{r}_1)\phi(\bar{r}_2)$ (and, should they exist, partner product functions such as $\psi'(\bar{r}_1)$ $\times \phi(\bar{r}_2)$ degenerate with it) also generate a representation. This latter set of matrices, however, is not necessarily irreducible.

Consider what occurs, for example, for C_{3v} if the $\psi(\bar{r}_1)$ and $\phi(\bar{r}_2)$ of interest should both happen to be E-type functions; then the pairs $\psi(\bar{r}_1)$ and its partner $\psi'(\bar{r}_1)$, and $\phi(\bar{r}_2)$ and $\phi'(\bar{r}_2)$, each span two-dimensional invariant subspaces of Hilbert space; but the four degenerate product functions

$$\{\psi(\bar{r}_1)\phi(\bar{r}_2), \psi'(\bar{r}_1)\phi(\bar{r}_2), \psi(\bar{r}_1)\phi'(\bar{r}_2), \psi'(\bar{r}_1)\phi'(\bar{r}_2)\} \tag{9.13a}$$

span a *four*-dimensional invariant subspace. This subspace is clearly reducible, since the irreducible representations of C_{3v} can be only one- (A_1 or A_2) or two- (E) dimensional. If we reduce the four-dimensional $E \times E$ matrices generated by $\psi(\bar{r}_1)\phi(\bar{r}_2)$ and its partners by the methods displayed in Section 8.8, we find that in block-diagonal form, each $E \times E$ matrix is composed of one A_1 block, one A_2 block, and one E block.

$$E \times E \to A_1 + A_2 + E \quad \text{for } C_{3v} \tag{9.13b}$$

Thus if each of two electrons on a C_{3v} molecule individually inhabits an E-state, then the compound wave function for the composite system can behave like A_1, A_2, or E. Which of these three possibilities actually occurs depends on other, symmetry-unrelated, factors. Note that product states do not necessarily enjoy as many options as in the $E \times E$ case; if the two single particle states were, say, A_2, then the compound state could *only* be A_1! For C_{3v}, the possibilities are as in Table 9.1.

Table 9.1

$A_1 \times A_1 = A_1$	$A_2 \times A_2 = A_1$
$A_1 \times A_2 = A_2$	
$A_1 \times E = E$	$A_2 \times E = E$
$E \times E = A_1 + A_2 + E$	

*PROBLEM 9.14 Verify Table 9.1.

PROBLEM 9.15 Must every one-electron molecule or molecular ion be allowed A_1 states?

We have introduced product spaces by constructing two-particle states. But the range of applicability of the approach is much broader than this exercise may suggest. For example, in Chapter 11 we shall demonstrate the use of product spaces in evaluating the matrix elements of importance to perturbation theory; the method reveals almost instantaneously whether or not a perturbing potential can break a degeneracy or cause transitions of the system between various quantum states.

9.6 Exchange Symmetry

Two important non-geometric symmetries which influence quantum mechanical calculations are the invariance of a system to the interchange of identical constituent parts (e.g. electrons), and to a change in the direction of the flow of time. We consider exchange symmetry first.

For an n-electron system, one might expect a general state to be, to a lowest order of approximation, the product of one-electron states:

$$\Psi(\bar{r}_1, \bar{r}_2, \ldots \bar{r}_n) = \psi_\alpha(\bar{r}_1)\psi_\beta(\bar{r}_2) \ldots \psi_\nu(\bar{r}_n) \tag{9.14a}$$

where we have assumed that we can follow the 'first' electron long enough to ascertain that it is in the αth state, etc. The problem lies with this assumption.

Electrons are completely indistinguishable, and an operator $\hat{P}_{ij}$ which interchanges the electrons at positions $\bar{r}_i$ and $\bar{r}_j$ is a symmetry operator for the system. Thus

$$\hat{P}_{12}\{\psi_\alpha(\bar{r}_1)\psi_\beta(\bar{r}_2) \ldots \psi_\nu(\bar{r}_n)\} = \psi_\alpha(\bar{r}_2)\psi_\beta(\bar{r}_1) \ldots \psi_\nu(\bar{r}_n)$$
$$= \psi_\beta(\bar{r}_1)\psi_\alpha(\bar{r}_2) \ldots \psi_\nu(\bar{r}_n) \tag{9.14b}$$

is as good a description of the system as is Eq. (9.14a). There are $n!$ such exchange operators (of which $n(n-1)$ affect only two particles at a time) corresponding to the $n!$ permutations of n objects, and they constitute the 'symmetric group' S_n discussed in Section 7.2.

So far we have considered the Wigner Theorem only with respect to a group of geometric symmetry operators such as rotations, reflections, and translations. The theorem, however, is sufficiently flexible that it applies also to a group of permutation operators as well. In particular, we can say that correct wave functions must display a behaviour under the interchange of particles compatible with one of the irreducible representations of S_n.

It can be shown that S_n for any n has exactly two distinct one-dimensional irreducible representations. Of these, one is the *fully symmetric*, generated by functions which are completely invariant to any exchange of particles; a function belonging to the *fully anti-symmetric* irreducible representation, on the other hand, changes sign with any pair-wise exchange of particles. Although

QUANTUM MECHANICAL STATES   163

other irreducible representations of higher dimensionality are allowed by the Wigner Theorem, and are mathematically meaningful, only fully symmetric and fully antisymmetric wave functions are found in nature. They describe, respectively, bosons and fermions, or integral-spin and half-integral-spin particles.

Consider a three-identical-particle system, with the particles in the αth, βth, and γth states. The permutation group on three objects S_3 happens to be isomorphic to C_{3v}, and has the same character table, etc. (See Problem 7.11.)

Table 9.2

S_3	[123]	[132] [213] [321]	[312] [231]
A_1	1	1	1
A_2	1	−1	1
E	2	0	−1

One can use projection operator techniques or trial and error to find linear combinations of terms such as $\psi_\alpha(\bar{r}_1)\psi_\beta(\bar{r}_2)\psi_\gamma(\bar{r}_3)$, $\psi_\beta(\bar{r}_1)\psi_\alpha(\bar{r}_2)\psi_\gamma(\bar{r}_3)$, ... etc., with suitable symmetry.

$$\Psi_S = \psi_\alpha(\bar{r}_1)\psi_\beta(\bar{r}_2)\psi_\gamma(\bar{r}_3) + \psi_\alpha(\bar{r}_1)\psi_\beta(\bar{r}_3)\psi_\gamma(\bar{r}_2)$$
$$+ \psi_\alpha(\bar{r}_2)\psi_\beta(\bar{r}_1)\psi_\gamma(\bar{r}_3) + \psi_\alpha(\bar{r}_2)\psi_\beta(\bar{r}_3)\psi_\gamma(\bar{r}_1) \quad (9.15a)$$
$$+ \psi_\alpha(\bar{r}_3)\psi_\beta(\bar{r}_1)\psi_\gamma(\bar{r}_2) + \psi_\alpha(\bar{r}_3)\psi_\beta(\bar{r}_2)\psi_\gamma(\bar{r}_1)$$

is clearly fully symmetric under any interchange of particles, and serves as a basis vector for generating A_1 of S_3. Similarly the Slater determinant

$$\Psi_A = \begin{vmatrix} \psi_\alpha(\bar{r}_1) & \psi_\beta(\bar{r}_1) & \psi_\gamma(\bar{r}_1) \\ \psi_\alpha(\bar{r}_2) & \psi_\beta(\bar{r}_2) & \psi_\gamma(\bar{r}_2) \\ \psi_\alpha(\bar{r}_3) & \psi_\beta(\bar{r}_3) & \psi_\gamma(\bar{r}_3) \end{vmatrix} \quad (9.15b)$$

is antisymmetric to the interchange of any two particles, and can describe a fermion.

PROBLEM 9.16 *The Pauli Exclusion Principle* states that two fermions cannot co-exist in the same state. Show that Ψ_A of Eq. (9.15b) incorporates this requirement.

†9.7 Time Reversal Symmetry

Consider the Schroedinger equation in which the potential is time independent and real:

$$\mathcal{H}\psi(t) = i\hbar \frac{\partial}{\partial t} \psi(t) \tag{9.16a}$$

Replacement of t by $-t$ yields the different equation

$$\mathcal{H}\psi(-t) = -i\hbar \frac{\partial}{\partial t} \psi(-t) \tag{9.16b}$$

But if we now take the complex conjugate of Eq. (9.16b), we return to our original Schroedinger equation, albeit with a different solution:

$$\mathcal{H}\psi^*(-t) = i\hbar \frac{\partial}{\partial t} \psi^*(-t) \tag{9.16c}$$

Equations (9.16a) and (9.16b) say that if $\mathcal{H}$ is real and invariant to a change of the sign of the time parameter, then $\psi^*(-t)$ is just as good a solution to the Schroedinger equation as $\psi(t)$ is. Thus if we filmed a quantum mechanical event described by $\psi(t)$, such as a collision, and played the film in reverse, then the time reversed sequence of events described by $\psi^*(-t)$ should also be a physically realizable occurrence, Fig. 9.2.

Time reversal symmetry can lead to degeneracies in addition to those which one would expect on the basis of geometric considerations alone. In particular, *Kramer's Theorem* says that in the absence of an externally applied magnetic field, the energy levels of any system with an odd number of electrons will be at least doubly (corresponding to spin-up and spin-down states) degener-

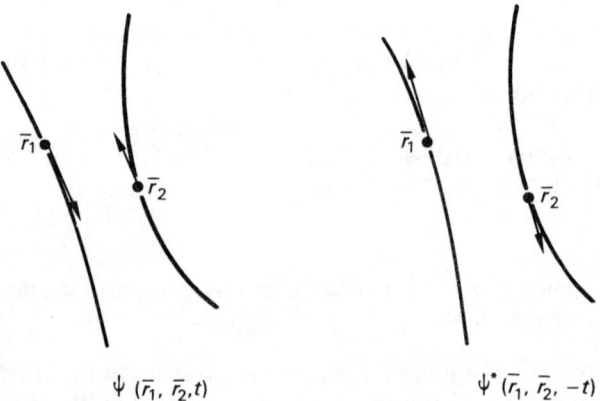

Figure 9.2.

QUANTUM MECHANICAL STATES

ate. This degeneracy has great practical importance: electron spin resonance studies usually involve the electron system's orbital ground state, in which spin state transitions are excited among the levels of a Kramer's multiplet whose degeneracy has been broken by an external magnetic field.

PROBLEM 9.17 Show that if $\psi(\bar{r}, t) = \psi_n(\bar{r}) \exp(-iE_n t/\hbar)$ is non-degenerate, then $\psi_n(\bar{r})$ is real, except for an arbitrary multiplicitive phase factor.

PROBLEM 9.18 Is $\bar{F} = m\bar{a}$ invariant to time reversal? Consider a charged particle moving in the field produced by a current loop.

*PROBLEM 9.19 You wouldn't expect the steam in the kitchen to return spontaneously to the kettle. Is this a counter-argument to the above discussion?

A word of warning: the *time reversal operator* $\hat{\Theta}$ is not linear like all the other dynamical and symmetry operators with which we ever deal; rather

$$\Theta c\psi(t) = c^*\psi^*(-t) = c^*\hat{\Theta}\psi(t) \tag{9.17a}$$

and $\hat{\Theta}$ is called *antilinear*. Moreover it is anti-unitary,

$$\langle \hat{\Theta}\psi | \Theta\phi \rangle = \langle \psi | \phi \rangle^* \tag{9.17b}$$

and possesses other irregular characteristics.

9.8 Molecular Orbitals†

Consider three ground state hydrogen atoms sitting far apart from one another, such that their three $|1s\rangle$ electrons are all of the same energy. Suppose we now bring them together to form a compact C_{3v} H_3 molecule‡. Such a thing does not appear to exist in nature in a bound state, but still it serves as a useful model system. The three previously degenerate 1s electrons must now inhabit A_1, A_2, or E-type composite states which are $1-$, $1-$, and 2-fold degenerate, respectively. Thus the original degeneracy is broken and the 's' label loses meaning as the separate atoms overlap to form a molecule.

*PROBLEM 9.20 *Without* using symmetry arguments, explain how the 1s degeneracy was broken; i.e. as the atoms were brought together, why weren't all the $|1s\rangle$ states distorted but simply shifted by the same amount of energy in the same direction?
PROBLEM 9.21 Why do 1s, 2s, 2p, etc. atomic states split into

†Tinkham,[39] pages 220–227, compares the molecular orbital and valence bond methods of calculating molecular states, and ionic versus covalent bonds.
‡Neglecting, as usual, the horizontal reflection $\hat{M}_h$.

energy bands when a gas condenses to form a solid? Why do the innermost (i.e. low principal quantum number) shells lead to the narrowest bands? Why do different bands respond differently (i.e. broaden and/or shift) to the application of hydrostatic pressure?

As a first quantitative attack on our molecule, we might try to solve the Schroedinger equation with a three-centred potential. But while there do exist analytic solutions for the two-centred potential, as for the H_2^+ ion, an approach like this has not borne much fruit with more complicated systems. Consequently one must look to approximation schemes.

A standard method of calculation of *molecular orbitals* (*MO*) is the *Linear Combination of Atomic Orbital* (*LCAO*) approximation. LCAO notes that the total distribution of charge on a molecule should be fairly much like that of a set of free atoms held in the same spatial configuration. Thus one might expect a state of the molecule of Fig. 9.3a to appear something like a superposition of the atomic states $|\phi_1\rangle, |\phi_2\rangle, |\phi_3\rangle$ of Fig. 9.3b, and we are thus tempted to construct and examine MO's which are linear combinations of these three.

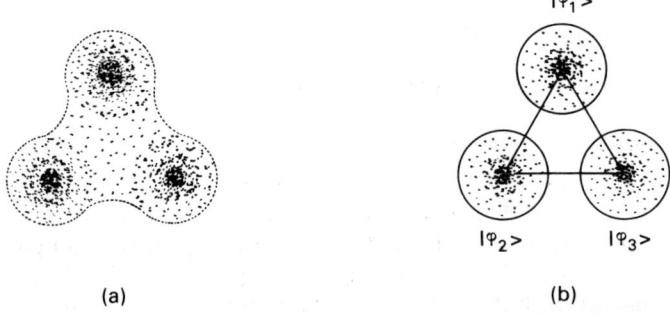

(a)   (b)

Figure 9.3.

A general LCAO involving only ϕ_1, ϕ_2, and ϕ_3 (if we have reason to believe that these three atomic states alone are sufficient for an adequate description of our molecule) is of the form

$$\psi = \alpha_1\phi_1(\beta_1) + \alpha_2\phi_2(\beta_2) + \alpha_3\phi_3(\beta_3) \tag{9.18a}$$

where the β's are internal constants parameterizing the ϕ's. The best molecular ground state of this form can be found by varying $\alpha_1, \alpha_2, \alpha_3$. $\beta_1, \beta_2,$ and β_3 until the expectation value of the energy,

$$\boxed{\langle \mathcal{H} \rangle = \langle \psi | \mathcal{H} | \psi \rangle / \langle \psi | \psi \rangle} \tag{9.18b}$$

QUANTUM MECHANICAL STATES   167

is minimized. This is just a straightforward application of the 'variational' method of approximately diagonalizing $\mathcal{H}$ by means of selecting trial functions with parameters most closely resembling those of the true eigenstates.

Although we could employ Eq. (9.18b) directly in Eq. (9.18a), much effort may be spared by choosing a trial function ψ of proper symmetry. This reduces to a small number the possible combinations of α_1, α_2, and α_3 which can appear in any LCAO describing the system. Thus we must find α-coefficients such that under rotations and reflections, ψ is of A_1, A_2, or E symmetry; no other LCAO's are allowed.

As an example, we shall assume ϕ_1, ϕ_2, and ϕ_3 to be s-like, and begin by determining which symmetrized LCAO's suitable for a C_{3v} molecule will be generated by such a set of s-states.

We begin by determining the effects of the six symmetry operations of C_{3v} upon our basis set $\{\phi_i\}$. Figure 9.4 shows, for example, that $\hat{M}_I$ interchanges ϕ_2 and ϕ_3, but leaves ϕ_1 alone. We denote these processes as $\hat{M}_I \phi_2 = \phi_3$, $\hat{M}_I \phi_1 = \phi_1$, etc.

The effects of the six elements of C_{3v} upon the three $\{\phi_i\}$ are recorded in Table 9.3:

Table 9.3

ϕ_i	$\hat{I}\phi_i$	$\hat{R}_{120}\phi_i$	$\hat{R}^{-1}_{120}\phi_i$	$\hat{M}_I\phi_i$	$\hat{M}_{II}\phi_i$	$\hat{M}_{III}\phi_i$
ϕ_1	ϕ_1	ϕ_3	ϕ_2	ϕ_1	ϕ_3	ϕ_2
ϕ_2	ϕ_2	ϕ_1	ϕ_3	ϕ_3	ϕ_2	ϕ_1
ϕ_3	ϕ_3	ϕ_2	ϕ_1	ϕ_2	ϕ_1	ϕ_3

We can now form a representation of C_{3v} with the $\{\phi_i\}$ basis set if we employ the approximation that

$$\langle \phi_i | \phi_j \rangle = \delta_{ij} \tag{9.18c}$$

This orthogonality condition is equivalent to the assumption that the atomic orbitals at different vertices of our triangle do not overlap (Why?). From Table 9.4 and Eq. (9.18c), we find, for example:

$$\hat{M}_I |\phi_1\rangle = |\phi_1\rangle$$

therefore

$$\left. \begin{array}{l} \langle \phi_1 | \hat{M}_I | \phi_1 \rangle = 1 \\ \langle \phi_2 | \hat{M}_I | \phi_1 \rangle = 0 \\ \text{etc.} \end{array} \right\} \tag{9.19a}$$

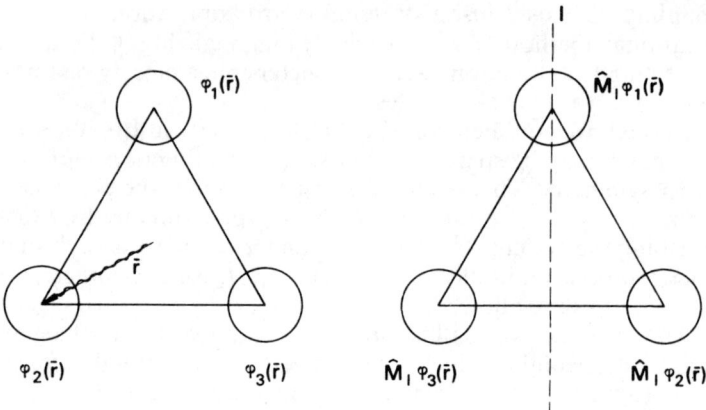

Figure 9.4.

Then, exactly as with Eqs (4.5) and (6.30), we obtain the following representation, which we shall call Γ^ϕ, of C_{3v} (see also Eq. (8.12))

$$\bar{\bar{\Gamma}}^\phi(\hat{M}_\mathrm{I}) = \begin{pmatrix} 1 & 0 & 0 \\ 0 & 0 & 1 \\ 0 & 1 & 0 \end{pmatrix}; \quad \chi^\phi(\hat{M}_\mathrm{I}) = 1$$

$$\bar{\bar{\Gamma}}^\phi(\hat{M}_\mathrm{II}) = \begin{pmatrix} 0 & 0 & 1 \\ 0 & 1 & 0 \\ 1 & 0 & 0 \end{pmatrix}; \quad \chi^\phi(\hat{M}_\mathrm{II}) = 1$$

$$\bar{\bar{\Gamma}}^\phi(\hat{M}_\mathrm{III}) = \begin{pmatrix} 0 & 1 & 0 \\ 1 & 0 & 0 \\ 0 & 0 & 1 \end{pmatrix}; \quad \chi^\phi(\hat{M}_\mathrm{III}) = 1 \qquad (9.19b)$$

$$\bar{\bar{\Gamma}}^\phi(\hat{I}) = \begin{pmatrix} 1 & 0 & 0 \\ 0 & 1 & 0 \\ 0 & 0 & 1 \end{pmatrix}; \quad \chi^\phi(\hat{I}) = 3$$

$$\bar{\bar{\Gamma}}^\phi(\hat{R}_{120}) = \begin{pmatrix} 0 & 1 & 0 \\ 0 & 0 & 1 \\ 1 & 0 & 0 \end{pmatrix}; \quad \chi^\phi(\hat{R}_{120}) = 0$$

$$\bar{\bar{\Gamma}}^\phi(\hat{R}_{240}) = \begin{pmatrix} 0 & 0 & 1 \\ 1 & 0 & 0 \\ 0 & 1 & 0 \end{pmatrix}; \quad (\hat{R}_{240}) = 0$$

Note that for every ϕ_i which is left undisturbed by a symmetry operation $\hat{R}$, there appears a '1' along the diagonal of $\bar{\bar{\Gamma}}^\phi(\hat{R})$ (why?),

QUANTUM MECHANICAL STATES

and 1 is added to $\chi(\hat{R})$. If ϕ_i is moved to a new position by $\hat{R}$, a corresponding 0 occurs in $\bar{\bar{\Gamma}}^\phi(\hat{R})$; and finally, if the sign of ϕ_i is reversed, as might happen to a p-state under reflection, we find a -1, Fig. (9.5a).

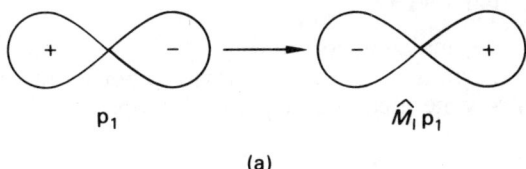

(a)

Figure 9.5(a).

We first determine what possible symmetries of LCAO will fall out of our s-like $\{\phi_i\}$ representation. From Eq. (9.19), we have the character table for Γ^ϕ of C_{3v}:

Table 9.4a

C_{3v}	$\hat{I}$	$\hat{R}_{120}, \hat{R}_{240}$	$\hat{M}_\text{I}, \hat{M}_\text{II}, \hat{M}_\text{III}$
Γ^ϕ	3	0	1

From Table 8.2, the character table for the irreducible representations of C_{3v} is

Table 9.4b

C_{3v}	$\hat{I}$	$\hat{R}_{120}, \hat{R}_{240}$	$\hat{M}_\text{I}, \hat{M}_\text{II}, \hat{M}_\text{III}$
A_1	1	1	1
A_2	1	1	-1
E	2	-1	0

Then, by Eq. (8.11)

$$\Gamma^\phi = \sum_v^3 a_v \Gamma^{(v)} \quad (9.20a)$$

$$a_v = \frac{1}{N} \sum \chi^{(v)}(\hat{R})^* \chi^\phi(\hat{R}) \quad (9.20b)$$

and, as in Eq. (8.14b)

$$\Gamma^\phi = A_1 + E \quad (9.20c)$$

We have established that a MO for C_{3v} constructed out of s-like atomic orbitals $\{\phi_i\}$ can only be of A_1 or E-type symmetry. If the $\{\phi_i\}$ had been more elaborate trial functions, we might have found A_2 MO's too. It now remains actually to construct the symmetrized LCAO's.

It is easily seen that the only linear combination of ϕ_1, ϕ_2, and ϕ_3 which is totally invariant to all elements of C_{3v} is

$$\psi^{A_1} = C(\phi_1 + \phi_2 + \phi_3) \tag{9.21}$$

where C is a normalizing constant.

States of E symmetry are not so obvious, but symmetrized LCAO can be generated via the projection operator of Eq. (9.9):

$$\psi_i^E = \frac{1}{3} \sum_R^N \chi^{(E)}(\hat{R})^* \hat{P}_R \phi_i \tag{9.22a}$$

We apply Eq. (9.9) to ϕ_1, ϕ_2, and then ϕ_3 individually. With the aid of the E line from the C_{3v} character table, we find $\psi_1^E = 2\phi_1 - 1\phi_2 - 1\phi_3 + 0 + 0 + 0$, etc., aside from the 1/3.

$$\begin{aligned} \psi_1^E &= 2\phi_1 - (\phi_2 + \phi_3) \\ \psi_{1'}^E &= 2\phi_2 - (\phi_1 + \phi_3) \\ \psi_{1''}^E &= 2\phi_3 - (\phi_1 + \phi_2) \end{aligned} \tag{9.22b}$$

PROBLEM 9.22 Determine Eq. (9.21) through the use of Eq. (9.9).

Equations (9.22b) are all properly symmetrized, but unfortunately there are too many of them, and they are not orthogonal. We need an E-type subspace spanned by *two* wave-vectors only. Our way out of this dilemma will be to choose linear combinations of ψ_1^E, $\psi_{1'}^E$, and $\psi_{1''}^E$ which yield two linearly independent basis vectors. This linear combination will still have the right symmetry, just as the sum of two even functions will still be even.

PROBLEM 9.23 Assume that ψ_1^E, alone can serve as one of these two. Show that ψ_1^E is even under $\hat{M}_1$, and that if it is to be one of our two basis vectors, then the other, ψ_2^E, must be odd under $\hat{M}_1$: $\hat{M}_1 \psi_2^E = -\psi_2^E$ [HINT: $\chi(\hat{M}_1) = 0$.] Then construct a linear combination of $\psi_{1'}^E$ and $\psi_{1''}^E$ which is odd under $\hat{M}_1$; this is our ψ_2^E.

PROBLEM 9.24 Use Eq. (9.11) and ψ_1^E to find ψ_2^E.

PROBLEM 9.25 Show that the three p-states of Fig. 9.5b generate an A_2 representation of C_{3v}.

In summary, three s-like atomic orbitals $\{\phi_i\}$ may be used in LCAO's so as to yield A_1 and E MO's.

$$A_1: \psi^{A_1} = \phi_1 + \phi_2 + \phi_3 \tag{9.23a}$$

QUANTUM MECHANICAL STATES

$$E: \begin{cases} \psi_1^E = 2\phi_1 - (\phi_2 + \phi_3) \\ \psi_2^E = \phi_2 - \phi_3 \end{cases} \quad (9.23b)$$

These are in a form suitable for the evaluation of matrix elements of the Hamiltonian, perturbation operators, etc. Other pairs of basis vectors can be found to span E-space, of course, and they, like the three vectors in Eq. (9.23), should be normalized.

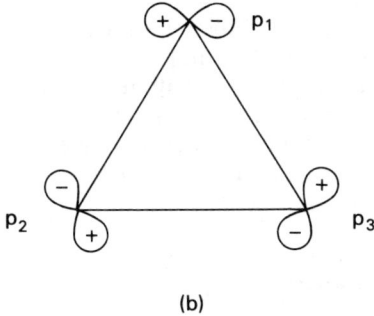

(b)

Figure 9.5(b).

PROBLEM 9.26 According to Table 9.1, if an A_1 electron and an E electron co-exist on a C_{3v} molecule, the compound two-electron state displays E-type symmetry. Will the E states of Eq. (9.24) be suitable for describing such a two-particle system? Will the product of the A_1 and one of the E-states?

PROBLEM 9.27 The Wigner Theorem tells us that for a symmetry operator $\hat{R}$

$$\mathcal{H}(\hat{R}\bar{r}) = \mathcal{H}(\bar{r}),$$

or equivalently

$$[\hat{P}_R, \mathcal{H}(\bar{r})] = 0$$

States, on the other hand, in general do not behave so simply; it is not necessarily true (except for A_1 states) that $\psi(\hat{R}\bar{r}) = \psi(\bar{r})$. Show, however, that for any state $\psi(\bar{r})$

$$\langle \psi(\hat{R}\bar{r}) | \psi(\hat{R}\bar{r}) \rangle = \langle \psi(\bar{r}) | \psi(\bar{r}) \rangle$$

Does $\psi^*(\hat{R}\bar{r})\psi(\hat{R}\bar{r}) = \psi^*(\bar{r})\psi(\bar{r})$? What would this mean physically? Does it agree with Table 9.1?

†PROBLEM 9.28 To the Dirac characters $\{\hat{\Omega}_C\}$ (See Problem 8.16)) there correspond function space operators $\{\hat{P}_C\}$. $\mathcal{H}$ and $\{\hat{P}_C\}$ form, for a given group, a commutating set, and consequently may be diagonalized simultaneously. What does this imply? How can the $\{\hat{P}_C\}$ obey eigenequations if the constituent $\{\hat{P}_R\}$ do not?

PROBLEM 9.29 Choosing suitable bases vectors, verify for C_{3v} the eigenequation

$$\hat{P}_C \psi_i^{(v)} = \Omega^{(v)}(C) \psi_i^{(v)} \tag{9.24a}$$

$$\Omega^{(v)}(C) = \frac{c\psi^{(v)}(\hat{R}')}{l^{(v)}}, \quad \hat{R}' \in C \tag{9.24b}$$

where $l^{(v)}$ is the degeneracy of the vth irreducible representation, and c the order of the class C. Compare with Eq. (9.4). Now find the possible eigenvalues of $\{\hat{\Omega}_C\}$ a different way, employing the multiplication table of the $\{\hat{\Omega}_C\}$ and the fact that the eigenvalues of a set of commuting operators obey any algebraic equation satisfied by the operators. Note that $\hat{\Omega}^{4}{}_1(I) = 1$.

9.9 Molecular Vibrations

It might appear that the Wigner Theorem applies solely to electronic states. But just as one can construct a Hamiltonian for an electron bound by Coulombic forces, so also can one do the same for atoms bound together by Hooke-law forces.

PROBLEM 9.30 Why are the elastic forces binding together a molecule or solid of the Hooke-law form, $\bar{F} = -k\bar{x}$? Or are they?

The solutions to the Schroedinger equation for a set of coupled harmonic oscillators can describe the vibrations of molecules, and Theorem 9.1 again applies. In this section we shall demonstrate by example that, as with electronic states, it is meaningful to talk about the symmetry properties of modes of molecular vibration. We shall examine our standard example of a planar C_{3v} system.

To describe the possible contortions that our molecule might undergo, we could attach to each vertex two orthogonal unit vectors, Fig. 9.6; any distortions would be expressible in terms of components $\{X_i'\}$ along these.

It is every bit as informative to use, instead, the six linearly independent generalized co-ordinates of Fig. 9.7. Any distortion noted before as the components $\{X_1' \ldots X_6'\}$ could equally well be reported as $\{Q_1, Q_2, \ldots Q_6\}$. The generalized co-ordinates $|Q_4\rangle$, $|Q_5\rangle$, and $|Q_6\rangle$ referring to translational and rotational motion are not of concern to us here; we do wish, however, to see whether or not $|Q_1\rangle$, $|Q_2\rangle$, and $|Q_3\rangle$ could be thought of as possessing A_1, A_2, or E-type symmetry properties.

QUANTUM MECHANICAL STATES

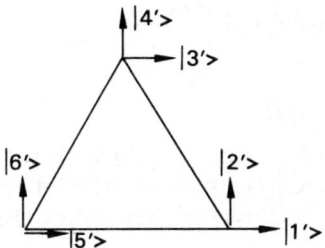

Figure 9.6.

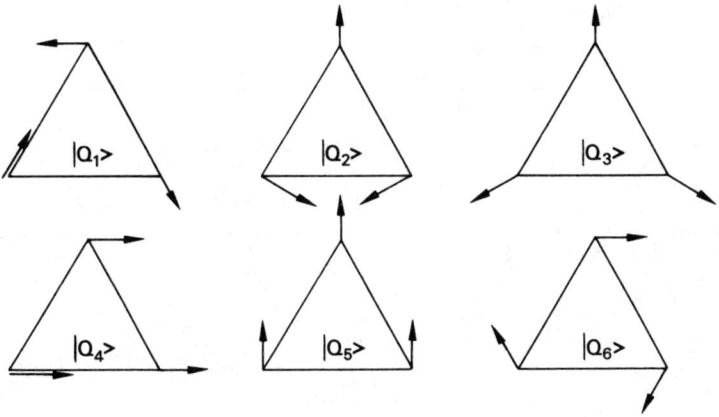

Figure 9.7.

$|Q_3\rangle$ is clearly invariant to any of the rotations or reflections of C_{3v}. Like $|3\rangle$ of ordinary three-space, then, $|Q_3\rangle$ is of A_1 symmetry. So far so good.

Neither $|Q_1\rangle$ nor $|Q_2\rangle$ is invariant to rotation by 120°, so neither belongs to A_1 or A_2. That means either that the two form an E-type basis set for C_{3v}, or that they are of some mixed symmetry. We shall now argue that, despite the fact that $|Q_1\rangle$ and $|Q_2\rangle$ bear little resemblance to the $|1\rangle$ and $|2\rangle$ of three-space, still they do span a subspace (of 'vibration mode' space) of E symmetry; $|Q_1\rangle$ and $|Q_2\rangle$ behave, under C_{3v}, like $|1\rangle$ and $|2\rangle$!

Consider first the images of $|1\rangle$ and $|2\rangle$ across the mirror plane I. The mirror image of $|1\rangle$ is $\hat{M}_I|1\rangle$, and similarly for $|2\rangle$. From Fig. 9.8a, $|1\rangle$ and $|2\rangle$ behave under $\hat{M}_I$ according to

$$|1\rangle + \hat{M}_I|1\rangle = 0$$
$$|2\rangle + \hat{M}_I|2\rangle = 2|2\rangle \qquad (9.25a)$$

On the other hand, from Fig. 9.8b

$$|Q_1\rangle + \hat{M}_I|Q_1\rangle = 0$$
$$|Q_2\rangle + \hat{M}_I|Q_2\rangle = 2|Q_2\rangle \qquad (9.25b)$$

Comparison of Eqs (9.25a) and (9.25b) leads us to the tentative assertion that $|Q_1\rangle$ and $|Q_2\rangle$ may behave, respectively, like $|1\rangle$ and $|2\rangle$ not only for $\hat{M}_I$, but under *all* the symmetry operations of C_{3v}.

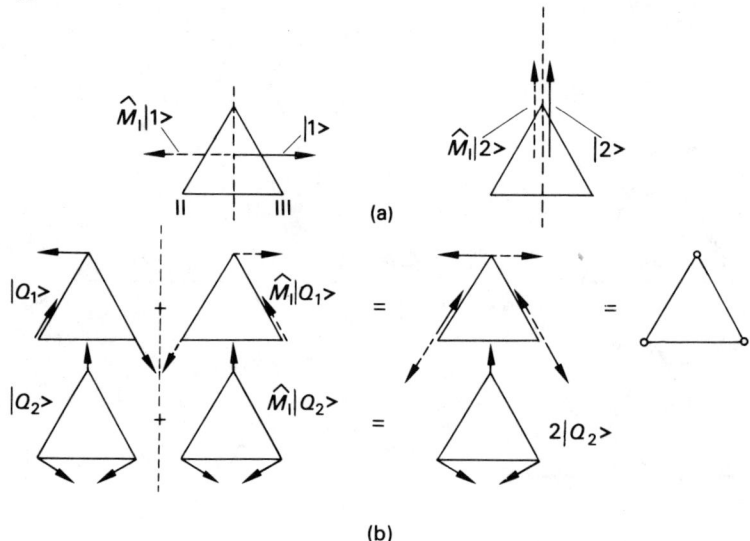

Figure 9.8.

We know from Chapter 4 how the other five operators of C_{3v} mix $|1\rangle$ and $|2\rangle$. For example

$$\hat{R}_{120}|1\rangle = -\tfrac{1}{2}|1\rangle + \tfrac{1}{2}\sqrt{3}\,|2\rangle \qquad (9.26a)$$

We might suspect, that as in Problem 9.3,

$$\hat{R}_{120}|Q_1\rangle = -\tfrac{1}{2}|Q_1\rangle + \tfrac{1}{2}\sqrt{3}\,|Q_2\rangle \qquad (9.26b)$$

Actual construction of a figure which looks like the right-hand side (so as to prove the equality) of Eq. (9.26b), is not difficult, and is left as an exercise for the artistically inclined. We shall offer here an alternative bit of evidence which itself should convince you that $\{|Q_1\rangle, |Q_2\rangle\}$ do indeed behave like $\{|1\rangle, |2\rangle\}$ under C_{3v}.

Figure 9.9a demonstrates the identity

$$\hat{R}_{240}|2\rangle = \sqrt{\tfrac{1}{3}}(|1\rangle + \hat{M}_{III}|1\rangle) \qquad (9.27a)$$

QUANTUM MECHANICAL STATES

If our $\{|1\rangle, |2\rangle\}$ to $\{|Q_1\rangle, |Q_2\rangle\}$ correspondence is valid, then

$$\hat{R}_{240}|Q_2\rangle = \sqrt{\tfrac{1}{3}}(|Q_1\rangle + \hat{M}_{\text{III}}|Q_1\rangle) \tag{9.27b}$$

Like the case of Eq. (9.25b), Eq. (9.27b) is easily illustrated, Fig. (9.9b).

Although no proof, hopefully the examples of Figs 9.8 and 9.9 have convinced the reader that it is meaningful to say that a C_{3v} system can undergo A_1 ('breathing mode') vibrations, and also E-type vibrations, or even some complex motion expressible as a linear combination of the two.

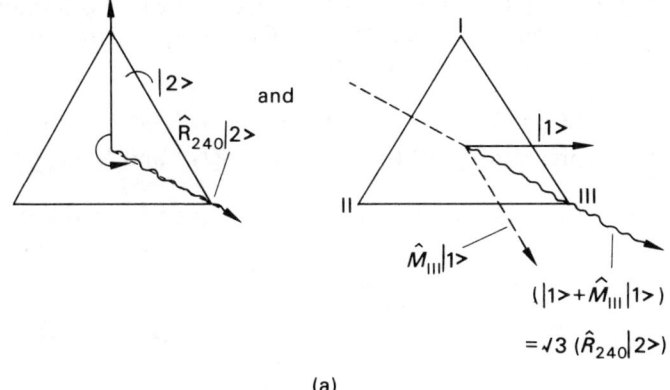

(a)

Figure 9.9(a).

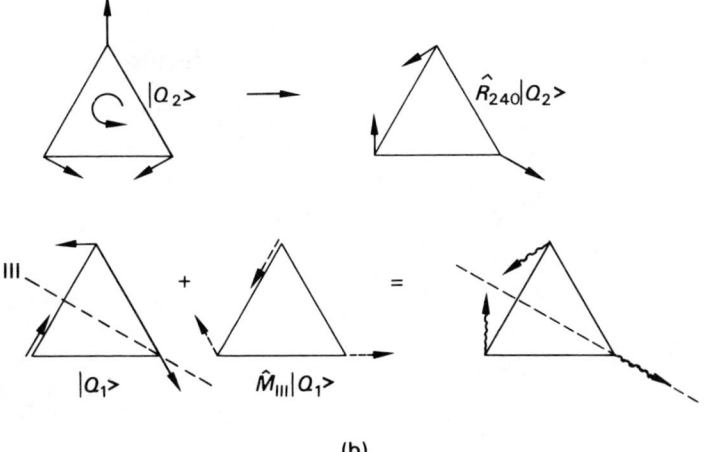

(b)

Figure 9.9(b).

We have not indicated the procedure by which one finds the generalized coordinates of Fig. 9.7. An educated guess, confirmed as in the above discussion of this section, will often lead to them; otherwise a little tedious but straightforward grinding is necessary:

†PROBLEM 9.31 Derive the shapes of the pair of E-modes $|Q_1\rangle$ and $|Q_2\rangle$. Since they should behave like $|1\rangle$ and $|2\rangle$ under $\hat{M}_{\text{III}}$, assume that $|Q_1^\circ\rangle$, $|Q_2^\circ\rangle$ of Fig. 9.10 is a good place to begin. By Eq. (9.26b), $\hat{R}_{120}|Q_1^\circ\rangle = -\frac{1}{2}|Q_1^\circ\rangle + (\sqrt{3}/2)|Q_2^\circ\rangle$, Fig. 9.26b. But Fig. 9.10c offers a different approach to $\hat{R}_{120}|Q_1^\circ\rangle$. Thus from Fig. 9.10b and 9.10c together, $\hat{R}_{120}|Q_1^\circ\rangle$ should look in part like $R_{120}|Q_1'\rangle$ of Fig. 9.10d; and $|Q_1'\rangle$ itself like Fig. 9.10e. Continue in this fashion until you have completely constructed $|Q_1\rangle$ and $|Q_2\rangle$. Now construct $|Q_3\rangle$. How would you use projection operators to carry out this process?

PROBLEM 9.32 Can $|Q_2\rangle$ and $\hat{R}_{240}|Q_2\rangle$ of Fig. 9.9b be used to generate a representation of C_{3v}? $|Q_2\rangle$, $\hat{R}_{120}|Q_2\rangle$, and $\hat{R}_{240}|Q_2\rangle$? Of C_3?

PROBLEM 9.33 Do the rotation and two translation generalized

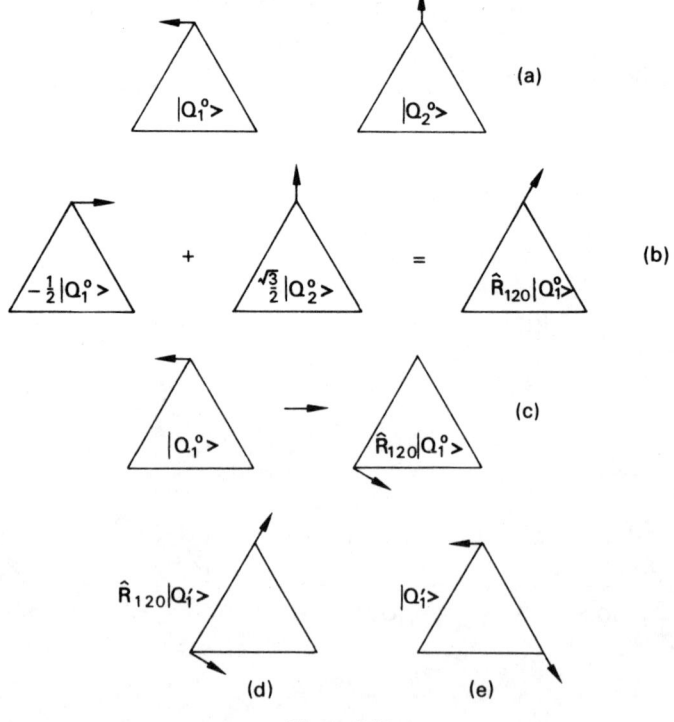

Figure 9.10.

QUANTUM MECHANICAL STATES   177

co-ordinates of Fig. 9.7 generate irreducible representations of C_{3v}?

PROBLEM 9.34 The Wigner Theorem was introduced for quantum mechanical states, not physical displacements (even if described in terms of generalized coordinates) in real space. Resolve this dilemma with a *quantum mechanical* argument.†

We shall discuss the vibrational and electronic states of molecules further when we examine the Jahn–Teller effect in Section 11.4.

†9.10 Classical Applications

We have been considering the Wigner Theorem only with respect to quantum mechanical systems, but the theorem may be applied to classical problems as well. If $\hat{W}(\bar{r})$ is an operator belonging to a system's totally symmetric irreducible representation ($\hat{W}(\bar{r})$ is unaffected by, and commutes with, all the symmetry operations),‡ then the Wigner Theorem can help solve the eigenequation

$$\hat{W}(\bar{r})|Q_\alpha\rangle = \lambda|Q_\alpha\rangle \tag{9.28}$$

if it happens to be of physical interest.

Figure 9.11a shows a C_{3v} system of equal masses joined by springs of equal length and elasticity k. We define six displacement coordinates x_i, $i = 1, \ldots 6$, along six basis unit vectors $|1'\rangle, \ldots |6'\rangle$, and shall examine the normal modes of vibrations through the use of Newton's Second Law

$$m\ddot{x}_j(t) = -\frac{\partial V}{\partial x_j} \tag{9.29}$$

In the last section, we attacked the same problem in a quantum mechanical spirit, but the treatment we now present is strictly classical. You should keep an eye open for areas of overlap and connections between the two approaches.

The potential of the system increases with a deviation of any of the three inter-particle separation distances from the equilibrium value l. If the two masses at the base are displaced as in Fig. 9.11b, the potential is increased by the amount

$$\tfrac{1}{2}k[\{(l+x_1-x_5)^2+(x_2-x_6)^2\}^{1/2}-l]^2 \cong \tfrac{1}{2}k(x_1-x_5)^2 \tag{9.30a}$$

The complete approximate potential is of the form

$$V \simeq \tfrac{1}{2}k[(x_1-x_5)^2 + \{-\tfrac{1}{2}(x_3-x_1)+\tfrac{1}{2}\sqrt{3}(x_4-x_2)\}^2$$
$$+ \{\tfrac{1}{2}(x_5-x_3)+\tfrac{1}{2}\sqrt{3}(x_6-x_4)\}^2] \equiv \tfrac{1}{2}k\sum V_{ij}x_ix_j \tag{9.30b}$$

†Schonland,[35] page 142.
‡See Eqs (5.25d) and (9.6).

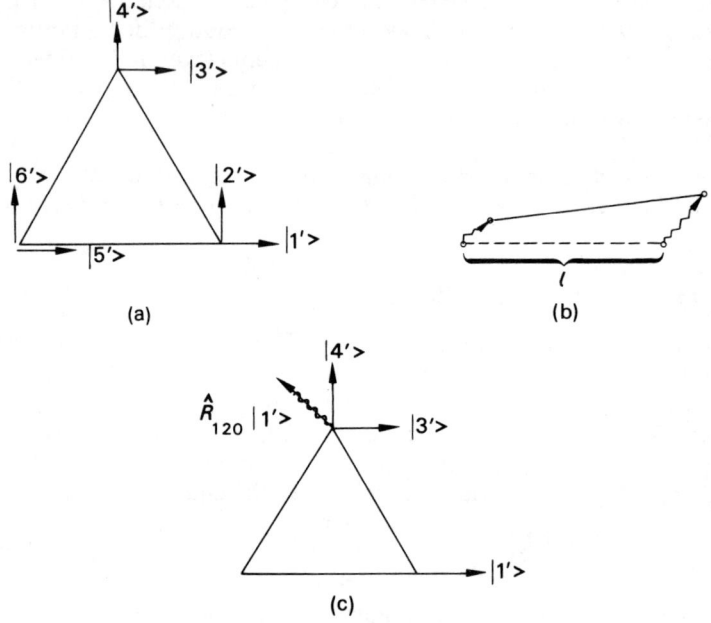

Figure 9.11.

PROBLEM 9.35 One can express Eq. (9.30b) as $(x_1 \ldots x_6) k \bar{\bar{V}} \begin{pmatrix} x_1 \\ \vdots \\ x_6 \end{pmatrix}$; find $\bar{\bar{V}}$. Show that

$$V_{ij} = \frac{\partial^2 V}{\partial x_i \partial x_j};$$

must

$$\frac{\partial^2 V}{\partial x_i \partial x_j} = \frac{\partial^2 V}{\partial x_j \partial x_i}?$$

Since in a normal mode the three bodies move synchronously, we employ the trial solutions

$$x_j(t) = x_j \exp(i\omega t) \tag{9.31}$$

in Eq. (9.29), and the classical (as opposed to the Schroedinger) equation of motion reduces to

$$\sum_i V_{ij} x_i = \lambda x_j \tag{9.32a}$$

$$\lambda = m\omega^2 / k$$

If we could transform from $\{|1'\rangle, \ldots |6'\rangle\}$ to a basis system

QUANTUM MECHANICAL STATES   179

$\{|Q_\alpha\rangle\}$ (the normal modes) in which $\bar{\bar{V}}$ is diagonal, Eq. (9.32a) would reduce, as in Eq. 9.28, to the uncoupled set of equations

$$V_{\alpha\alpha}Q_\alpha = \lambda Q_\alpha \tag{9.32b}$$

and give us the values of λ; thus the problem is to diagonalize $\bar{\bar{V}}$. Normally this would be done by solving the secular equation $|V_{ij} - \lambda\delta_{ij}| = 0$, with the $\bar{\bar{V}}$ of Problem 9.29. With more than a few particles, however, this can become a messy process.

We make use, therefore, of the formal equivalence of Eq. (9.32a) to the Hamiltonian eigenequation, Eq. (5.19), of quantum mechanics,

$$\sum_i \mathscr{H}_{ij}\psi_{n_j} = E_n\psi_{n_j} \tag{9.33}$$

and work by analogy.

Recall that Eq. (9.33) was obtained by employing a suitable complete basis set $\{|a_i\rangle\}$ to form a representation of $\mathscr{H}$, and that the eigensolutions are found by diagonalizing $\mathscr{H}$, or solving the secular equation $|\mathscr{H}_{ij} - E\delta_{ij}| = 0$. If we had initially chosen the $\{|a_i\rangle\}$ to belong to the irreducible representations of the group of the Schroedinger equation, then the matrix $\mathscr{H}$ would automatically be in block form; one can thus obtain preliminary information on the system without diagonalizing $\mathscr{H}$, by reducing the representation of the symmetry group generated by $\{|a_i\rangle\}$. For a C_{3v} molecule, $\Gamma^{(i)} \to a_{A,1}A_1 + a_{A,2}A_2 + a_E E$, as in Eq. (9.20a); we thus discover, by the Wigner Theorem, the types of energy levels and their degeneracies.

The above observation leads us in our current classical problem to build and reduce the $\{|1'\rangle, \ldots |6'\rangle\}$ representation of C_{3v}, Fig. 9.11a. This should tell us something of the eigensolutions of Eq. (9.32), and therefore of the normal modes of vibration of the system, without diagonalizing $\bar{\bar{V}}$.

Consider the effect of $\hat{R}_{120}$ upon $|1'\rangle$, Fig. 9.11c: $\hat{R}_{120}|1'\rangle = -\frac{1}{2}|3'\rangle + \sqrt{3/2}|4'\rangle$. In the usual fashion, and setting $\langle i'|j'\rangle = \delta_{i'j'}$ (legal?) we can build up 6×6 representations of the elements of C_{3v}. Thus

$$\bar{\bar{\Gamma}}(\hat{R}_{120}) = \frac{1}{2}\begin{pmatrix} 0 & 0 & 0 & 0 & -1 & -\sqrt{3} \\ 0 & 0 & 0 & 0 & \sqrt{3} & -1 \\ -1 & -\sqrt{3} & 0 & 0 & 0 & 0 \\ \sqrt{3} & -1 & 0 & 0 & 0 & 0 \\ 0 & 0 & -1 & -\sqrt{3} & 0 & 0 \\ 0 & 0 & \sqrt{3} & -1 & 0 & 0 \end{pmatrix} \tag{9.34}$$

and so on.

PROBLEM 9.36 Find $\overline{\overline{\Gamma}}(\hat{R})$ for the six elements of C_{3v}. You need find only *one* reflection matrix, and then use the isomorphism of $\{\overline{\overline{\Gamma}}(\hat{R})\}$ to $\{\hat{R}\}$, whose multiplication table is known, for the others. Why is $\{|i'\rangle\}$ of Fig. 9.11a to be considered complete?

The use of characters lets us break down this six-dimensional representation into

$$A_1 + A_2 + 2E \tag{9.35}$$

Immediately we conclude, without having solved the equations of motion, that the normal modes are either one or two-fold degenerate, and that they display A_1, A_2, or E-type symmetry. We have used the Wigner theorem, but not quantum mechanics, to get this result.

One of the E-pairs and the A_2 correspond, respectively, to translational and rotational modes, and are of frequency $\omega = 0$. But without our group theoretic arguments, we would have had no reason *a priori* to think that two of our modes of vibration should be of the same frequency, or have A_1 or E-type symmetry.

One can employ group theoretic tricks and find

$$\lambda = m\omega^2/k = \begin{cases} 3, & A_1 \\ \frac{3}{2}, & E \end{cases} \tag{9.36}$$

without ever solving the secular determinant. The interested reader is referred to Mathews and Walker,[29] page 448.

9.11 Summary

1. According to the Wigner Theorem, the orthonormalized degenerate eigenvectors of the Hamiltonian operator must behave like the generators of an irreducible representation of the group of the Schroedinger equation. The properties of such generators, however, are known from group theoretic (as opposed to quantum mechanical) considerations; thus once one learns the symmetry group of a system, one immediately knows much about the properties of possible energy states.

2. The Wigner Theorem applies not only to geometric symmetry operators such as rotations and reflections, but also to time reversal and particle exchange symmetry operators. But in the latter case, other constraints demand that of all the irreducible representations of S_n for an n-particle system, only fully symmetric (boson) and fully antisymmetric (fermion) states be found in nature.

3. Given a group $\{\hat{R}\}$, the operator

$$\hat{\mathcal{P}}^{(\mu)} = \sum_R \chi^{(\mu)}(\hat{R})^* \hat{P}_R$$

will project out of a function that portion which displays the symmetry of the μth irreducible representation of the group. The closely related

$$\hat{\mathscr{P}}_{ik}^{(\mu)} = \sum_R^N \frac{n_\mu}{N} \Gamma_{ik}^{(\mu)}(\hat{R})\hat{P}_R$$

operating on the basis vector $|\psi_i^{(\mu)}\rangle$ will yield the partner function $|\psi_k^{(\mu)}\rangle$.

CHAPTER 10

More Examples of the Use of the Wigner Theorem: Continuous Groups

We present several worked-out examples of the use of Wigner's Theorem. First the theorem is employed in a derivation of the Bloch Theorem for a one-dimensional crystal. Then, as the inter-lattice-site distance shrinks to 0, this becomes the case of a system with an axis of complete rotational symmetry, for which group theoretic arguments alone fully determine the allowed angular dependences of spin-0 quantum states. This leads to a discussion of the odd-dimensional irreducible representations of the symmetry groups of the sphere, and to spinor-generated irreducible representations and double groups. Finally, we consider briefly a scheme for classifying the symmetry properties of operators, as opposed to those of states.

10.1 Bloch Functions

Crystalline solids, and the one-dimensional lattice of Fig. 10.1, display translational symmetry. To avoid difficulties with end or boundary effects, we imagine a perfect lattice of infinite extent, where each atom is separated from its nearest neighbour by the distance a. Clearly a displacement of the system by na, where n is any positive or negative integer, will leave its appearance unchanged; a system of periodicity a is invariant to displacements by na.

Periodicity places severe constraints upon the form of the wave

EXAMPLES OF THE USE OF THE WIGNER THEOREM 183

function of any electron in a crystal; as an example of the power of the Wigner Theorem, we shall prove *Bloch's Theorem* for a one-dimensional periodic lattice:

THEOREM 10.1: The Bloch Theorem. If the potential felt by an electron is periodic of period a,

$$\boxed{\hat{V}(x) = \hat{V}(x+a)} \quad (10.1\text{a})$$

then possible wave functions are of the form

$$\boxed{\begin{aligned}\psi^{(k)}(x) &= u^{(k)}(x) \exp(ikx) \\ u^{(k)}(x) &= u^{(k)}(x+a)\end{aligned}} \quad (10.1\text{b})$$

Such functions are called *Bloch functions*.

While there are simple arguments which lead to Eq. (10.1b), a group theoretic approach is appropriate and instructive here. We define $\hat{R}_n$ by an extension of Eq. (6.7) to be the geometric operator which causes a shift of the lattice by the distance na:

$$\hat{R}_n x = x + na; \quad \hat{R}_n = (\hat{R}_1)^n, \quad \text{integer } n \quad (10.2\text{a})$$

With each $\hat{R}_n$, we associate an operator $\hat{P}_n$ which performs in function space in accordance with Eq. (6.8):

$$\hat{P}_n \psi(x) = \psi(x - na) = (\hat{P}_1)^n \psi(x) \quad (10.2\text{b})$$

$\{\hat{R}_n\}$ form the group of the Schroedinger equation, and $[\mathscr{H}, \hat{P}_n] = 0$. The mathematics remains most manageable if we demand that

$$\hat{R}_N = \hat{R}_0 \quad (10.3\text{a})$$

and

$$\psi(x + Na) = \psi(x), \quad N \gg 1 \quad (10.3\text{b})$$

This imposition of *periodic boundary conditions* is equivalent to modifying the model of Fig. 10.1 slightly; instead of a straight, infinite chain, our system now consists of N particles in a single long closed loop, Fig. 10.2. With a sufficiently large N, there should be no noticeable difference from our original picture of an infinite chain. But Eq. (10.3a) has converted the infinite-order group $\{\hat{R}_n\}$ into a cyclic group of order N, C_N.

The Wigner Theorem says that the possible wave functions $\{\psi_i(x)\}$ of the system generate the irreducible representations of

Figure 10.1.

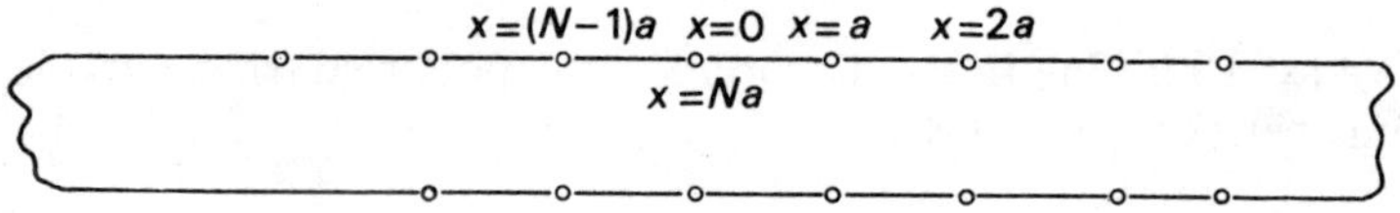

Figure 10.2.

the group $\{\hat{R}_n, n=1,\ldots N\}$, and thus

$$\hat{P}_n \psi_i^{(\nu)}(x) = \sum_j \Gamma_{ji}^{(\nu)}(\hat{R}_n) \psi_j^{(\nu)}(x) \tag{10.4a}$$

We make use of the fact, proven as Problem 8.5, that the irreducible representations of an abelian group are all one-dimensional; Eq. (10.4a) becomes

$$\hat{P}_n \psi^{(\nu)}(x) = \Gamma^{(\nu)}(\hat{R}_n) \psi^{(\nu)}(x) \tag{10.4b}$$

which tells us of the coarse, long range spatial dependence of $\psi^{(\nu)}(x)$. It says nothing, however, of the behaviour of $\psi^{(\nu)}(x)$ in the immediate vicinity of any particular value of x; that can be learned only by solving the Schroedinger equation exactly.

Equation (10.4b), however, reveals two significant points about the $\{\psi^{(\nu)}(x)\}$. First, the eigenstates of an electron in our simple, one-dimensional crystal are non-degenerate. And, second, these states can be distinguished from one another, and labelled meaningfully in terms of the representation index ν, by the spatial behaviour described in Eq. (10.4b).

An important particular case of Eq. (10.4b) is that of $\hat{P}_N$:

$$\hat{P}_N \psi^{(\nu)} = (\hat{P}_1)^N \psi^{(\nu)} = \{\Gamma^{(\nu)}(\hat{R}_1)\}^N \psi^{(\nu)}, \quad \text{any } \nu \tag{10.5a}$$

but

$$\hat{P}_N \psi^{(\nu)} = \hat{P}_0 \psi^{(\nu)} = \psi^{(\nu)} \tag{10.5b}$$

therefore

$$(\Gamma^{(\nu)}(\hat{R}_1))^N = 1, \quad \text{any } \nu \tag{10.5c}$$

Then, for the different ν's,

$$\Gamma^{(\nu)}(\hat{R}_1) = 1^{1/N} = \exp(2\pi i \mu / N) \quad \mu = 1, \ldots, N \tag{10.5d}$$

There are N values of ν (why?), and similarly N values of μ. Is there any reason to associate each ν with a μ in a one-to-one fashion? We can see that there is when we consider what happens if two non-equivalent irreducible representations, with $\nu = \alpha$ and

EXAMPLES OF THE USE OF THE WIGNER THEOREM

$\nu = \beta$, respectively, use the same μ, namely μ_0:

$$\Gamma^{(\alpha)}(\hat{R}_1) = \exp(2\pi i \mu_0/N)$$

$$\Gamma^{(\beta)}(\hat{R}_1) = \exp(2\pi i \mu_0/N)$$

Then $\psi^{(\alpha)}(x)$ and $\psi^{(\beta)}(x)$, by Eq. (10.4b), have exactly identical transformational properties, in contradiction to our supposition that they generate non-equivalent representations. Thus the N different irreducible representations of Eq. (10.5d) must be associated with N different values of μ. Or, more simply

$$\Gamma^{(\nu)}(\hat{R}_1) = \exp(2\pi i \nu/N), \qquad \nu = 1, \ldots, N \qquad (10.6a)$$

This tells us that $\psi^{(\nu)}(x)$ generates the νth one-dimensional representation of the group $\{\hat{R}_n\}$ such that

$$\hat{P}_1 \psi^{(\nu)}(x) = \exp(2\pi i \nu/N)\, \psi^{(\nu)}(x) \qquad (10.6b)$$

$$\hat{P}_n \psi^{(\nu)}(x) = \exp(2\pi i n \nu/N)\, \psi^{(\nu)}(x) \qquad (10.6c)$$

Calling the length of the chain $Na = L$, and defining $k = \nu(2\pi/L)$, Eq. (10.6b) becomes

$$\boxed{\psi^{(k)}(x+a) = \exp(ika)\, \psi^{(k)}(x) \qquad k = \frac{2\pi}{L}, \frac{4\pi}{L}, \ldots, \frac{2\pi}{a}} \qquad (10.7)$$

*PROBLEM 10.1 Show that a function which obeys Eq. (10.7) must be of the form of Eq. (10.1b). The theorem is proven.
*PROBLEM 10.2 A real crystal may display not only translational, but also rotational, etc. symmetries. How will this affect the state functions?
PROBLEM 10.3 Construct the character table for an N-point lattice.
PROBLEM 10.4 Is it meaningful to talk of a Bloch form for vibrational states in a crystal?

10.2 Systems of Axial Symmetry

A main result of the preceeding section was Eq. (10.6c), rewritten here, in slightly different form, as

$$\hat{P}_n \psi^{(m)}(x) = \Gamma^{(m)}(\hat{R}_n) \psi^{(m)}(x) \qquad (10.8a)$$

$$\Gamma^{(m)}(\hat{R}_n) = \exp(m\, 2\pi i n a/Na), \qquad m, n = 1, \ldots, N \qquad (10.8b)$$

Let us imagine our closed-loop lattice to form a circle of radius one unit; then na runs between a and $Na=2\pi$. If we allow the separation a between points to decrease indefinitely, but cause N to increase correspondingly such that $Na=2\pi$ still, then the lattice points will form a quasi-continuum paramaterized by the angle ϕ, where $0 \leq (\phi = na) \leq (Na = 2\pi)$.

In the limit† $N \to \infty$

$$\Gamma^{(m)}(\phi) = \exp(im\phi) \qquad m = 0, \pm 1, \pm 2, \ldots \tag{10.9a}$$

The system we are now describing is one with complete axial symmetry, such as that of a right cylinder or cone, for which all rotations (even infinitesimal) about the axis are symmetry operations.

We conclude via the Wigner Theorem and Eq. (10.9a) that any system with full axial symmetry must have wave-functions of the form

$$\boxed{\psi(\bar{r}) = f(r, \theta) \exp(im\phi) \qquad m = 0, \pm 1, \pm 2, \ldots} \tag{10.9b}$$

Not surprisingly, the complex-valued spherical harmonics are of this sort.

We can derive Eq. (10.9a) in a different fashion as follows. The group of rotations about a single axis is abelian, and therefor $\hat{P}_\phi \psi = \Gamma(\phi)\psi$ for any ψ. If two rotations are performed in sequence,

$$\hat{P}_{\phi_1} \hat{P}_{\phi_2} \psi = \Gamma(\phi_1)\Gamma(\phi_2)\psi = \hat{P}_{(\phi_1 + \phi_2)}\psi = \Gamma(\phi_1 + \phi_2)\psi$$

therefore

$$\Gamma(\phi_1)\Gamma(\phi_2) = \Gamma(\phi_1 + \phi_2) \tag{10.10a}$$

Given a reasonable-seeming boundary condition which prevents our state from being a multiple-valued function of ϕ (and see Eq. (10.3b))

$$\Gamma(2\pi) = \Gamma(0) = 1 \tag{10.10b}$$

Eq. (10.10) can be satisfied only by the function

$$\Gamma(\phi) = \exp(im\phi) \qquad m = 0, \pm 1, \pm 2, \ldots \text{ Q.E.D.} \tag{10.9a}$$

The m of this equation has a definite physical meaning. A system for which $|3\rangle$ is an axis of full rotational symmetry should have a well-defined component of angular momentum J_3; then

---

†$m = 0, -1, -2, \ldots$ arise with a boundary condition slightly different from Eq. (10.3b). What is it?

EXAMPLES OF THE USE OF THE WIGNER THEOREM   187

$$\hat{J}_3\psi(\bar{r})=\hat{J}_3\{f(r,\theta)\exp(im\phi)\}=f(r,\theta)\left\{-i\hbar\frac{\partial}{\partial\phi}\exp(im\phi)\right\}$$
$$=m\hbar f(r,\theta)\exp(im\phi)=m\hbar\psi(\bar{r}) \qquad (10.11)$$

and our system is in the mth eigenstate of the operator $\hat{J}_3$.

Since $\hat{J}_3\psi=m\hbar\psi$, we are tempted to examine the *operator* formed by replacing m with $-\hat{J}_3/\hbar$ in Eq. (10.9a) (where the minus sign is thrown in for convenience)

$$\exp(im\phi)\rightarrow\exp(-i\hat{J}_3\phi/\hbar)\equiv\hat{P}_\phi \qquad (10.12a)$$

Now consider the effect, for $\phi=\phi_0$, of this proposed operator upon the arbitrary state $\psi(\phi)=\exp(im'\phi)$ of our axially symmetric system:

$$\hat{P}_{\phi_0}\psi(\phi)=\exp(-i\hat{J}_3\phi_0/\hbar)\{\exp(im'\phi)\}=\exp\{im'(\phi-\phi_0)\}$$
$$=\psi(\phi-\phi_0) \qquad (10.12b)$$

Equation (10.12b), however, is what we should expect of a function-space $\hat{P}$-operator (Eq. (6.4))

$$\hat{P}_{\phi_0}\psi(\hat{R}_{\phi_0}\bar{r})=\psi(\bar{r})$$

or

$$\hat{P}_{\phi_0}\psi(\phi+\phi_0)=\psi(\phi)$$

We therefore assert that a generator of finite rotations about an axis of rotation $\bar{n}$ is our new $\hat{P}_\phi$:

$$\boxed{\hat{P}_{\phi,\bar{n}}=\exp(-i\hat{J}\cdot\bar{n}\phi/\hbar)} \qquad (10.13)$$

This is, of course, the operator we constructed out of the generators of infinitesimal rotations in Eq. (6.16).

PROBLEM 10.5 Why do we find exponential operators whenever we consider the generators of continuous transformations in space or time, Eq. (5.16), (6.11), (10.13)?

*PROBLEM 10.6 Will the operator $\hat{P}_{\phi,\bar{n}}$ transform the states of a C_{3v} system properly for an arbitrary ϕ? For $\phi=120°$?

One can show that a linear molecule, of symmetry group $C_{\infty v}$, has two one-dimensional irreducible representations, and an infinite number of two-dimensional varieties. Since for any ϕ, $\{\hat{R}_\phi,\hat{R}_{-\phi}\}$ constitute a class, the character table for $C_{\infty v}$ is as in Table 10.1. The spectroscopic notation $\Sigma, \Pi, \Delta, \Phi\ldots$ (versus $s, p, d, f\ldots$) is normally used for linear molecules.

Table 10.1

$C_{\infty v}$	$\hat{I}$	$\hat{R}_\phi$	$\hat{M}$
(A_1) Σ^+	1	1	1
(A_2) Σ^-	1	1	-1
(E) π	2	$2\cos\phi$	0
(E') Δ	2	$2\cos 2\phi$	0
(E'') Φ	2	$2\cos 3\phi$	0

PROBLEM 10.7 Use Eq. (10.9a) to validate Table 10.1. Which representations of $C_{\infty v}$ are produced by real or complex d-states?

PROBLEM 10.8 According to Eqs (9.4) and (10.9a), rotation of a linear molecule by 120 degrees about its axis of symmetry only causes a change of phase of its state. By Eqs (7.16) through (7.18), however, rotation of a C_{3v} molecule by the same amount can cause the admixture of states. Why the difference? See Problem 6.8.

10.3 Full Rotational Group

On several occasions we have employed real valued forms of the spherical harmonics of order l to generate $(2l+1)$-dimensional representations of various geometric operators. We have seen, for example, that p-states and d-states lead to block-diagonal representations of the rotation operator for any angle about $|3\rangle$, Eq. (6.31), and similarly for vertical mirror plane operators. They give the reducible three- and five-dimensional representations of C_{3v}, in particular, of which we have made much use.

These same spherical harmonics of order l can, as well, be used in generating a $(2l+1)$ dimensional representation of the full rotation group, which consists of all rotations about all axes passing through a fixed point. This can be done by repeating the procedure of Section 6.6, or by employing Eq. (10.13) $\hat{P}_{\phi,\bar{n}} = \exp(-i\hat{J}\cdot\bar{n}\phi/\hbar)$. By suitable generalizations of the methods of Chapter 8 (for example, the sum over $\hat{R}$ in the Great Orthogonality Theorem must be converted to an integral), one finds that the representations so produced are *irreducible* representations of the full rotation group.

In Section 5.11 we showed that the spherical harmonics provide the exact angular dependences of the eigensolutions for a spherical Hamiltonian. It should not be surprising, then, by Wigner's Theorem, that they generate irreducible representations of the symmetry group of the sphere. What is perhaps less obvious is that they constitute the *only* set of functions to do so; no $(2l+1)$ functions other than the $\{Y_l^m\}$, or linear combinations thereof, will produce the lth odd-dimensional irreducible representation of the full rotational group. The spectroscopic labels $s, p, d, f. \ldots$ thus not only refer to atomic states of angular momentum $|\bar{L}| = \hbar\sqrt{l(l+1)}$, $l = 0, 1, 2, 3, \ldots$, but also serve to identify the $1, 3, 5, 7 \ldots$ dimensional irreducible representation of the atom's symmetry group!

Representation theory for the group of the sphere can become quite elaborate, and rather than pursue it, we shall simply indicate a few of the many ways in which it is used. The *Vector Model* for the addition of electronic angular momenta on an atom, for example, has a group theoretic foundation. If we have two atomic electrons in states with well-defined angular momentum quantum numbers l_1 and l_2, then we can use product states of the form $\{Y_{l_1}^{m_1}(\bar{r}_1) Y_{l_2}^{m_2}(\bar{r}_2)\}$, with l_1 and l_2 fixed, and $l_1 \geq m_1 \geq -l_1$, $l_2 \geq m_2 \geq -l_2$, to generate a representation of the full rotation group. This representation is found by the method of product spaces to be reducible, and is *always* decomposable as

$$\Gamma^{(l_1)} \times \Gamma^{(l_2)} = \Gamma^{(l_1+l_2)} + \Gamma^{(l_1+l_2-1)} + \cdots + \Gamma^{(|l_1-l_2|)} \qquad (10.14)$$

Thus the compound state of the two-electron system can have as an orbital angular momentum quantum number any integer between and including $l_1 + l_2$ and $|l_1 - l_2|$.

Likewise, the commutation relations of the angular momentum operators $\hat{L}_1, \hat{L}_2, \hat{L}_3$

$$[\hat{L}_1, \hat{L}_2] = i\hbar \hat{L}_3, \text{ etc.} \qquad (10.15)$$

follow from (or rather, are embedded within and can be made to materialize via) the operator $\exp(-i\hat{J} \cdot \bar{n}\phi/\hbar)$ of Eq. (10.13), which is valid for any alignment of $\bar{n}$. Consider what happens when we try to rotate a real-space vector of unit length $|x\rangle$ about various axes by very small angles. Assume, Fig. 10.3a, that $|x\rangle$ lies initially along the $|1\rangle$ axis. Rotating $|x\rangle$ through the angle θ_2 about the $|2\rangle$ axis via the $\hat{R}_2$ operator will leave the arrowhead of $\hat{R}_2|x\rangle$ a distance θ_2 below that of $|x\rangle$, Fig. 10.3b. Rotating $\hat{R}_2|x\rangle$ then through θ_1 about $|1\rangle$ moves the head another $\theta_2\theta_1$, finally yielding $\hat{R}_1\hat{R}_2|x\rangle$, Fig. 10.3c. We now attempt the same result but with the order of rotations reversed. Figure 10.4b shows that beginning with $\hat{R}_1$ has no effect at all on $|x\rangle$. And $\hat{R}_2\hat{R}_1|x\rangle$ of Fig. 10.4c is clearly not the same as $R_1R_2|x\rangle$ of Fig. 10.3c. One can, however, apparently mix in an additional rotation by the angle $\theta_1\theta_2$ about $|3\rangle$ to catch up; thus

$$\hat{R}_1\hat{R}_2|x\rangle = \hat{R}_3\hat{R}_2\hat{R}_1|x\rangle \qquad (10.16a)$$

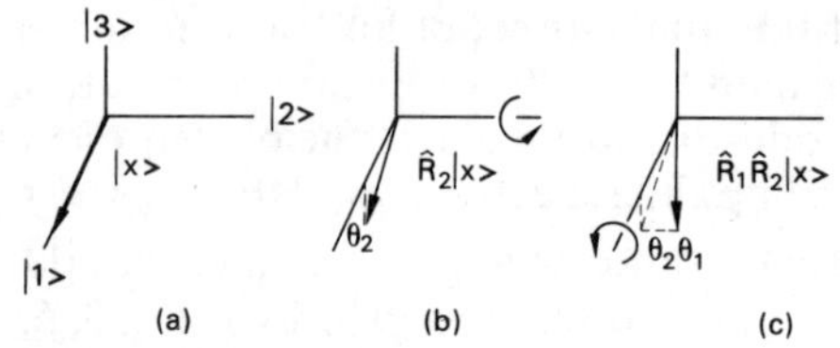

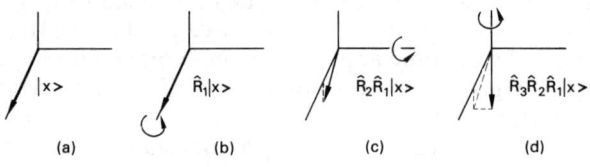

Figure 10.3.

Figure 10.4.

Transferring this information to function space, which is allowed since $\{\hat{P}_R\}$ is isomorphic to $\{\hat{R}\}$, we find

$$\exp(-i\hat{L}_1\theta_1/\hbar)\exp(-i\hat{L}_2\theta_2/\hbar) = \exp(-i\hat{L}_3\theta_1\theta_2/\hbar)$$
$$\times \exp(-i\hat{L}_2\theta_2/\hbar)\exp(-i\hat{L}_1\theta_1/\hbar) \qquad (10.16b)$$

or, Taylor expanding and keeping terms no higher than quadratic in the small angles θ, we arrive at the ubiquitous angular momentum commutation relations:

$$[\hat{L}_1, \hat{L}_2] = i\hbar\hat{L}_3 \qquad (10.15)$$

Finally, in contradistinction to all of the above, we again stress that if we deal with a finite-order group such as C_{3v} instead of a Lie (continuous) group, we no longer can automatically find the exact angular dependences of the solutions to the Hamiltonian by symmetry alone; one can only say that correct functions must satisfy certain transformational requirements (they must behave in a certain fashion under the six operations of C_{3v}).

PROBLEM 10.9 Why is there not a good orbital angular momentum quantum number for an electron on a C_{3v} molecule? [HINT: consider a large, near-circular orbit in the plane of the C_{3v} frame.] Is any angular momentum of the system conserved?

The several standard formal approaches to the full rotational group are complex, and treated at various levels of sophistication

EXAMPLES OF THE USE OF THE WIGNER THEOREM 191

in the books listed in the bibliography. These yield not only the odd-dimensional irreducible representations generated by the spherical harmonics, but also the spinor-generated, even-dimensional representations. In the next section, we shall only briefly introduce the two-dimensional representation of the full rotation group produced by spin-$\frac{1}{2}$ states, and the useful 'double group' construction.

†10.4 Spin and Double Groups

In a magnetic field, a free† electron's spin will align either with or against the field, and the particle will inhabit an eigenstate, $|\alpha\rangle$ or $|\beta\rangle$, common to $\hat{\mathcal{H}}$, $\hat{S}$, and $\hat{S}_z$. Even when the external field is reduced to zero, $\{|\alpha\rangle, |\beta\rangle\}$ can still serve as a complete orthonormal basis set for describing the arbitrary spin state $|\psi\rangle$:

$$|\psi\rangle = c_\alpha |\alpha\rangle + c_\beta |b\rangle \qquad (10.17a)$$

c_α and c_β may, of course, be functions of $\bar{r}$ or other parameters. In this system, the respresentations of $|\alpha\rangle$, $|\beta\rangle$, and $|\psi\rangle$ are, respectively,

$$|\alpha\rangle \to \begin{pmatrix} 1 \\ 0 \end{pmatrix}$$

$$|\beta\rangle \to \begin{pmatrix} 0 \\ 1 \end{pmatrix} \qquad (10.17b)$$

$$|\psi\rangle \to \begin{pmatrix} c_\alpha \\ c_\beta \end{pmatrix}$$

Also, since $|\alpha\rangle$ and $|\beta\rangle$ are eigenstates of $\hat{S}_z$,

$$\bar{\bar{S}}_z = \frac{\hbar}{2} \begin{pmatrix} 1 & 0 \\ 0 & -1 \end{pmatrix} \qquad (10.17c)$$

The problem naturally arises of having to find the transformational properties of $|\psi\rangle$ and $\begin{pmatrix} c_\alpha \\ c_\beta \end{pmatrix}$.

Immediately we see that we cannot use the methods of Section 6.7, since $|\alpha\rangle$ and $|\beta\rangle$ are not necessarily functions of $\bar{r}$! We resort, therefore, to the explicit form of $\hat{P}_{\phi,\bar{n}}$ of Eq. (10.13):

$$\hat{P}_{\phi,n} = \exp(-i\hat{J} \cdot \bar{n}\phi/\hbar) \qquad (10.13)$$

---

†Or an atomic or molecular electron, when the spin-orbit interaction is negligible. The notation of Section 5.12 is used here.

But Eq. (10.13) was itself derived either (i) by making the assumption that $\hat{J} = \hat{L} = \hat{r} \times \hat{p}$, a function of $\bar{r}$, or (ii) by demanding that $\Gamma(2\pi) = \Gamma(0) = 1$ in Eq. (10.10b). The first method does not apply to the situation $\hat{J} = \hat{S}$, and, as we shall soon see, for half-integral spin states it is no longer true that $\hat{P}_{2\pi}\psi = \psi$; our ultimate justification for adopting Eq. (10.13) must come from formal analysis of the full rotation group, which unfortunately is beyond the scope of this text.

The $\{|\alpha\rangle, |\beta\rangle\}$ representation of $\exp(-i\hat{S}\cdot\bar{n}\phi/\hbar)$ presents no problem in the specific case for which $\bar{n} = |3\rangle$; by Eq. (10.17c), this is just

$$\bar{\bar{\Gamma}}(\hat{R}_\phi) = \exp(-i\phi\bar{\bar{S}}_z/\hbar) = \exp\left\{-\frac{i\phi}{2}\begin{pmatrix}1 & 0\\ 0 & -1\end{pmatrix}\right\}$$

$$= \cos\frac{\phi}{2}\begin{pmatrix}1 & 0\\ 0 & 1\end{pmatrix} - i\sin\frac{\phi}{2}\begin{pmatrix}1 & 0\\ 0 & -1\end{pmatrix} \quad (10.18a)$$

If the coordinate system or physical system is rotated by ϕ about $|3\rangle$, Eq. (6.19) becomes, for the particular example of $|\psi\rangle = |\alpha\rangle$,

$$\hat{P}_R|\alpha\rangle = \exp(-i\phi/2)|\alpha\rangle \quad (10.18b)$$

*PROBLEM 10.10 Verify Eq. (10.18). What does a rotation by ϕ about $|3\rangle$ do to $\begin{pmatrix}c_\alpha\\ c_\beta\end{pmatrix}$? Why does $\hat{P}_R$ admix no $|\beta\rangle$ into $|\alpha\rangle$ in Eq. (10.18b)?

*PROBLEM 10.11 What does a rotation by 2π do to $|\alpha\rangle$? To $|\psi\rangle$?

Rotations about an arbitrarily aligned axis, however, are not so simple. The process involves finding commutation relations for spin operators, $[\hat{S}_1, \hat{S}_2] = i\hbar\hat{S}_3$, etc., in analogy to $[\hat{L}_1, \hat{L}_2] = i\hbar\hat{L}_3$ of Eq. (10.15), inventing raising and lowering operators as in Eq. (5.53), and finally establishing a convention (usually in terms of Euler angles) for describing precisely any rotation in question.† The net result of all this is that our $\{|\alpha\rangle, |\beta\rangle\}$ representation of $\hat{P}_{\phi,\bar{n}}$ becomes

$$\bar{\bar{\Gamma}}^{(1/2)}(\hat{R}) = \exp(-i\phi\bar{\bar{\sigma}}\cdot\bar{n}/2) \quad (10.19a)$$

$$\bar{\bar{\sigma}}_1 = \begin{pmatrix}0 & 1\\ 1 & 0\end{pmatrix}$$

$$\bar{\bar{\sigma}}_2 = \begin{pmatrix}0 & -1\\ 1 & 0\end{pmatrix} \quad (10.19b)$$

$$\bar{\bar{\sigma}}_3 = \begin{pmatrix}1 & 0\\ 0 & -1\end{pmatrix}$$

†Merzbacher,[31] Chapter 12.

where σ_i are the well-known Pauli spin matrices. The superscript on $\overline{\overline{\Gamma}}^{(1/2)}$ serves as a reminder that the basis set $\{|\alpha\rangle, |\beta\rangle\}$ is appropriate for a spin-$\frac{1}{2}$ quantum system.

$\hat{P}_{\phi,n} = \exp(-i\hat{S} \cdot \bar{n}\phi/\hbar)$ and $\{|\alpha\rangle, |\beta\rangle\}$ lead to a two-dimensional irreducible representation of the full rotation group. In a similar fashion, sets of spin 3/2, 5/2, 7/2, ... states lead to 4-, 6-, 8-, ... dimensional irreducible representations of this group. These can be handled as above.

Equation (10.18b) and Problem 10.11 reveal, however, a paradox with spinor representations not occurring in those generated by spherical harmonics: under a 2π rotation about $|3\rangle$, which should return a physical object or coordinate system to its initial condition, $|\alpha\rangle$ becomes $-|\alpha\rangle$! This double-valuedness of spinors† leads to no physically measurable inconsistencies, because probability densities and expectation values always involve the product of two such states, and the awkward -1 phase factors cancel out; but it does affect our scheme for classifying states by their transformational properties.

Noting the double-valued nature of a spinor under $\hat{R}_{2\pi}$, but that no factor of -1 appears with a rotation by 4π, Hans Bethe suggested in 1929 that for a system with half-integral spin, a group such as C_{3v} be expanded to include a distinctly new kind of symmetry operation: rotation by 2π. One assumes the *fiction* that $\hat{R}_{2\pi} \neq \hat{I}$, but rather $\hat{R}_{2\pi}^2 = \hat{R}_{4\pi} = \hat{I}$. Thus a marked triangle is supposedly not brought back into original alignment under a 2π rotation, but only after 4π! The *double group* of C_{3v}, for example, now includes operations $\hat{R}_{360}$, $\hat{R}_{480}$, and $\hat{R}_{600}$; and similarly, $\hat{R}_{2\pi}\hat{M}_I = \hat{M}_I\hat{R}_{2\pi}$, $\hat{R}_{2\pi}\hat{M}_{II}$, and $\hat{R}_{2\pi}\hat{M}_{III}$ are elements. C_{3v} is doubled in size, and three totally new classes are added to the three ordinary classes. There are thus a total of six irreducible representations of the double group, of dimensions 2, 2, 1, 1, 1, 1. The character table is not hard to work out, and is presented as Table 10.2. The new additional irreducible representations can be generated by states of half-integral angular momentum; these are distinct from our original C_{3v} irreducible representations obtained using, say, real-valued spherical harmonics (integral angular momentum states).

The character table of C_{3v} itself is included as a portion of that of the double group, a normal feature of double groups. The fact that for C_{3v} the number of classes, and therefore of irreducible representations, is also doubled, however, is not a general result. For example, the symmetry group of the cube has five irreducible representations, of dimensions 3, 3, 2, 1, 1; the corresponding

---

†See also Goldstein[16] on the Cayley–Klein parameters used in analysing the rotational motions of classical bodies, page 109.

Table 10.2

C_{3v} double group	$\hat{I}$	$\hat{R}_{120}$ $\hat{R}_{240}$	$\hat{M}_{\rm I}$ $\hat{M}_{\rm II}$ $\hat{M}_{\rm III}$	$\hat{R}_{2\pi}$	$\hat{R}_{480}$ $\hat{R}_{600}$	$\hat{R}_{2\pi}\hat{M}_{\rm I}$ $\hat{R}_{2\pi}\hat{M}_{\rm II}$ $\hat{R}_{2\pi}\hat{M}_{\rm III}$
A_1	1	1	1	1	1	1
A_2	1	1	-1	1	1	-1
E	2	-1	0	2	-1	0
Γ_4	1	1	i	-1	-1	$-i$
Γ_5	1	1	$-i$	-1	-1	i
Γ_6	2	-1	0	-2	1	0

double group has only three more irreducible representations, of dimensions 2, 2, and 4.

For finite-order symmetry groups, the problem of the transformation properties of electron wave functions which include both orbital and spin angular momenta can be messy; but the double group character table may be used in the labelling of such states and in the evaluation of perturbation matrix elements in the normal fashion, and hence the great value of the construct.

*PROBLEM 10.12 Do $\{|\alpha\rangle, |\beta\rangle\}$ form a basis for an irreducible representation of the double group of C_{3v}?

PROBLEM 10.13 Show that the generators of representations Γ_4 and Γ_5 of C_{3v} must be states of half-integral angular momentum.

PROBLEM 10.14 Are the double group and the colour group of C_{3v} isomorphic?

†10.5 Irreducible Spherical Tensor Operators

Until now we have been concerned primarily with the transformational properties of quantum states. It was suggested, however, in our discussion of projection operators, Eq. (9.6), and earlier, Eq. (5.25d), that *any* function of $\bar{r}$ may display symmetry behaviour, even an operator. Here we shall pursue that idea a bit further. We shall keep in mind a physical system which appears invariant to all the rotations of the sphere, $\{\hat{R}\}$; then as in Eq. (4.15), a state $|\psi(\bar{r})\rangle$ and an operator $\hat{A}(\bar{r})$ should transform simultaneously under any such rotation as

$$|\psi(\bar{r})\rangle \to |\psi'(\bar{r})\rangle \equiv \hat{P}_R|\psi(\bar{r})\rangle \tag{10.20a}$$

$$\hat{A}(\bar{r}) \to \hat{A}'(\bar{r}) \equiv \hat{P}_R \hat{A}(\bar{r}) \hat{P}_R^{-1} = \hat{P}_R \hat{A}(\bar{r}) \hat{P}_R^+ \tag{10.20b}$$

EXAMPLES OF THE USE OF THE WIGNER THEOREM

Under a rotation of real space, such that $|\psi(\bar{r})\rangle$ changes into $|\psi'(\bar{r})\rangle = \hat{P}_R|\psi(\bar{r})\rangle$, the operator equation

$$\hat{A}(\bar{r})|\psi(\bar{r})\rangle = |\phi(\bar{r})\rangle \tag{10.20c}$$

becomes

$$\hat{P}_R \hat{A}(\bar{r}) \hat{P}_R{}^+ \hat{P}_R |\psi(\bar{r})\rangle = \hat{P}_R |\phi(\bar{r})\rangle \tag{10.20d}$$

and $\hat{A}(\bar{r})$ changes into $\hat{A}'(\bar{r}) = \hat{P}_R \hat{A}(\bar{r}) \hat{P}_R{}^+$; hence Eq. (10.20b).

*PROBLEM 10.15 Show that the expectation value of $\hat{A}(\bar{r})$ in the state $|\psi(\bar{r})\rangle$ equals the expectation value of $\hat{A}'(\bar{r}) \equiv \hat{P}_R \hat{A}(\bar{r}) \hat{P}_R{}^+$ in the state $|\psi'(\bar{r})\rangle = \hat{P}_R|\psi(\bar{r})\rangle$. Should $\langle\psi(\bar{r})|\hat{A}(\bar{r})|\psi(\bar{r})\rangle = \langle\psi'(\bar{r})|\hat{A}(\bar{r})|\psi'(\bar{r})\rangle$?

Consider first the system's Hamiltonian. Since $\mathcal{H}$ commutes with all $\hat{P}_R$ in the group of the Schroedinger equation, and $\mathcal{H}(\hat{R}\bar{r}) = \mathcal{H}(\bar{r})$, then under $\hat{R}$,

$$\mathcal{H} \to \mathcal{H}' \equiv \hat{P}_R \mathcal{H} \hat{P}_R{}^+ = \mathcal{H} \tag{10.21a}$$

This is reminiscent of the behaviour of an s-state, for which

$$|s(\bar{r})\rangle \to |s'(\bar{r})\rangle \equiv \hat{P}_R|s(\bar{r})\rangle = \Gamma^{(s)}(\hat{R})|s(\bar{r})\rangle = |s(\bar{r})\rangle \tag{10.21b}$$

for all $\hat{P}_R$. Returning to $\mathcal{H}$, one could emphasize the analogy by writing

$$\mathcal{H} \to \mathcal{H}' \equiv \hat{P}_R \mathcal{H} P_R{}^+ = \Gamma^{(s)}(\hat{R})\mathcal{H} = \mathcal{H} \tag{10.21c}$$

In terms of the projection operator theorem, Eq. (9.6), one says that $\mathcal{H}$ transforms as the fully symmetric irreducible representation of the symmetry group, or more simply, that $\mathcal{H}$ is a *scalar operator*. Another example of a scalar operator was discussed in Section (9.10). $\hat{J}^2$ is a third.

PROBLEM 10.16 There exist alignments of function space such that not only are the representations of all $\hat{P}_R$ for a finite group in block-diagonal form, but also $\mathcal{H}$ is *fully* diagonal. Does this depend on $\mathcal{H}$ being a scalar operator?

Now let us look at the angular momentum operator $\hat{L}(\bar{r}) \equiv \hat{L}$. The classical three-vector $\bar{L}$ is proportional to an observable vector entity, the magnetic moment $\bar{\mu}$, and we certainly would want the physically measurable *expectation values* of the *operators* $\hat{L}$ and $\hat{\mu}$ also to behave like real vectors in three-space. Suppose a rotation of the real-space coordinate system takes every state $|\psi(\bar{r})\rangle$ into $|\psi'(\bar{r})\rangle = \hat{P}_R|\psi(\bar{r})\rangle$. The expectation values of the components of $\hat{L}$, $\langle \hat{L}_i \rangle$, should behave exactly like the components of any other three-vector under a rotation:

$$\langle L_i \rangle \to \sum_j^3 R_{ij} \langle L_j \rangle = \sum_j^3 R_{ij} \langle \psi(\bar{r}) | \hat{L}_j | \psi(\bar{r}) \rangle \tag{10.22a}$$

On the other hand, under the transformation $|\psi(\bar{r})\rangle \to |\psi'(\bar{r})\rangle$, the expectation value of operator $\hat{L}_i$ should change also as

$$\langle \psi(\bar{r}) | \hat{L}_i | \psi(\bar{r}) \rangle \to \langle \psi'(\bar{r}) | \hat{L}_i | \psi'(\bar{r}) \rangle = (\langle \psi(\bar{r}) | \hat{P}_R^+) \hat{L}_i (\hat{P}_R | \psi(\bar{r}) \rangle) \tag{10.22b}$$

Eq. (10.22a) and (10.22b) can hold simultaneously for all $\psi(\bar{r})$ only if

$$\hat{P}_R^+ \hat{L}_i \hat{P}_R = \sum_j^3 R_{ij} \hat{L}_j \tag{10.22c}$$

The entities $\langle L_1 \rangle$, $\langle L_2 \rangle$, and $\langle L_3 \rangle$ are thus not only the expectation values of three related operators, but also the components of an ordinary three-vector $\overline{\langle L \rangle}$.

*PROBLEM 10.17 Show that Eq. (10.22c) can be re-written as

$$\hat{L}_i \to \hat{L}_i' \equiv \hat{P}_R \hat{L}_i \hat{P}_R^+ = \sum_j R_{ji} \hat{L}_j \tag{10.22d}$$

Comparison of Eqs (10.22d) and (10.20b) suggests a useful interpretation of the transformation properties of the $\{\hat{L}_i\}$. Under a rotation of real space such that $\hat{L}_i \to \hat{P}_R \hat{L}_i \hat{P}_R^+$, the operators $\hat{L}^i$ mix among themselves like three *basis vectors* spanning the function space which contains $\hat{L}(\bar{r})$. The $\langle \hat{L}_i \rangle$, as noted above, behave differently, and more like the components of some vector $\overline{\langle L \rangle}$.

We are tempted to pursue the analogy noted in Eq. (10.21c) between the transformational properties of states and those of operators. Recall the equivalence of $\overline{\overline{R}}$, the representation of $\hat{R}$ generated by $\{|1\rangle, |2\rangle, |3\rangle\}$, to $\overline{\overline{\Gamma}}^{(p)}$, the representation of $\hat{P}_R$ generated by real-valued p-states $\{|p_1\rangle, |p_2\rangle, |p_3\rangle\}$, where

$$\hat{P}_R | p_i(\bar{r}) \rangle = \sum_j^3 \Gamma_{ji}^{(p)}(\hat{R}) | p_j(\bar{r}) \rangle \tag{10.23a}$$

If $\overline{\overline{R}} = \overline{\overline{\Gamma}}^{(p)}$, then by Eq. (10.22d) the three operators $\{\hat{L}_i\}$ appear to transform exactly like three real-valued p-states. And we can emphasize this by writing

$$\hat{L}_i \to \hat{L}_i' = \sum_j^3 \Gamma_{ji}^{(p)}(\hat{R}) \hat{L}_j \tag{10.23b}$$

An operator which behaves in this fashion is called a *vector operator*. Scalar and vector *operators*† thus transform under rotations exactly like real-valued s- and p-*states*, respectively.

In Eqs (10.21b) and (10.23a), we have employed the representations of the full rotational group generated by the real-valued

†Also called Cartesian tensor operators of ranks zero and one, respectively.

EXAMPLES OF THE USE OF THE WIGNER THEOREM

forms of the spherical harmonics, Table 5.4. If we had, instead, used the more natural (for systems of spherical symmetry) complex forms, Table 5.3, we would have needed the equivalent (i.e. related by a similarly transformation) representations

$$D^{(0)} \equiv \Gamma^{(s)}, D^{(1)} \equiv \Gamma^{(p)}, D^{(2)} \equiv \Gamma^{(d)}, \text{ etc.} \quad (10.24a)$$

The 'D' notation is conventional for dealing with the complex-valued form of the spherical harmonic basis vectors:

$$\hat{P}_R Y_l^m(\theta, \phi) = \sum_{m'=-l}^{l} D_{m'm}^{(l)}(\hat{R}) Y_l^{m'}(\theta, \phi) \quad (10.24b)$$

In particular, the states $|Y_0^0\rangle \equiv |s\rangle$ and $\{|Y_1^0\rangle, |Y_1^{\pm 1}\rangle\} \equiv \{|p_3\rangle, (|p_1\rangle \pm i|p_2\rangle)/\sqrt{2}\}$ transform as

$$\hat{P}_R |Y_0^0(\theta, \phi)\rangle = D^{(0)}(\hat{R})|Y_0^0\rangle = |Y_0^0(\theta, \phi)\rangle \quad (10.24c)$$

$$\hat{P}_R |Y_1^m(\theta, \phi)\rangle = \sum_{m'=-1}^{+1} D_{m'm}^{(1)}(\hat{R}) |Y_1^{m'}(\theta, \phi)\rangle \quad (10.24d)$$

Now reconsider operators. Equation (10.21c) transcribes to the new notation as

$$\mathscr{H} \to \mathscr{H}' \equiv \hat{P}_R \mathscr{H} \hat{P}_R^+ = D^{(0)}(\hat{R})\mathscr{H} = \mathscr{H} \quad (10.25a)$$

And just as one can transform from the real-value states $\{|p_1\rangle, |p_2\rangle, |p_3\rangle\}$ to the complex-valued $\{|Y_1^0\rangle, |Y_1^{\pm 1}\rangle\}$, so also we define three new operators: $\hat{L}_0 = \hat{L}_3$, $\hat{L}_{\pm 1} = (\hat{L}_1 \pm i\hat{L}_2)/\sqrt{2}$, which behave as

$$\hat{L}_m \to \hat{L}'_m \equiv \hat{P}_R \hat{L}_m \hat{P}_R^+ = \sum_{m'=-1}^{1} D_{m'm}^{(1)}(\hat{R}) \hat{L}_{m'} \quad (10.25b)$$

Equations (10.25) say that $\mathscr{H}$ and $\{\hat{L}_m\}$ act under $\{\hat{R}\}$ exactly like the complex-valued spherical harmonics Y_0^0 and $\{Y_1^m\}$, respectively, which are the generators of the one- and three-dimensional irreducible representations of the group of the sphere.

There is great advantage in generalizing this idea:

DEFINITION 10.1 A set of $(2l+1)$ operators $\{\hat{A}_l^{-l}, \hat{A}_l^{-l+1}, \ldots \hat{A}_l^l\}$ which transform under $\{\hat{R}\}$ as

$$\hat{A}_l^m \to \hat{A}_l'^m = \hat{P}_R \hat{A}_l^m \hat{P}_R^+ = \sum_{m'=-l}^{l} D_{m'm}^{(l)}(\hat{R}) \hat{A}_l^{m'} \quad (10.26)$$

is called an *irreducible spherical tensor operator of rank l*.

The $D^{(l)}$ matrices are basically the same as the $\Gamma^{(l)}$ we have been using all along, only generated by complex-valued spherical harmonics rather than the real-valued variety.

Equation (10.26) defines an irreducible spherical tensor operator of rank l by saying, in essence, that $\{\hat{A}_l^m\}$ is something which behaves like $\{Y_l^m\}$. The $\{Y_l^m\}$, on the other hand, are eigenvectors of $\hat{L}$ and $\hat{L}_z$; the components of $\hat{L}$ (which, as we have seen, can themselves be put in the form of an irreducible spherical operator of rank 1) obey well-defined commutation relations, Eq. (10.15), as do the associated raising and lowering operators (which happen to be, of course, none other than $\hat{L}_{\pm 1}$ of Eq. (10.25b)!) Rigorous pursuit of this chain of associations leads[†] to the equivalent definition:

DEFINITION 10.2 A set of $(2l+1)$ operators $\{\hat{A}_l^{-l}, \hat{A}_l^{-l+1}, \ldots \hat{A}_l^l\}$ whose commutation relations with the angular momentum operators $\{\hat{L}_0, \hat{L}_{\pm 1}\}$ are

$$[\hat{L}_0, \hat{A}_l^m] = m\hbar \hat{A}_l^m \tag{10.27a}$$

$$[\hat{L}_{\pm 1}, \hat{A}_l^m] = \sqrt{\{(l \mp m)(l \pm m + 1)\}}\hbar A_l^{m \pm 1} \tag{10.27b}$$

constitutes an irreducible spherical tensor operator of rank l.

Since $\{\hat{L}_0, \hat{L}_{\pm 1}\}$ is itself an irreducible spherical tensor operator of rank 1, as such it could be written $\{\hat{L}_1^{\,0}, \hat{L}_1^{\,+1}, \hat{L}_1^{\,-1}\}$.

PROBLEM 10.18 What can one learn about $\hat{L}$ from Eq. (10.27) and the assumption that $\{\hat{L}_0, \hat{L}_{\pm 1}\}$ should be a tensor operator of rank 1?

The value of all of the above becomes apparent when one attempts to evaluate involved matrix elements of a difficult operator. If the operator can be expressed as a linear combination of irreducible spherical tensor operators, then the Wigner–Eckart Theorem leads to great simplifications.

PROBLEM 10.19 Show that in a cubic crystal, the electrical conductivity tensor $\bar{\bar{\theta}}$,

$$\bar{I} = \bar{\bar{\theta}} \bar{E}$$

reduces to a scalar. [HINT: If $\hat{R}$ is a symmetry operator, must the symmetry transformation $\theta' = R\bar{\bar{\theta}}R^{-1}$ leave $\bar{\bar{\theta}}$ invariant, $\bar{\bar{\theta}}' = \bar{\bar{\theta}}$?] This is an example of a situation where one could apply *Neumann's principle*: Any directly measurable physical property of a crystal must display A_1 symmetry (i.e. belong to the totally invariant representation of its point group).

PROBLEM 10.20 Graphite consists of carbon atoms, each of which is trigonally bonded to its three nearest neighbours, and only

---

[†] Rose[34], p. 82.

EXAMPLES OF THE USE OF THE WIGNER THEOREM   199

weakly attracted to C atoms on other planes. Can thermal conduction be isotropic in graphite?

PROBLEM 10.21 Did we use a form of Neumann's principle in our proof of the Wigner Theorem, Theorem 9.1?

†**PROBLEM 10.22** $\{Y_l^m\}$ are eigenvectors to $\hat{J}_z$ and the scalar operator $\hat{J}^2$. Does this automatically follow from the fact that for a system of spherical symmetry, $\{Y_l^m\}$ span the subspaces invariant with respect to the non-commuting vector operators $\hat{J}_x, \hat{J}_y, \hat{J}_z$?

†**PROBLEM 10.23** For a spherically symmetric system, the $\{\hat{L}_i\}$ are the generators of infinitesimal rotations, and the eigenvectors of $\mathcal{H}$ must also be eigenvectors to $\{\hat{L}^2, \hat{L}_3\}$. Is either $\hat{L}^2$ or $\hat{L}_3$ a Dirac character for the system?

†10.6 Clebsch-Gordan Coefficients and the Wigner-Eckart Theorem

In this final section, we offer a very brief introductory discussion first of Clebsch–Gordan coefficients, and then of the related Wigner–Eckart Theorem.

The study of Clebsch–Gordan coefficients takes us back to the direct-product spaces of Sections 5.12 and 9.5. Consider two sub-systems, denoted by numbers 1 and 2, each of which inhabits states of well-defined angular momentum $\bar{J}$, and of well-defined component $J_3 = \bar{J} \cdot |3\rangle$. The sub-systems will be in states $\{|j_1 m_1\rangle\}$ and $\{|j_2 m_2\rangle\}$, respectively, which are eigenvectors of both $\bar{J}$ and $\hat{J}_3$; for integral spin, the $|j_i m_i\rangle$ will be spherical harmonics. If the sub-systems interact, then the compound system may be described by the product states $\{|j_1 m_1\rangle|j_2 m_2\rangle\}$. Since m_i can assume integer values ranging between j_i and $-j_i$, there are $(2j_1 + 1)(2j_2 + 1)$ such degenerate† and orthonormal product states on the manifold of fixed j_1 and j_2. The problem is to find $(2j_1 + 1)(2j_2 + 1)$ degenerate, orthonormal states $\{|j_3 m_3\rangle\}$ for which the quantum numbers j_3 and m_3 refer to the total angular momentum of the *compound* system.

Assuming that $\{|j_3 m_3\rangle\}$ do exist, we expand a typical element $|j_3 m_3\rangle$ from this set as (with j_1 and j_2 fixed)

$$|j_3 m_3\rangle = \sum_{m_1 m_2} (|j_1 m_1\rangle |j_2 m_2\rangle)(\langle j_1 m_1|\langle j_2 m_2|)|j_3 m_3\rangle \quad (10.29\text{a})$$

$$= \sum_{m_1 m_2} C(j_1 j_2 j_3; m_1 m_2 m_3)|j_1 m_1\rangle|j_2 m_2\rangle$$

where the *Clebsch–Gordan* or *Wigner coefficients*

$$C(j_1 j_2 j_3; m_1 m_2 m_3) \equiv (\langle j_1 m_1|\langle j_2 m_2|)|j_3 m_3\rangle \quad (10.29\text{b})$$

†In the absence of external magnetic fields.

provide the needed unitary transformation in terms of some suitable scalar product. The evaluation of the coefficients for given j_1 and j_2 is straightforward (complete tables are also available), and discussed in the references mentioned below. It is easily shown, for example, that these angular momentum coupling coefficients vanish unless

$$m_3 = m_1 + m_2 \tag{10.29c}$$

$$j_3 = j_1 + j_2, \ j_1 + j_2 - 1, \ldots, |j_1 - j_2| \tag{10.29d}$$

where the latter equation is the 'triangle inequality' of the atomic vector model, Eq. (10.14). The coefficients may be normalized so as to be real, from which it follows that

$$C(j_1 j_2 j_3; m_1 m_2 m_3) = \langle j_3 m_3 | (|j_1 m_1\rangle | j_2 m_2 \rangle) \tag{10.29e}$$

They also display index-permutation symmetry properties such as

$$C(j_1 j_2 j_3; m_1 m_2 m_3) = (-1)^{j_1 + j_2 - j_3} C(j_2 j_1 j_3; m_2 m_1 m_3) \tag{10.29f}$$

PROBLEM 10.24 Find the transformation inverse to Eq. (10.29a).

Now to the Wigner–Eckart Theorem, which concerns itself with the evaluation of integrals such as

$$\langle j'm' | \hat{A}_j^m | j''m'' \rangle \tag{10.30}$$

Since the irreducible spherical tensor operator of rank j, $\{\hat{A}_j^m\}$, was defined in the first place as being composed of functions which transform like $\{Y_j^m\}$, Eq. (10.26), it should come as no surprise that the components $\hat{A}_j^m$ can themselves be used to create product spaces, and that these also obey

$$\hat{A}_{j_3}^{m_3} = \sum_{m_1 m_2} C(j_1 j_2 j_3; m_1 m_2 m_3) \hat{A}_{j_1}^{m_1} \hat{A}_{j_2}^{m_2} \tag{10.31a}$$

in direct analogy to Eq. (10.29a). The functions involved happen to be operators rather than states, but their transformation properties are the same; so the admixture coefficients $C(j_1 j_2 j_3; m_1 m_2 m_3)$ should be also.

Equation (10.31a) tempts us to press our luck and consider a hybrid space containing products of $\hat{A}_j^m$ functions and $|jm\rangle$ functions; the relevant coupling coefficients should still be the same as

$$\langle j_3 m_3 | (|j_1 m_1\rangle | j_2 m_2 \rangle) = C(j_1 j_2 j_3; m_1 m_2 m_3) \tag{10.31b}$$

If we replace $|j_1 m_1\rangle$ by $\hat{A}_{j_1}^{m_1}$, we are left with

$$\langle j_3 m_3 | \hat{A}_{j_1}^{m_1} | j_2 m_2 \rangle = C(j_1 j_2 j_3; m_1 m_2 m_3)) K \tag{10.31c}$$

where the K arises because $\hat{A}_{j_1}^{m_1}$ were (unlike the $\{|jm\rangle\}$) never normalized to unity. Interchanging the indices '1' and '2', and then making use of Eq. (10.29f), we arrive finally at the *Wigner–Eckart Theorem*:

$$\langle j_3 m_3 | \hat{A}_j^{m_2} | j_1 m_1 \rangle = C(j_1 j_2 j_3; m_1 m_2 m_3) \langle j_3 \| \hat{A}_{j\,2} \| j_1 \rangle \qquad (10.32)$$

The normalization constant of proportionality in Eq. (10.31c) becomes the *reduced matrix* element $\langle j_3 \| \hat{A}_{j,2} \| j_1 \rangle$, which is independent of m_1, m_2, or m_3! The entire 'geometry' of the problem is contained solely within the Clebsch–Gordan coefficients; the beauty of the theorem is that if one has a large number of integrals like $\langle j_3 m_3 | \hat{A}_j^{m_2} | j_1 m_1 \rangle$ to evaluate, with fixed j_1, j_2, and j_3, but numerous combinations of m's, one need find $\langle j_3 \| \hat{A}_{j2} \| j_1 \rangle$ only once, and let tables of Clebsch–Gordan coefficients do the rest of the work!

For a deeper and more rigorous examination of all the topics of this chapter, the reader is referred to the books of Tinkham[39], Joshi[22], Brink and Satchler[5], Edmonds[11], Merzbacher[31], and Rose[34]. All of these provide worked examples of tensor operators, and the Wigner–Eckart Theorem, as does Slichter[36]. Lipkin[28] presents a most readable account of the closely related topic of Lie algebras.

†PROBLEM 10.25 By the above analyses, one might expect that

$$|j_3 m_3 \rangle = \sum_{m_1 m_2} C(j_1 j_2 j_3; m_1 m_2 m_3) \hat{A}_j^{m_1} | j_2 m_2 \rangle \qquad (10.33)$$

is meaningful, and that the triangle inequality still obtains. In particular, let $\{\hat{A}_j^{m_1}\} = \{\hat{L}_1{}^0, \hat{L}_1{}^{+1}, \hat{L}_1{}^{-1}\}$; then j_3 should be allowed the values $j_2 + 1, j_2$, and $|j_2 - 1|$. As raising and lowering, operators in Eq. (10.33), however, $\hat{L}_1{}^{-1}$ and $\hat{L}_1{}^{+1}$ can alter m_2 but not j_2; consequently $j_3 = j_2$ only. What is the source of this restriction?

PROBLEM 10.26 Is Eq. (6.33) an example of the Wigner–Eckart Theorem?

10.7 Summary

1. The non-degenerate wave function $\psi^{(k)}(x)$ of an electron on a linear periodic lattice generates the kth one-dimensional irreducible representation of the translation group $\{\hat{R}_n\}$, and must be of the Bloch form.

2. The wave function of an electron in a potential of axial symmetry must be of the form $\psi(\vec{r}) = f(r, \theta) \exp(im\phi)$. Rotation of the system about the $\bar{n}$ axis is brought about via the operator $\hat{P}_{\phi,\bar{n}} = \exp\{-i\hat{J} \cdot \bar{n}\phi/\hbar\}$. In the case of *full* rotational symmetry, the spherical harmonics multiplied by appropriate radial functions span the invariant subspaces of odd dimensionality in Hilbert space.

3. Spinors generate the even-dimensional irreducible representations of the full rotation group. Calculations involving spin states and symmetry groups of finite order often are approached most conveniently by means of double groups.

4. Operators, as well as wave functions, may display symmetry properties. In particular, the concepts of scalar and vector operator lead into the useful generalization of irreducible spherical tensor operators, and then to the Wigner–Eckart theorem, Eq. (10.32).

CHAPTER 11

Perturbations Static and and Dynamic

One of the most useful applications of group representation theory to quantum mechanics is in the evaluation of the matrix elements of importance to perturbation theory calculations. Although group theory cannot reveal the magnitudes of such elements, still it can tell us when they must, for symmetry reasons alone, vanish. The way in which a static perturbation of low symmetry effects the degenerate states of a highly symmetric zero-th order Hamiltonian is illustrated with a C_{3v} example.

The level crossing theorem describes the conditions under which two energy eigenvalues may become degenerate and then cross as a parameter in the Hamiltonian, such as magnetic field or interatomic spacing, is varied.

We close with a worked example of the Jahn–Teller effect, in which a degenerate electronic level of a C_{3v} molecule couples with E-type vibrational modes, the net result being a spontaneous distortion of the molecule to a configuration of lower symmetry.

11.1 Why Integrals Vanish

The study of small time-independent and time-dependent perturbations centres on the evaluation of integrals such as

$$\int \psi_i^*(\bar{r}) F(\bar{r}) \psi_j(\bar{r}) \, d^3\bar{r} \tag{11.1}$$

Thus given a function $f(\bar{r})$, itself perhaps a product of other functions associated with a physical system, we wish to be able to predict when the integral $\int f(\bar{r}) \, d^3\bar{r}$ must, for symmetry reasons alone, vanish.

Let us consider an object with symmetry group $\{\hat{R}\}$, and denote by $f^{(\mu)}(\bar{r})$ a function having the symmetry corresponding to (behaving under $\hat{P}_R$ like) a basis vector for the μth irreducible representation of $\{\hat{R}\}$. According to Eq. (9.6) *any* function can be decomposed into such functions of specified symmetry, and for our $f(\bar{r})$ of interest

$$f(\bar{r}) = \sum_{\mu} f^{(\mu)}(\bar{r}) \tag{9.6}$$

Let $g(\bar{r})$ be another function in our space; the product $\langle g(\bar{r})|f(\bar{r})\rangle$ is a scalar quantity, invariant to any symmetry (and thus unitary) transformation $\hat{P}_R$:

$$\langle g(\bar{r})|f(\bar{r})\rangle = (\langle g(\bar{r})|P_R^+)(\hat{P}_R|f(\bar{r})\rangle) \tag{11.2a}$$

If we choose $g(\bar{r})$ to be a totally symmetric unit function, such that $g(\bar{r}) = \hat{P}_R g(\bar{r}) = 1$, then Eq. (11.2a) becomes

$$\int f(\bar{r}) \, d^3\bar{r} = \int \hat{P}_R f(\bar{r}) \, d^3\bar{r} \tag{11.2b}$$

Summing this over all $\hat{R}$ and interchanging sum and integral yields

$$\int f(\bar{r}) \, d^3\bar{r} = \frac{1}{N} \int \left(\sum_R^N \hat{P}_R\right) f(\bar{r}) \, d^3\bar{r} \tag{11.2c}$$

By Eq. (9.9), however,

$$\frac{1}{N} \sum_R^N \hat{P}_R = \mathscr{P}^{A_1} \tag{11.2d}$$

is the projection operator which precipitates out the totally symmetric portion of any function, and annihilates the rest. Therefore

$$\int f(\bar{r}) \, d^3\bar{r} = \int \mathscr{P}^{A_1}(f(\bar{r})) \, d^3\bar{r} = \int \mathscr{P}^{A_1}\left(\sum_{\mu} f^{(\mu)}(\bar{r})\right) d^3\bar{r} \tag{11.2e}$$

and if none of the $f^{(\mu)}$ belongs to A_1, then the integral vanishes. We have proven

THEOREM 11.1 The integral of $f(\bar{r})$ survives

$$\boxed{\int f(\bar{r}) \, d^3\bar{r} \neq 0} \tag{11.3}$$

only if $f(\bar{r})$ is the sum of parts at least one of which belongs to the totally symmetric irreducible representation A_1 of the group of the physical system involved.

PERTURBATIONS STATIC AND DYNAMIC 205

This is an immensely useful result; consider an integral such as

$$\langle \psi_i^{(\nu)}(\bar{r})^* | h(\bar{r}) | \psi_k^{(\mu)}(\bar{r}) \rangle = \int \psi_i^{(\nu)}(\bar{r})^* h(\bar{r}) \psi_k^{(\mu)}(\bar{r}) \, d^3\bar{r}$$

We examine the function $f(\bar{r}) = \psi_i^{(\nu)}(\bar{r})^* h(\bar{r}) \psi_k^{(\mu)}(\bar{r})$ by means, for example, of the projection operator of Eq. (9.9), and determine whether or not it contains components of totally symmetric form. If not, the integral vanishes. And of course this sort of thing can be done almost automatically via product spaces, Section 9.5. Given the functions $\{f^{(\alpha)}(\bar{r}), g^{(\beta)}(\bar{r}), h^{(\gamma)}(\bar{r}), i^{(\delta)}(\bar{r}) \ldots\}$, belonging to the various irreducible representations of a group, the method of product groups immediately gives the symmetry types of all products such as $f^{(\alpha)}(\bar{r})g^{(\beta)}(\bar{r})$ or $f^{(\alpha)}(\bar{r})g^{(\beta)}(\bar{r})h^{(\gamma)}(\bar{r})$. In the process, it reveals which of such products contain totally invariant A_1 components; the integrals of these particular products may be (but are not necessarily) non-vanishing.

PROBLEM 11.1 Electronic transitions are excited by time-varying electric or magnetic fields. To calculate electric and (neglecting spin) magnetic dipole transition rates, one must evaluate, respectively, integrals of the form $\langle \psi_i | \hat{r} \cdot \bar{n} | \psi_j \rangle$ and $\langle \psi_i | \hat{L} \cdot \bar{n} | \psi_j \rangle$, where $\bar{n}$ is a unit vector in real space pointing in the direction of the E or B field polarization, $\hat{r}$ and $\hat{L}$ are position and angular momentum (vector) operators, and $|\psi_i\rangle$ and $|\psi_j\rangle$ are electronic states. Consider a molecule of C_{3v} symmetry. What transitions are forbidden if (a) $\bar{n}$ is parallel to one side of the base triangle, (b) $\bar{n}$ is normal to one side of the triangle, (c) $\bar{n}$ is normal to the plane of the triangle?
[HINT : $\langle \psi_i | \hat{r} \cdot \bar{n} | \psi_j \rangle = \sum_k^3 n_k \langle \psi_i | x_k | \psi_j \rangle$, and see Problem 9.11.]

11.2 Perturbations

In Section 5.13, we examined the influence of the perturbation $\Delta \hat{V}$ upon the eigenstates $\{|k°\rangle\}$ and energies $\{E_k°\}$ of an unperturbed Hamiltonian $\mathcal{H}°$; the states and eigenenergies are shifted and/or split by $\Delta \hat{V}$, becoming eigensolutions $\{|k\rangle\}$ and $\{|E_k\rangle\}$ to $\mathcal{H} = \mathcal{H}° + \Delta \hat{V}$. In this section, we shall explore the connection between the influence of a perturbation upon a system, and the symmetries of the perturbation and states involved.

For small $\Delta \hat{V}$, the approximations Eq. (5.61) are valid:

$$|k\rangle = |k°\rangle - \sum_{n \neq k} \frac{\langle n° | \Delta \hat{V} | k° \rangle}{E_n° - E_k°} |n°\rangle + \cdots \quad (11.4a)$$

$$E_k = E_k° + \langle k° | \Delta \hat{V} | k° \rangle - \sum_{n \neq k} \frac{|\langle n° | \Delta \hat{V} | k° \rangle|^2}{E_n° - E_k°} + \cdots \quad (11.4b)$$

In case $|k°\rangle$ and $|m°\rangle$ are degenerate, then the singularities are removed from these expressions by diagonalizing $\Delta\hat{V}$ in the $k°-m°$ manifold—i.e. by rotating this subspace until off-diagonal terms vanish. In the process, $\Delta\hat{V}$ will break the degeneracy according to Eq. (5.64b) *unless* it should happen that

$$\langle m°|\Delta\hat{V}|m°\rangle = \langle k°|\Delta\hat{V}|k°\rangle \tag{11.4c}$$
$$\langle m°|\Delta\hat{V}|k°\rangle = \langle k°|\Delta\hat{V}|m°\rangle = 0 \tag{11.4d}$$

Thus the perturbation can split the level if it is able to distinguish physically between $|k°\rangle$ and $|m°\rangle$; this cannot happen if $\Delta\hat{V}$ has the full symmetry of $\mathcal{H}°$ (i.e. is of itself A_1 symmetry).

Similarly, in examining rates of perturbation-induced transitions between states, or in following the time development of perturbed systems, one must again consider integrals of the form

$$\langle k°|\Delta\hat{V}|n°\rangle \tag{11.4e}$$

The frequent use of Fermi's Golden Rule† provides a case in point.

The problem boils down, for all of the above, to determining the magnitude of integrals like $\langle k°|\Delta\hat{V}|n°\rangle$. Group theoretic arguments can tell us effortlessly when such matrix elements must, for symmetry reasons alone, vanish. And because it is sufficient in many situations simply to know whether or not a transition is forbidden, or whether or not a certain type of perturbation can never break a given degeneracy, the group approach may provide a total answer to the problem. In any case, the method can greatly reduce the labour involved by indicating in advance which integrals must vanish by symmetry requirements alone.

Consider, for example, a C_{3v} molecule with electronic A_1, A_2, and E states. Suppose we wish to determine if the particular matrix element $\langle\psi_1|\Delta\hat{V}|\psi_2\rangle$, of significance to some part of Eq. (11.4), vanishes when, say, $\langle\psi_1|$ is of type A_1, $|\psi_2\rangle$ belongs to E, and the perturbation is pure A_2. According to Table 11.1, reproduced here for convenience from Table 9.1, a product of A_1, E, and A_2 functions is itself $A_1 \times A_2 \times E = (A_1 \times A_2) \times E = A_2 \times E = E$. Or in shorthand notation, $\langle A_1|A_2|E\rangle = E$.

Table 11.1

$$C_{3v}: \begin{array}{l} A_1 \times A_1 = A_1 \\ A_2 \times A_2 = A_1 \\ A_1 \times A_2 = A_2 \\ A_1 \times E = E \\ A_2 \times E = E \\ E \times E = A_1 + A_2 + E \end{array}$$

†Merzbacher,[31] page 475.

Table 11.2

	$\langle\psi_1\|$	$\|\psi_2\rangle$	A_1	$\Delta\hat{V}$ A_2	E
I	A_1	A_1	×		
	A_2	A_2	×		
	A_1	A_2		×	
	A_2	A_1		×	
	A_1	E		▨	×
	A_2	E			×
	E	A_1			×
	E	A_2			×
II	E	E	×	×	×

C_{3v}

The integrand in $\int \psi_1^* \Delta\hat{V} \psi_2 d^3\bar{r}$ is thus of pure E symmetry, containing no A_1 component. By the results of the proceeding section. the integral *must* vanish! It's that simple!

In Table 11.2, we use the contents of Table 11.1 to evaluate in this fashion all possible triple products for the symmetrical functions of C_{3v}. The second column may be thought of as being the initial or unperturbed state in a matrix, in the sense of Eq. (11.4), the first as the final or admixed state, and the three right-hand columns as the three possible symmetry components of $\Delta\hat{V}$; while the states must be of pure symmetry, the total $\Delta\hat{V}$ need not be. We have indicated with an '×' the possible combinations of states and potentials which will yield a triple product consisting of or containing an A_1 part; and, we have also noted the case $A_1 \times A_2 \times E$ of our above $\langle\psi_1|\Delta\hat{V}|\psi_2\rangle$ example, which does *not* contain an A_1 contribution. The order in which the products are taken is immaterial and the triple product is associative, so the table could be put in more condensed form; as it stands, it contains much redundant information.

Table 11.2 separates nicely into two portions, denoted I and II. According to Eq. (11.4d), and the table, the only situation where an E state degeneracy is broken is where matrix elements such as $\langle E|\Delta\hat{V}|E\rangle$ are non-vanishing; in all other situations one need not worry about the degeneracy question. Therefore Part I of Table 11.2 is concerned with the effects of various perturbations upon (or transitions from) states when degeneracy-breaking is not an issue; Part II is discussed later.

We can now turn to specific problems involving time-dependent and time-independent perturbation theory. In Fig. 11.1 we illus-

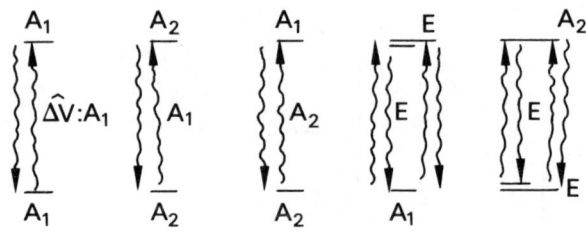

Figure 11.1.

trate for C_{3v} the possible transitions allowed to or from a non-degenerate level, and the symmetry of perturbation required for them to occur. It is important to note that a transition need not occur simply because it is not disallowed by symmetry considerations! Similarly, electric-dipole transitions from (to) an A_1 level need not occur at the same rates to (from) the $|x\rangle$-like and $|y\rangle$-like states in an E level.

As with transition probabilities, the effect of static perturbations upon non-degenerate levels presents no problems. Suppose we are interested in the effects of perturbations of various sorts upon an A_1 state $|A_1\rangle$. Table 11.2 and Eq. 11.4b say an A_1 $\Delta \hat{V}$ may cause a first order energy shift in $|A_1\rangle$; an A_2 potential, on the other hand can cause only a second-order shift, and in the process admixes into $|A_1\rangle$ states of A_2 symmetry. This may at first be worrysome—after all, shouldn't the states of a C_{3v} molecule be pure A_1, A_2, or E, and not mixed?

*PROBLEM 11.2 Resolve this dilemma. [HINT: What sort of state can an A_1 potential such as $\mathcal{H}^\circ$ admix with $|A_1\rangle$, $|A_2\rangle$, and $|E\rangle$, respectively, and why?].

Things can get a bit more complex when we consider the effects of time-independent perturbations upon degenerate states. With regard to Part II of Table 11.2, we must distinguish between two cases: either the two E states in question belong to different levels, Fig. 11.2a, or to the same level, Fig. 11.2b.

The first case is handled as above, and we need concern ourselves only with the second. By Table 11.2, an A_1, A_2, or E-type perturbation should be allowed to give non-vanishing $\langle k^\circ | \Delta \hat{V} | m^\circ \rangle$, and hence to split the level. And A_2 and E perturbations in reality may do so. But what about an A_1 $\Delta \hat{V}$, which displays the full symmetry of the unperturbed Hamiltonian $\mathcal{H}^\circ$? If such a $\Delta \hat{V}$ fails to lower the symmetry of our system, surely it cannot break a degeneracy, even though Table 11.2 says it is allowed to!

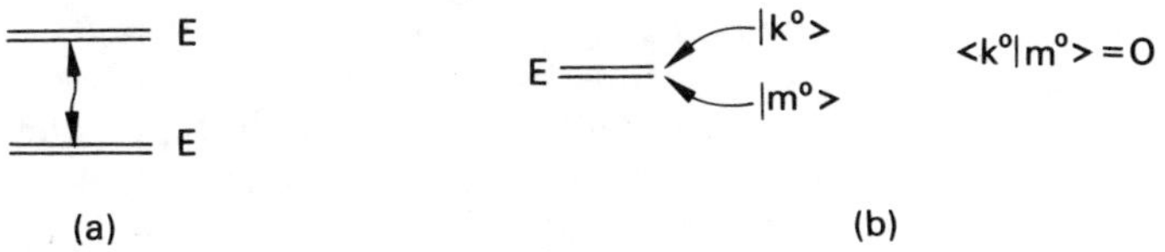

PERTURBATIONS STATIC AND DYNAMIC

```
         E                        ___ |k°>
    ⊥                    E ═══ ⟨                  <k°|m°> = 0
         E                        ‾‾‾ |m°>

        (a)                          (b)
```
Figure 11.2.

*PROBLEM 11.3: Resolve this dilemma also. [HINT: Why does $\langle k°|\mathscr{H}°|m°\rangle = 0$? Will a symmetric perturbation break the E-level degeneracy?

We shall pursue further the problems of degeneracy and symmetry reductions in the following sections. First we follow the splitting of levels of a highly symmetrical (spherical) system with the application of a lower (C_{3v}) perturbation; then we examine the conditions under which two levels can cross accidentally as a parameter in the Hamiltonian (such as magnetic field) varies; and finally we see how electron-phonon coupling can lead to a spontaneous reduction of symmetry and degeneracy-breaking in a C_{3v} molecule, by means of the so-called Jahn–Teller effect.

*PROBLEM 11.4 Show that if $|\psi^{(\mu)}\rangle$, $|\psi^{(\eta)}\rangle$, and $F^{(\nu)}$ are functions belonging, respectively, to the irreducible representations $\Gamma^{(\mu)}$, $\Gamma^{(\eta)}$, and $\Gamma^{(\nu)}$ of some group, then†

$$\langle \psi^{(\mu)}|F^{(\nu)}|\psi^{(\eta)}\rangle = 0$$

if $\Gamma^{(\nu)} \times \Gamma^{(\eta)}$ does not contain $\Gamma^{(\mu)}$.

11.3 Reduction of Symmetry

In the last section, we discussed a method of examining the influence of a small $\Delta \hat{V}$ upon a system, based upon study of the matrix elements of perturbation theory. Here we outline a more general attack, valid even when $\Delta \hat{V}$ is comparable in magnitude to $\mathscr{H}°$. This method stems directly from the Wigner Theorem, rather than upon the matrix element theorem; one should bear in mind, however, that the two approaches are closely related, and that in many situations, both are applicable. But the type of information obtainable is not exactly the same.

We shall examine the consequences of adding a C_{3v} perturbation to a system whose Hamiltonian was initially spherically symmetric. Our development bears a close resemblance to the subject

---

†If $|\psi^{(\mu)}\rangle$ belongs to $\Gamma^{(\mu)}$, then $\langle \psi^{(\mu)}|$ belongs, strictly speaking, to the complex conjugate representation $\Gamma^{(\mu)*}$. In general $\Gamma^{(\mu)} = \Gamma^{(\mu)*}$, and a distinction must be made only for certain one-dimensional irreducible representations of complex-valued character. See, for example, Schonland,[35] page 99.

matter of Chapter 6, namely the formation of a reducible representation of C_{3v} with spherical harmonic basis vectors—the philosophy, however, is quite different. Here we wish to learn how a lower symmetry perturbation influences the properties of an electron originally in an eigenstate of a Hamiltonian which happens to display high (spherical) symmetry.

† PROBLEM 11.5 Consider a $(2l+1)$-dimensional irreducible representation of the full rotation group of the sphere. Show that the character corresponding to a rotation through ϕ about any axis is

$$\chi^{(l)}(\phi) = \sum_{m=-l}^{l} \exp(im\phi) \tag{11.5}$$

(see Problem 8.15 and Eq. (10.13)). Use this to find the characters for the representation of C_{3v} generated by d-states; does it matter whether we use real or complex-valued d-states? Recall $\hat{R}_p = \hat{M}\hat{R}_{180}$, Problem 1.4.

Let us consider the seven f-state functions, spherical harmonics of order three, $\{Y_3^m, m = -3, -2, \ldots +3\}$. These form a set of mutually orthonormal vectors which generate one particular irreducible representation of the full rotation group, and mix freely among themselves (but with spherical harmonics of no other order!) under the operations of that group. What would happen to these seven functions if the spherically symmetric object they describe were placed in an environment of reduced symmetry? For example, what would happen if a spinless Bohr atom with $l = 3$ were placed at the centre of an equilateral triangle of three equal charges?

C_{3v} is known to have only three kinds of irreducible representations, two one-dimensional types, A_1 and A_2, and one two-dimensional representation, E. Thus all seven of the original functions must somehow be separated into, or be modified so as to assume the symmetry types of, the three new types of invariant subspaces. The altered functions now transform in accordance with the three newly allowed symmetries.

How this happens can be seen as follows. We use the seven Y_3^m vectors to form a seven-dimensional representation of the group of the sphere, the symmetry group of the unperturbed system. All rotations through the same angle ϕ (but about any axis) are in the same class; there will be, therefore, an infinite number of such seven-by-seven matrices $\bar{\bar{\Gamma}}^{(f)}(\hat{R}_\phi)$ for the sphere, one for each axis and value of ϕ. From this set, and the analogous set representing reflections, we extract six matrices corresponding to the six particular elements of C_{3v}. But this six-matrix representation, now

PERTURBATIONS STATIC AND DYNAMIC

describing C_{3v}, must be reducible, since the irreducible representations of the triangle are only one-, one-, and two-dimensional, respectively. (The group of the sphere, of course, *can* have a seven-dimensional irreducible representation.) By the normal procedure of reducing a representation (rotating the seven-dimensional f-state function subspace into an optimal orientation which leaves a block-diagonal form), we find that each of the six seven-by-seven C_{3v} matrices is broken into one one-dimensional sub-block of type A_1, two one-dimensional blocks of type A_2, and two E blocks

$$\Gamma^{(f)} \to A_1 + 2A_2 + 2E \qquad (11.6)$$

as, for example, in Fig. 11.3a.

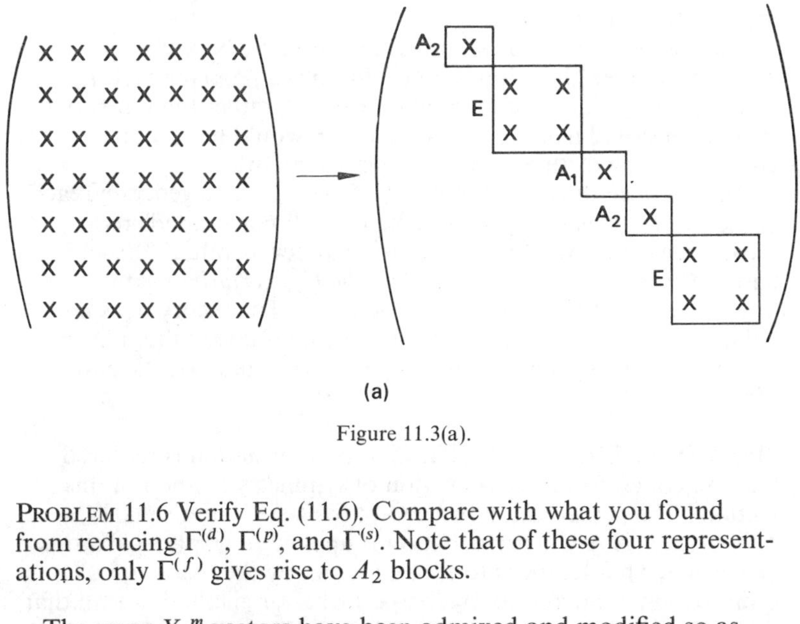

(a)

Figure 11.3(a).

PROBLEM 11.6 Verify Eq. (11.6). Compare with what you found from reducing $\Gamma^{(d)}$, $\Gamma^{(p)}$, and $\Gamma^{(s)}$. Note that of these four representations, only $\Gamma^{(f)}$ gives rise to A_2 blocks.

The seven Y_3^m vectors have been admixed and modified so as to become the generators of: one A_1 totally symmetric, one-dimensional invariant subspace, two A_2 one-dimensional spaces of a lower symmetry, and two sets of doubly-degenerate E states (Fig. 11.3b). Thus the process of reducing the seven-dimensional representation is equivalent to rotating and separating Y_3-space into five new subspaces of three general categories of symmetry, such that each new subspace is invariant under the symmetry operations of C_{3v}.

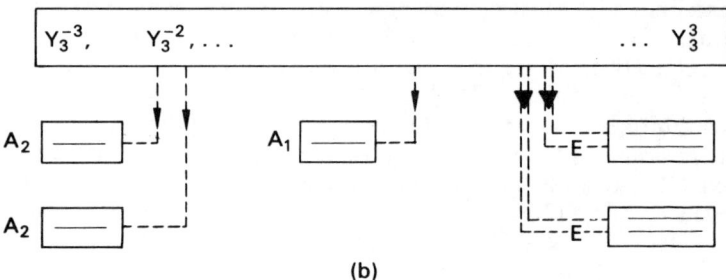

(b)

Figure 11.3(b).

Two related questions arise immediately: are the four E states, say, degenerate, and do they mix under $\{\hat{P}_R\}$ of C_{3v}? After all, they transform similarly under C_{3v}, and they arose originally from seven degenerate states, the $\{Y_3{}^m\}$, which *do* mix together under *any* rotation or reflection.

The answer to both questions is an emphatic NO. If they were mixed, then the admixture coefficients $\Gamma_{ij}^{(\mu)}(\hat{R})$ for each $\hat{R}$ would reside within one four-by-four block, rather than in two two-dimensional matrices; this, however, would be completely contrary to the dictate of the Wigner Theorem!

And so, excluding the possibility of accidental degeneracy, each of the five C_{3v} basis sets derived from $\{Y_3{}^m\}$ corresponds to a separate energy level. Our seven original degenerate $Y_3{}^m(\theta, \phi)$ basis vectors are *split and altered by the C_{3v} perturbation* into five levels, two of which are doubly degenerate. The theory does *not* tell us the ordering of the states or the magnitude of the splitting, only the possible general kinds of perturbed state, A_1, A_2, and E, of C_{3v}.

*PROBLEM 11.7 In general, what kind of information is required for, and comes from, the reduction of symmetry method of this section? The perturbation method of the last section?

*PROBLEM 11.8 A C_{3v} perturbation is applied to a spherical atom. Determine what happens to $l=1$ states (a) by the reduction of symmetry method and (b) by the perturbation method. Recall that real and complex p-states generate equivalent irreducible representations of C_{3v} and of the group of the sphere. What happens to the symmetries of states as the C_{3v} potential reduces to zero? What happens to the p-states if in addition to the C_{3v} perturbation, an electric field is applied along $|3\rangle$? Along $|1\rangle$?

PROBLEM 11.9 In the development of Chapter 6 and what followed, we made use of the fact that atomic orbitals corresponding to different angular momentum quantum number are orthogonal. Why does this orthogonality occur?

PERTURBATIONS STATIC AND DYNAMIC

*PROBLEM 11.10 What happens physically to $\{|p_1\rangle, |p_2\rangle, |p_3\rangle\}$ states when a hydrogen atom is placed in an octahedral (symmetry of a cube) electrostatic field? If the cube is squashed a bit along one four-fold axis of rotation? If the cube is compressed along a three-fold axis?

*PROBLEM 11.11 Atomic d-states are split into a triplet and a doublet by an octahedral field. Why? (Draw a diagram of the orbitals and the environment.) The energy splitting is sometimes called '10 Dq'.

PROBLEM 11.12 What happens to a C_{3v} molecule when placed in a magnetic field aligned along $|3\rangle$? Does the polarity of the field reverse under reflection? If so, would this affect the Hamiltonian?

PROBLEM 11.13 Figure 11.4 reveals that C_{3v} is a subgroup of 0, the symmetry group of the cube (see Problem 7.17); a $\hat{R}_{120}$ axis and a mirror plane shared by both groups are indicated. Suppose three degenerate electronic states which transform like x, y, z (i.e. like p_x-, p_y- and p_z-states) form an irreducible representation of 0, named T_1. Characters and typical generating functions for T_1, and the other irreducible representations of 0, are presented in Table 11.3; $\hat{R}_{90}$ and $\hat{R}'_{180}$ refer to classes of symmetry operators found in 0 but not C_{3v}. The first three characters in the T_1 row define a representation of C_{3v}, but this happens to be reducible. Show that when a cubic crystal is compressed along a body diagonal (i.e. along a (111) direction), then a T_1 level of a centrally trapped ion is split into an A_2 and an E level of C_{3v}. What happens to a T_2 level? to E? to any nondegenerate state such as A_2?

Table 11.3

0	$\hat{I}$	$\hat{R}_{120}$	$\hat{M}_I$	$\hat{R}'_{180}$	$\hat{R}_{90}$	
A_1	1	1	1	1	1	$(x^2+y^2+z^2)$
A_2	1	1	−1	1	−1	
E	2	−1	0	2	0	$(x^2-y^2, 3z^2-r^2)$
T_1	3	0	−1	−1	1	(x, y, z)
T_2	3	0	1	−1	−1	(xy, yz, zx)

†11.4 Level Crossing Theorem†

The solid curve in Fig. 11.5 is an enlargement of that portion of Fig. 9.1 in the region of point C, where the levels of states $|\gamma\rangle$ and $|\delta\rangle$ cross. In some situations, as a physically significant parameter (in this case the magnetic field H) and consequently some portion

†See also Landau and Lifshitz,[26] pages 277–282, and Davydov,[8] pages 175–178.

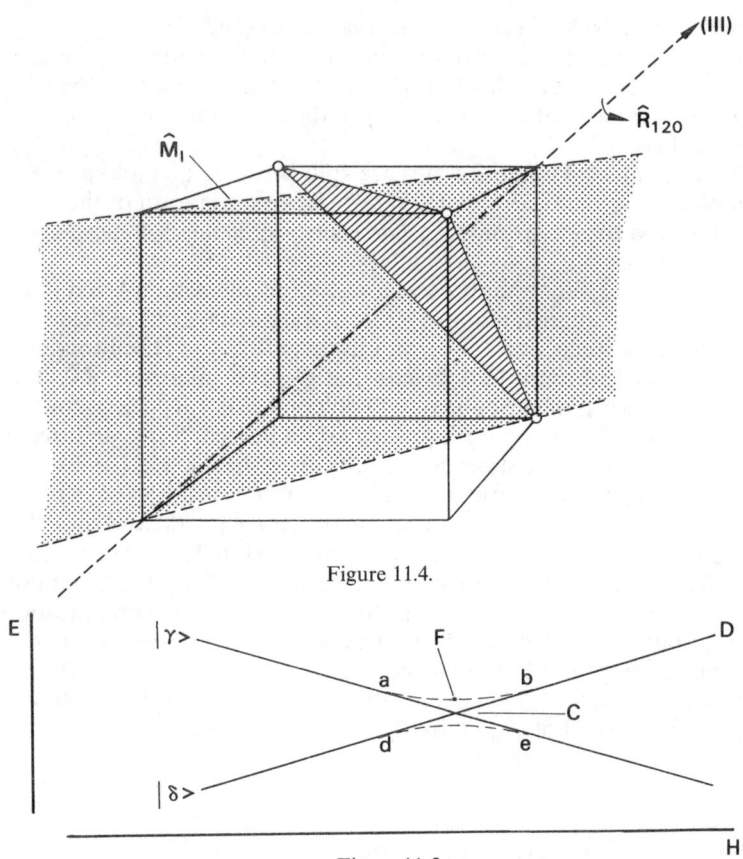

Figure 11.4.

Figure 11.5.

of the Hamiltonian varies, the energies of two levels will cross, and we have an accidental degeneracy. But if the two levels in question, $|\gamma\rangle$ and $|\delta\rangle$, are of the same symmetry type, then the *Level Crossing Theorem* states that no crossing will occur; the lines *a–e* and *b–d* of Fig. 11.5 will be replaced by the 'mutually repelling' broken curves *a–b* and *d–e*. This occurs because $|\gamma\rangle$ and $|\delta\rangle$ have no *a priori* reason to be orthogonal; thus nothing prevents terms in the Hamiltonian from having non-vanishing off-diagonal matrix elements between these states of the same symmetry, and $\langle\gamma|\mathcal{H}|\delta\rangle \neq 0$. Equation (11.4) says that this situation lifts the degeneracy at *C*, and if two levels cannot become degenerate, they clearly cannot cross. Thus energy levels of the same symmetry do not cross with variation of a physical parameter such as *H*.

It may even be that levels corresponding to states of different symmetry are not allowed to cross. Suppose there exists a small *normally ignored* term $\Delta\mathcal{H}$ of low symmetry in the Hamiltonian due, for example, to elastic distortions of the system. Then the matrix element $\langle \delta | \Delta \mathcal{H} | \gamma \rangle$ may be non-vanishing even with $|\delta\rangle$ and $|\gamma\rangle$ of different symmetry types, and the system will be described by the curves a–b and d–e, rather than cross.

But a warning. Occasionally, the non-crossing rule breaks down. For example, let the parameter H of Fig. 11.5 be replaced by r, the interatomic separation in a di-atomic molecule; Fig. 11.5 is now called a *configuration-coordinate diagram*. If the molecule is vibrating sufficiently rapidly back and forth through the C region, such that the kinetic energy of the system far exceeds the expected energy level splitting at C, then the system may follow the paths a–e or b–d, even though such a crossing should by symmetry considerations be forbidden.

PROBLEM 11.14 Assume that $|\gamma\rangle$ and $|\delta\rangle$ of Fig. 11.5 are prevented by symmetry arguments from crossing. Will a state at the point D have the symmetry of $|\gamma\rangle$ or of $|\delta\rangle$? At the point F?

11.5 Born–Oppenheimer Approximation

Before discussing the possibility of coupling the electronic and vibrational modes of a system, we shall offer a brief account of the Born–Oppenheimer, or 'adiabatic' approximation. As a specific example, we consider a planar C_{3v} molecule.

The total Hamiltonian for a C_{3v} molecule is of the form

$$\mathcal{H} = \sum_i (\mathrm{KE})_i + \sum_\alpha (\mathrm{KE})_\alpha + \sum_{i,\alpha} V(Q_\alpha, \bar{r}_i) \tag{11.7}$$

where i in the first summation labels the ith electron at position $\bar{r}_i$, and the second sum represents the total kinetic energy (KE) arising from the motions of all the nuclei in the system (i.e. the vibrational kinetic energy of our molecule). The third sum is the net electron-electron, nuclear-nuclear, and electron-nuclear interaction energy, a function both of the $\bar{r}_i$ and of $\{Q_\alpha\}$, the system's generalized-coordinate components, Fig. 11.6; this portion of the Hamiltonian accounts for any changes in the system's potential energy arising from slight molecular distortions, and $\bar{Q} = \sum_\alpha^3 Q_\alpha |Q_\alpha\rangle$ describes such distortions in terms of the normal modes. Unless noted otherwise, we shall assume these distortions to be static rather than oscillatory.

Throughout our development, we have been making the tacit assumption that a molecule or solid is a static system with a well-defined and permanent symmetry. We have essentially been assuming the Born–Oppenheimer approximation, which says that

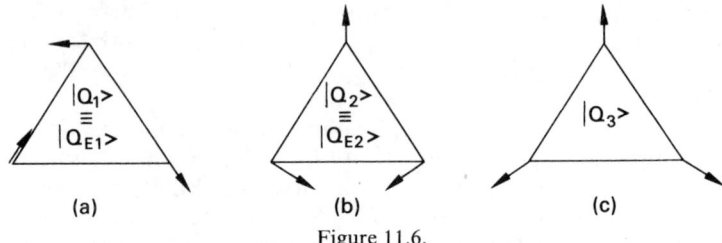

Figure 11.6.

because the nuclei are massive, molecular vibrations are small and slow; the kinetic energy of the nuclei and deviations from perfect geometric symmetry may therefore be neglected, to a first approximation, in a consideration of electronic processes. The electrons can be thought of as constituting a separate system subject to a slowly varying external nuclear potential; the nuclei in turn experience an effective potential which depends on the averaged-out motion of the electrons.

The assumption is usually valid, and the normal starting point for molecular or solid state calculations of electronic properties. The Hamiltonian of Eq. (11.7) reduces to

$$\mathcal{H}_e = \sum_i (\text{KE})_i + V(\bar{Q}, \bar{r}_i) \tag{11.8a}$$

which corresponds to a molecule with the nuclei clamped in a configuration described by generalized coordinate components $\{Q_1, Q_2, Q_3\}$. The solutions to the eigenequation of this essentially electronic Hamiltonian are of the form

$$\psi_e(\bar{r}_i, \bar{Q}), \quad E_e(\bar{Q}) \tag{11.8b}$$

The eigenvector $\psi_e(\bar{r}_i, \bar{Q})$ describes the state of the electron at $\bar{r}_i$, but is, like $E_e(\bar{Q})$, parameterized by fixed values of the components of the normal coordinates Q_α. Thus one chooses a particular set $\{Q_1, Q_2, Q_3\}$, solves the electronic eigenequation of Eq. (11.8a), chooses another set of values $\{Q_\alpha\}$, solves the eigenequation anew, and so on.

Having found $\psi(\bar{r}_i, \bar{Q})$ and $E_e(\bar{Q})$, one can then turn to the related problem of nuclear motion. Using electron-nuclear product states

$$\psi_e(\bar{r}_i, \bar{Q})\psi_n(\bar{Q}) \tag{11.9a}$$

as trial solutions in the eigenequation of the complete Hamiltonian Eq. (11.7), and with suitable manipulations and approximations, there drops out an equation describing *nuclear* motion (i.e. vibrational states) in terms of the electronic eigenenergies which we have, presumably, already determined:

$$\sum_\alpha (\text{KE})_\alpha + E_e(\bar{Q})\psi_n(\bar{Q}) = E_n \psi_n(\bar{Q}) \tag{11.9b}$$

Thus $E_e(\bar{Q})$ represents not only an electronic energy level for the static molecule, but also an effective elastic potential binding the molecule together. This is reasonable, since both $E_e(\bar{Q})$ and the binding energy are determined by the same interactions of electrons and nuclei. For small variations from equilibrium, the elastic potential of Eq. (11.9b) can be Taylor expanded as

$$E_e(\bar{Q}) = E(0) + \tfrac{1}{2} \sum_\alpha^3 K_\alpha Q_\alpha^2 \tag{11.9c}$$

using three vibrational normal coordinates in the case of C_{3v}. With any distortion, Hooke law forces elevate the net potential energy of the molecule in accordance with Eq. (11.9c).

The neglect of the nuclear kinetic energy in Eq. (11.8a), and then the use of the separable $\psi_e \psi_n$ product state to give Eq. (11.9b), are the basic ingredients in the Born–Oppenheimer method; and this completes our discussion of it. But before turning to the Jahn–Teller effect, we can say a bit more about the form assumed by the potential $\sum_i V(\bar{Q}, \bar{r}_i)$. For a C_{3v} case in which only a single electron at $\bar{r}$ is of interest, we Taylor expand the potential about equilibrium, $Q_\alpha = 0$ for all α; such an expansion is valid for sufficiently small Q_α.

$$V(\bar{Q}, \bar{r}) = V_0(\bar{r}) + \sum_\alpha^3 \frac{\partial V}{\partial Q_\alpha} Q_\alpha + \cdots \tag{11.10a}$$

$V_0(\bar{r})$ does not concern us, and will henceforth be ignored. The $\partial V/\partial Q_\alpha$ are functions of $\bar{r}$, and it will be of value to us to learn their transformational properties.

As a first approach, we might argue that $V(\bar{Q}, \bar{r})$, as a part of the Hamiltonian, has full C_{3v} symmetry. But $\bar{Q}_\alpha \neq 0$ means (except for Q_3, the fully symmetric or 'breathing' mode) precisely that the molecule is *distorted*, and no longer of C_{3v} symmetry! One can only say that $V(Q, \bar{r})$ has C_{3v} symmetry to lowest order. Let us assume, for the moment, that this is adequate.

We know from Section 9.8 and Problem 9.25 that the $|Q_\alpha\rangle$ transform under C_{3v} like the generators of A_1 and E irreducible representations. The totally symmetric A_1 vibrational mode does not reduce the symmetry of the molecule, cannot break any degeneracy of electronic states, and is not of further interest to us. To emphasize the fact that $|Q_1\rangle$ and $|Q_2\rangle$ correspond to a pair of degenerate E-type vibrational modes, we shall relabel them as $|Q_{E1}\rangle$ and $|Q_{E2}\rangle$. $\{|Q_{E1}\rangle, |Q_{E2}\rangle\}$ transform like the generators of an irreducible representation of C_{3v}; thus they must behave under the six transformations of C_{3v} like $\{x, y\}, \{x^2 - y^2, 2xy\} \ldots$, etc. Although we may not know which of such function pairs is most suitable for representing the vibrational modes of a particular

C_{3v} molecule, we can rest assured that any of them at least conforms with the constraints imposed by symmetry requirements.

Suppose, by way of example, that we choose to associate $\{|Q_{E1}\rangle, |Q_{E2}\rangle\}$ with $\{x, y\}$, as in the development of Section 9.8. The requirement that the potential

$$V(\bar{Q}, \bar{r}) = \frac{\partial V}{\partial Q_{E1}} Q_{E1} + \frac{\partial V}{\partial Q_{E2}} Q_{E2}$$

$$= \frac{\partial V}{\partial Q_{E1}} x + \frac{\partial V}{\partial Q_{E2}} y \qquad (11.10b)$$

be invariant to C_{3V} means that $\partial V/\partial Q_{E1}$ and $\partial V/\partial Q_{E2}$ themselves transform something like x and y; for then $V(\bar{Q}, \bar{r}) \cong (x^2 + y^2)$, which is indeed invariant to C_{3v}. Thus $V(\bar{Q}, \bar{r}) = c(xQ_{E1} + yQ_{E2})$ displays suitable symmetry behaviour.

Exactly as above, but assuming vibrational states $\{|Q_{E1}\rangle, |Q_{E2}\rangle\}$ which look like $\{x^2 - y^2, -2xy\}$ instead of $\{x, y\}$, then

$$V(\bar{Q}, \bar{r}) = c'(x^2 - y^2)Q_{E1} + c'(-2xy)Q_{E2} \qquad (11.10c)$$

is appropriate. In Section 11.7, it will be convenient to use this particular *approximation*, Eq. (11.10c), for $V(\bar{Q}, \bar{r})$ in Eq. (11.8a). $\mathcal{H}$ containing this $V(\bar{Q}, \bar{r})$ is not the *exact* Hamiltonian; the best we could do was to find a $V(\bar{Q}, \bar{r})$ which at least displays the correct symmetry properties under C_{3v}. This, however, will be sufficient!

PROBLEM 11.15 Find the transformational properties of the $\partial V/\partial Q_\alpha$ by examining $V(\bar{Q}, \bar{r})$ in the limit $\bar{Q} \to 0$. Note that $\partial V/\partial Q_\alpha$ is to be evaluated at $Q_\alpha = 0$ in the Taylor series, and that $\partial V/\partial Q_\alpha$ is independent of the magnitude of Q_α.

PROBLEM 11.16 What happened to the $\bar{r}$ dependence of $V(\bar{Q}, \bar{r})$ in Eq. (11.10c)? (i.e. When we say that $|Q_{E1}\rangle$ and $|Q_{E2}\rangle$ transform like $\{x, y\}$, x and y do *not* necessarily refer to the coordinates of the electron.)

PROBLEM 11.17 $E_e = E_e(\bar{Q})$ because of the existance of $V(\bar{Q}, \bar{r})$ in Eq. (11.8a). In Eq. (11.9c), the Taylor series expansion of $E_e(\bar{Q})$ must include the second order term; why then, can we get away with only first order terms in the expansion Eq. (11.10a) of $V(\bar{Q})$?

†11.6 The Jahn–Teller Theorem

THEOREM 11.2 *The Jahn–Teller Theorem*. A non-linear molecule tends to be unstable in a degenerate electronic state, simple Kramer's degeneracies excepted.

Consider a situation in which symmetry arguments and estimates of the ordering of energy levels suggest that a system should

PERTURBATIONS STATIC AND DYNAMIC

be in a degenerate (but not a Kramers spin-degenerate) electronic state. Then according to the Jahn-Teller Theorem, unless the system is a linear molecule, this electronic state will couple with vibrational modes in such a fashion that the system will spontaneously distort in shape. The molecule, now distorted, will be described by a new (and smaller) group of symmetry operators. The Wigner Theorem will no longer apply to the original, undistorted group, but only to the new system of lower symmetry, and our state of interest should no longer be degenerate. If the electronic-vibrational coupling is sufficiently strong, a Jahn-Teller instability will lead to experimentally observable effects.

In the next section we work through a C_{3v} example in some detail, but we illustrate the effect qualitatively here with the square planar system of symmetry D_{4h}, Fig. 11.7a. Suppose we have a molecule consisting of a central ion to which is bound a single p_x-electron and a square frame of four neutral atoms which *do not interact* with the p_x-electron. In Fig. 11.8a we plot the potential energy of the frame plus the total energy of the central ion as a function of the magnitude of a B_1 type distortion, Fig. 11.7b. This distortion can be a natural consequence of vibrations,† or caused by external static forces; note that for $Q<0$, the directions of the arrows in Fig. 11.7b are reversed. The curvature in Fig. 11.8a is a consequence of the Hooke-law forces binding the molecule (in particular, the frame) together.

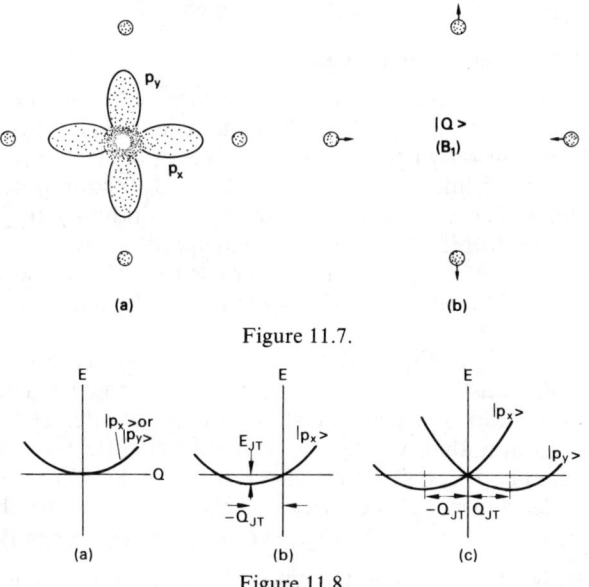

Figure 11.7.

Figure 11.8.

†In which case we subtract off and forget the nuclear kinetic energy.

Now consider a case when each of the atoms of Fig. 11.7a becomes negatively ionized. We define, again, the zero of energy to correspond to $Q=0$. With a $Q>0$ distortion, the overlap of a p_x-electron with neighbouring ions will increase, and consequently the electrostatic potential energy of the system increases also. The minimum in the energy curve will not be at $Q=0$, but rather at some $Q_{JT}<0$, Fig. 11.8b. Thus if the system is sitting initially at $Q=0$, and if there exists a mechanism for the p_x- electron to dissipate energy, the system will spontaneously settle into a Jahn-Teller distorted configuration, lowering the total energy by the amount E_{JT}, the Jahn-Teller distortion energy.† If the system contains a p_y-electron instead of one in $|p_x\rangle$, the energy minimum is shifted to $Q_{JT}>0$. The two possibilities are displayed together in Fig. 11.8c which shows that in a system sitting statically with Q fixed at either $+|Q_{JT}|$ or $-|Q_{JT}|$, the p-degeneracy is broken.

After a spontaneous Jahn-Teller distortion, which lowers the symmetry of the system from that of a square to that of a rectangle, the p_x- and p_y-states become, respectively, $|p'_x\rangle$ and $|p'_y\rangle$. If E_{JT} is small, the system can tunnel from one state to the other, and, at finite temperatures, jump back and forth via thermal activation; for large E_{JT}, the distortion is frozen permanently at either $+|Q_{JT}|$ or $-|Q_{JT}|$.

PROBLEM 11.18 $|p'_x\rangle$ centred at $-|Q_{JT}|$ and $|p'_y\rangle$ at $+|Q_{JT}|$ have the same energy. Is this an accidental degeneracy?

† 11.7 A C_{3v} Jahn–Teller Example

Here we work through a very simple example in which electron-phonon coupling in a C_{3v} system leads to a Jahn-Teller distortion. We shall draw heavily upon the results of Section 11.5, and in particular, we consider the way in which the distortion potential $V(\bar{Q}, \bar{r})$ affects electronic states. But instead of finding $\psi_e(\bar{r}, \bar{Q})$ by solving the electronic Hamiltonian eigenequation with various fixed values of the $\{Q_\alpha\}$, we shall rather use $V(Q, \bar{r})$ and the states $\psi_e(\bar{r}, 0) \equiv \psi(\bar{r})$ of the undistorted system in a perturbation calculation.

Like the $\partial V/\partial Q_\alpha$ of Eq. (11.10), the *zero*th order electronic wave functions $\psi(\bar{r})$ must conform to symmetry requirements. The non-degenerate A_1 and A_2 electronic states are not of interest to us, and so we assume that our system is in a degenerate E-state whose angular dependence is, say, that of $|x\rangle$ or $|y\rangle$ under C_{3v}. In employing first-order degenerate perturbation theory to evaluate the shift in such states due to the $V(\bar{Q}, \bar{r})$ of Eq. (11.10a), we must

†Equivalently, when we 'turn on' the negative charges of the frame, two ions will be repelled by the p_x-electron, and the system will find a new equilibrium at some $Q_{JT}<0$.

PERTURBATIONS STATIC AND DYNAMIC

diagonalize the matrix representation of $\partial V/\partial Q_\alpha$ generated by $|x\rangle$ and $|y\rangle$. To generate this perturbation matrix with our degenerate electronic states in the first place, by Eq. (11.10c) we must evaluate terms such as† $\langle x|(x^2-y^2)|x\rangle$ and $\langle x|(-2xy)|y\rangle$; this, of course, is why we sought out the symmetries of $V(\bar{Q}, \bar{r})$ and the electronic states. Converting to polar coordinates, $x = r \cos \phi$ and $y = r \sin \phi$, then

$$\langle x|x^2-y^2|x\rangle = f(r)\int_0^{2\pi} \sin^2\phi \,(\cos^2\phi - \sin^2\phi)\,d\phi = \frac{\pi}{2}f(r)$$

$$\langle x|-2xy|y\rangle = -2f(r)\int_0^{2\pi} \cos^2\phi \sin^2\phi \,d\phi = \frac{\pi}{2}f(r) \qquad (11.11a)$$

etc.

where the radial factor has been separated out; the matrix representation of $V(\bar{Q}, \bar{r})$ assumes the form (constant β)

$$V = \beta Q_{E1}\begin{pmatrix} 1 & 0 \\ 0 & -1 \end{pmatrix} + \beta Q_{E2}\begin{pmatrix} 0 & 1 \\ 1 & 0 \end{pmatrix} \qquad (11.11b)$$

Diagonalization of the secular determinant

$$\begin{vmatrix} \beta Q_{E1} - \Delta E & \beta Q_{E2} \\ \beta Q_{E2} & -\beta Q_{E1} - \Delta E \end{vmatrix} = 0 \qquad (11.11c)$$

leads to the energy splitting of $|x\rangle$ and $|y\rangle$ brought about by $V(\bar{Q}, \bar{r})$:

$$\Delta E = \pm \beta (Q_{E1}^2 + Q_{E2}^2)^{1/2} \qquad (11.12a)$$

Defining *abstract* coordinates X and Θ by means of

$$Q_{E1} = X \cos \Theta$$
$$Q_{E2} = X \sin \Theta \qquad (11.12b)$$

the total energy of the system, from Eq. (11.9c) and (11.12a), is

$$E_\pm = E(0) + \tfrac{1}{2} K X^2 \pm \beta X \qquad (11.13a)$$

$E_\pm$ are plotted as functions of X in Fig. 11.9a; since Eqs (11.13a)

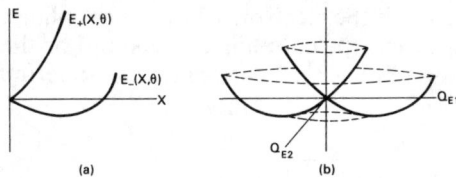

Figure 11.9.

†Bear in mind that here we are performing a model calculation where we have chosen typical, rather than general, functions for electronic states and $\partial V/\partial Q_\alpha$. In either case, the manipulations involved can be simplified through the use of the Wigner–Eckart Theorem.

are isotropic in $X-\Theta$ space, $E_\pm$ can also be pictured as the 'Mexican hat', of Fig. 11.9b. The physical meaning of X and Θ is not obvious; the fact that X increases with any E-type distortion but that the energies $E_\pm$ are independent of Θ suggests that they are remotely, but only remotely, connected with real-space polar coordinates. But a precise interpretation of X and Θ is not crucial; the important point is that *any* distortion of magnitude $X = (Q_{E1}^2 + Q_{E2}^2)^{1/2} = |\beta|/K = |Q_{JT}|$ results in a minimum in E_-, and a lowering of the system's energy by an amount $|\beta|^2/2K = E_{JT}$ below its $Q=0$ value. Thus the system, in choosing to reside at an equilibrium point of lowest energy, will prefer to sit in a distorted configuration, and the originally-expected degenerate E states will in fact be split; the actual state of lower energy is not $|x\rangle$ or $|y\rangle$, but rather of the form

$$\psi_- = \sin\frac{\Theta}{2}|x\rangle - \cos\frac{\Theta}{2}|y\rangle \qquad (11.13b)$$

Thus even if our C_{3v} system wants to be in a degenerate E-state, electron-phonon coupling still will cause a Jahn–Teller distortion which breaks the E-degeneracy. Note, however, that Θ can assume any value between 0 and 2π, and in this sense, the system is now infinitely degenerate.

PROBLEM 11.19 What is the state of higher energy? Does the Born–Oppenheimer approximation, and Eq. (11.9a) in particular, still apply for a system in the state ψ_- of Eq. (11.13b)?

The Jahn-Teller effect has been seen in many systems through X-ray and neutron scattering, magnetic resonance, and optical and infrared absorption studies. In practice the effect usually occurs in one of two forms; either the distortion is *static*, and the system is frozen into a contorted geometric state, or the system jumps continuously from one distorted spatial configuration to another, in a *dynamic* fashion, Fig. 11.10. Most of the discussion above and in Section 11.6 is applicable primarily to the static case, which occurs if the electron-vibration coupling is strong; for the dynamic case, the coupling is weak and/or the system is at a high temperature. One can sometimes determine the

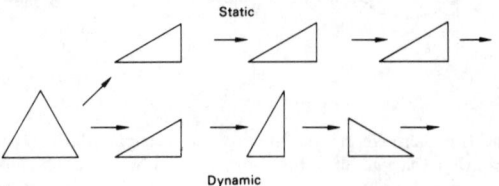

Figure 11.10.

strength of the Jahn–Teller coupling of electronic and vibrational states by following the effect as a function of temperature; experimental evidence of a switch from the static to the dynamic case appears roughly when T rises to about $T_{JT} = |\beta|^2 2Kk_B$, where k_B is the Boltzmann constant.

One can use group theory to discuss the Jahn–Teller effect even when the electron-phonon coupling is too strong to allow the above simple perturbation treatment, or in cases where the Born-Oppenheimer approximation itself breaks down and the separation of Eq. (11.9a) is not allowed.

PROBLEM 11.20 Instead of Eq. (11.10c), try using $V(\bar{Q}) = xQ_{E1} + yQ_{E2}$. Explain your results.

PROBLEM 11.21 What can group theory tell us about Eq. (11.11a)? Place as much of the above as possible in purely group theoretic language (i.e. do not choose specific forms for $V(\bar{Q})$ and electronic and vibrational states).

PROBLEM 11.22 Under what conditions might a static-to-dynamic change occur at $T_{JT} = |\beta|^2/2Kk_B$? What sorts of experiments might display this change, and how would they do so?
in perturbation calculations, we examine briefly a model system

†11.8 Pseudo-Jahn-Teller Effect

As a final example of the usefulness of symmetry arguments in perturbation calculations, we examine briefly a model system displaying a *pseudo–Jahn–Teller effect*; this is like the Jahn–Teller effect, but involves levels which are close but not strictly degenerate.

Consider a one-electron atom bound to the walls of a well by means of two very weak springs, Fig. 11.11a.

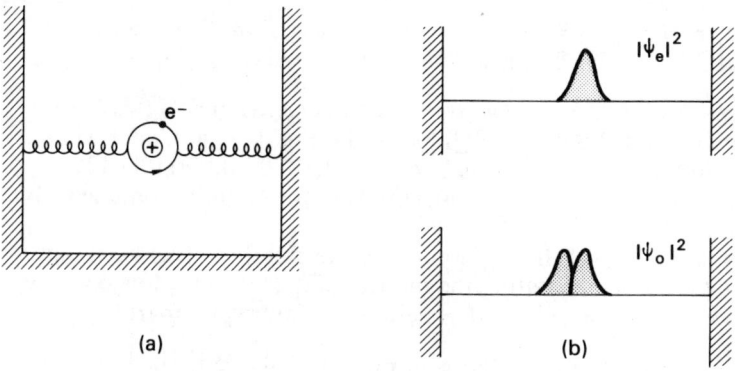

Figure 11.11.

The system displays inversion symmetry, and as such, its electronic states are even and odd (corresponding to the two irreducible representations of the symmetry group). An even state $\psi_e(\bar{r})$ and an odd $\psi_o(\bar{r})$ thus cannot be degenerate (unless accidentally), which is also to be expected on physical grounds since the probability distributions of charge must be dissimilar, Fig. 11.11b.

The electron at $\bar{r}$ experiences a potential of the form $V(\bar{r}, Q)$, where Q is to be some measure of the amplitude of vibration of the system; we shall take $+Q$ to be the displacement of the atom from its equilibrium position to the right. By inversion symmetry $V(\bar{r}, Q) = V(-\bar{r}, -Q)$, which Taylor expands about $Q=0$ as

$$V_0(\bar{r}) + V_1(\bar{r})Q + V_2(\bar{r})Q^2 + \cdots$$
$$= V_0(-\bar{r}) + V_1(-\bar{r})(-Q) + V_2(-\bar{r})(-Q)^2 + \cdots \quad (11.14\text{a})$$

Term by term comparison yields

$$\begin{aligned} V_0(\bar{r}) &= V_0(-\bar{r}) \\ V_1(\bar{r}) &= -V_1(-\bar{r}) \\ V_2(\bar{r}) &= V_2(-\bar{r}) \end{aligned} \quad (11.14\text{b})$$

The important point is that $V_1(\bar{r}) \equiv \partial V/\partial Q$ is an odd function of $\bar{r}$, and therefore a matrix representation of a $V_1(\bar{r})$ perturbation in the $\{\psi_e, \psi_o\}$ basis system will have only off-diagonal matrix elements.

Suppose that the two particular states ψ'_e and ψ'_o are *nearly* accidentally degenerate, and separated by the small energy difference Δ; alternatively, suppose they were at one time degenerate, but somehow split apart by a small, previously ignored low-symmetry contribution to the Hamiltonian. Then the representation of the complete Hamiltonian of the entire system is

$$\mathcal{H} = \begin{pmatrix} \Delta/2 & \beta'Q \\ \beta'Q & -\Delta/2 \end{pmatrix} + \tfrac{1}{2} m\omega_0^2 Q^2 \begin{pmatrix} 1 & 0 \\ 0 & 1 \end{pmatrix} + \cdots \quad (11.14\text{c})$$

where $\beta' = \langle \psi'_e | (\partial V/\partial Q) | \psi'_o \rangle$, m is the mass of the atom, and ω_0 its natural angular frequency of oscillation when $\beta' = 0$.

PROBLEM 11.23 Find eigenvectors and eigenvalues of this Hamiltonian, and show that $E(Q)$ looks like Fig. 11.12a. Under what conditions does $E(Q)$ look like Fig. 11.12b? Like Fig. 11.12c? Compare with Fig. 11.8 and 11.9. Will the system spontaneously distort?

PROBLEM 11.24 Why can we ignore the system's other even and odd states in generating the matrices Eq. (11.14c)? How could ψ'_e and ψ'_o have once been degenerate and later split apart?

Let us compare the terms in Eq. (11.11b) and in the left-hand matrix of Eq. (11.14c) which depend upon low-symmetry distor-

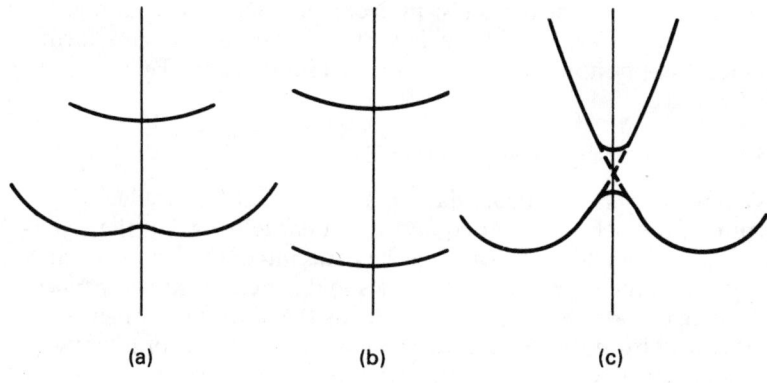

Figure 11.12.

tions. First, in both cases, the coupling is linear in the Q's. This is a consequence of our using only the lowest non-vanishing order of perturbation theory and (which is not the same thing!) only the lowest order term in the Taylor expansion of $V(\bar{Q}, \bar{r})$, Eq. (11.10a). If we had retained $\sum_{\alpha,\beta} (\partial^2 V/\partial Q_\alpha \partial Q_\beta) Q_\alpha Q_\beta$ there as well, the results would have been qualitatively different; for example, the Mexican hat of Fig. 11.9b would not have been cylindrically symmetric, but rather would have displayed three absolute minima in X–Θ, or Q_{E1}–Q_{E2}, space.†

Secondly, the perturbation of the Jahn–Teller example mixed two degenerate states belonging to the same irreducible representation, and its matrix representation can have diagonal and/or off-diagonal elements. The pseudo-Jahn–Teller perturbation mixed non-degenerate (albeit close) states of different irreducible representations, and the matrix can contain only off-diagonal linear terms. In essence, we can use first order perturbation theory in one case, but must go to second order in the other.

And finally, in our pseudo-Jahn–Teller example, the magnitude of Q provides a measure of the extent to which the perfect inversion symmetry of the system in real space is broken, and a measure also of the extent to which such a distortion admixes with a wavefunction of well-defined symmetry (odd, say) contributions of dfferent (even) symmetry. X plays an analogous role for the Jahn–Teller example.

PROBLEM 11.25 Relate the arguments of the last two paragraphs to one another.

†M. D. Sturge[38], page 118.

PROBLEM 11.26 Consider a C_{3v} molecule in which an A_1 and an A_2 electronic state are nearly accidentally degenerate. Can any form of electron-phonon interaction cause a pseudo Jahn–Teller distortion?

11.9 What Group Theory Cannot Do

We must emphasize the fundamental point that knowledge of the group of the Schroedinger equation can *not* tell us what the eigenenergies of a system are, or even the ordering of the levels; it can only reveal the possible degeneracies and general symmetry types of the eigenstates. Group theory tells us the allowed symmetries of states, and from this we can decide whether integrals of interest must necessarily vanish; on the other hand, the theory tells us nothing about the relative sizes of various non-vanishing matrix elements in question. It will not tell us, for example, that a system favours a certain mode of decay from an excited to the ground state; it only tells us whether the mode is allowed or disallowed to any order. The theory will not reveal how much a particular type of externally applied perturbation (electric field, uniaxial stress, etc.) will split a degenerate set of states, but only *if* it can cause a splitting.

11.10 Summary

1. An integral $\int f(\bar{r}) \, d^3 \bar{r}$, where $f(\bar{r})$ is associated with a physical object whose symmetry group is $\{\hat{R}\}$, will vanish unless $f(\bar{r})$ contains a component invariant (of A_1 symmetry) under all the operations $\{\hat{P}_R\}$. This theorem finds use in time-dependent and time-independent perturbation theory; in particular, the integral $\langle \psi^{(\mu)}(\bar{r}) | F^{(\nu)}(\bar{r}) | \psi^{(\eta)}(\bar{r}) \rangle$ must vanish unless $\Gamma^{(\mu)} \times \Gamma^{(\nu)} \times \Gamma^{(\eta)}$ contains A_1, or equivalently, unless $\Gamma^{(\nu)} \times \Gamma^{(\eta)}$ contains $\Gamma^{(\mu)}$.

2. The reduction of symmetry method allows one to ascertain what degeneracies will be broken with the introduction of a new term to the Hamiltonian even when perturbation theory is inapplicable. We have considered the example of a set of spherical harmonics of order l, $\{Y_l^m\}$, which span a subspace of Hilbert space invariant under the operations of the full rotation group. Their $(2l+1)$-fold degeneracy is lifted, and the states no longer all mix together under symmetry operations, if the system in question is placed in an environment of lower symmetry—i.e. if to the Hamiltonian there is added a lower-symmetry perturbation.

3. In a solid or non-linear molecule, the coupling of degenerate electronic states with vibrational modes may lead to a stable or dynamic physical deformation of the system. This Jahn–Teller effect has held the interest of theoreticians since its proposal in 1937, but was not seen experimentally until 1952.

CHAPTER 12

Symmetry and Conservation Laws

Until now, we have been concerned with the way in which a quantum system's geometric symmetries govern the transformational properties of its state functions, and with the subsequent problem of vanishing matrix elements. In this very short chapter, however, we shall change course altogether, and briefly examine the dynamical entities which are conserved as a consequence of symmetry. Invariance of a system to displacements in space, for example, demands that its linear momentum be conserved; invariance to rotations and the passage of time lead, respectively, to the laws of conservation of angular momentum and energy.

These three laws are associated with the insensitivity of a system to variation in the continuous parameters of space-time. Other symmetries involving properties distributed discretely, rather than continuously, such as left- and right-handedness, charge, and the direction of the flow of time, lead to less familiar, though equally important, conservation laws.

Some of the most exciting physics of the past few decades has followed from the discovery that these latter conservation laws are almost always valid, but not unfailingly so. Lee and Yang's prediction of 1956 of the *non*-conservation of parity, verified a year later by Wu, for example, first demonstrated that one of the fundamental interactions of nature displays an inherent asymmetry.

More advanced treatments of particle physics make much use of group theory and general algebraic techniques. This chapter introduces some of the relevant symmetry arguments at a most elementary level.

12.1 Conservation of Linear Momentum

The equations of motion of a classical system, in the Hamiltonian formalism, are

$$\dot{x} = \frac{\partial \mathcal{H}(p, x)}{\partial p} \tag{12.1}$$

$$\dot{p} = -\frac{\partial \mathcal{H}(p, x)}{\partial x}$$

where x and p are a coordinate and its conjugate momentum, and the second of these equations is equivalent to Newton's $F = ma$. If the system is invariant to displacement by an arbitrary amount x'—that is, if displacement by any x' is a symmetry operation—then by the second of these equations, $\dot{p} = 0$, and the momentum remains constant over time.

In the following quantum mechanical treatment, we shall show that if the operator which generates finite displacements in space commutes with the Hamiltonian, then linear momentum is conserved.

In Eq. (6.9) we defined the generator of displacements by the distance a in real space

$$\hat{R}_a x = x + a \tag{12.2a}$$

and in Eq. (6.11), a corresponding operator $\hat{P}_a$ in function space such that

$$\hat{P}_a \psi(\hat{R}_a x) = \psi(x) \tag{12.2b}$$

In Eq. (6.15), we found an explicit expression for $\hat{P}_a$:

$$\hat{P}_a = \exp(-i\hat{p}a/\hbar) \tag{12.2c}$$

where $\hat{p}$ is the linear momentum operator.

If $\hat{R}_a$ is a symmetry operator, then

$$[\hat{P}_a, \mathcal{H}] = 0 \tag{12.3a}$$

from which it follows, by Eq. (12.2c), that

$$[\hat{p}, \mathcal{H}] = 0 \tag{12.3b}$$

By Eq. (5.17), however,

$$i\hbar \frac{d}{dt} \langle \hat{p} \rangle = \langle [\hat{p}, \mathcal{H}] \rangle = 0 \tag{12.3c}$$

which completes the proof that $\langle \hat{p} \rangle$ is a constant over time, and momentum is conserved.

SYMMETRY AND CONSERVATION LAWS

PROBLEM 12.1 What, if anything, is conserved as a consequence of periodicity in a lattice? Compare the state of a free particle, $\psi = \exp(ikx)$, $k = p/\hbar$, with the Bloch functions.

PROBLEM 12.2 Is linear momentum necessarily conserved if $\mathscr{H}$ is invariant to infinitesimal displacements in space?

PROBLEM 12.3 Prove that if a system is invariant to rotations about an axis, the component of angular momentum along the axis is conserved.

PROBLEM 12.4 Prove that if a Hamiltonian is time independent, then the energy of the system is conserved.

PROBLEM 12.5 What is conserved if a system's Hamiltonian is invariant to the Lorentz transformation?

12.2 Broken Symmetry: Parity

Very few inaminate things distinguish between left and right. Fig. 12.1, for example, shows two tetrahedrally bonded molecules related to one another by the inversion operation; almost invariably the two will be found in equal natural abundances.† And if *ABCDE* is made in the laboratory by normal means, the two stereo-isomers are formed at the same rate. A solution of the compound, as a consequence, will not be optically active (will not rotate the plane of polarization of linearly polarized light).

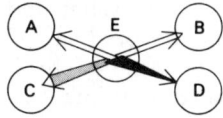 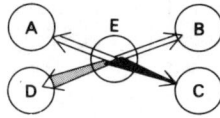

Figure 12.1.

It was long thought that the fundamental physical *forces* of nature also make no distinction between left and right, and display symmetry under the parity operation. Coulomb's law, for example, does not change if the vector $\bar{r}$ joining two charges is replaced by $-\bar{r}$.

PROBLEM 12.6 Show that the Lorentz force $F = e\bar{B} \times \bar{v}$ is invariant to the parity operation. [HINT: what produces B?]

In the experiment of Fig. 12.2, cobalt-sixty nuclei are oriented (by cooling a sample of ^{60}Co to $T \sim 0.01$ K and placing it in a

†Notable exceptions to this rule are the amino acids in proteins, and other biological molecules. Several billion years ago the binding properties of molecules and the evolutionary process combined to favour the existence of only a single handedness among the constituent parts of living organisms; the choice, however, could have gone either way, and in all likelihood nothing would be any different if proteins were composed of D-amino acids instead of the left-hand L variety, the other biomolecules being rebuilt accordingly.

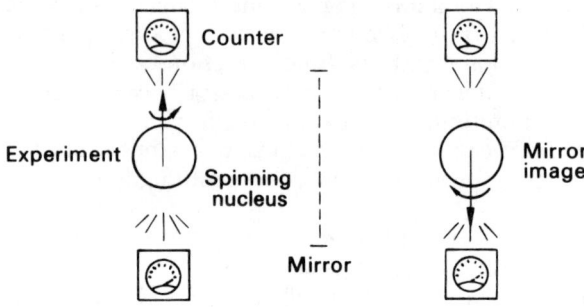

Figure 12.2.

strong, vertically aligned magnetic field) with their axes of spin upward and parallel to a mirror. (If the right hand curls with the spinning motion, the thumb points in the direction of the spin vector, by definition.) The rate of beta decay events is measured as a function of polar angle, and more decay products are counted per second below the sample, in the direction opposite to that of spin, than above. In the mirror system, one again sees more electrons coming off downward, but now *along* the direction of the spin vector; this, however, is in disagreement with the real experiment. The only way in which the physical system and the mirror image could possibly appear the same would be if identical decay rates were found above and below the sample. This is known experimentally *not* to be the case. Thus the mirror image of the physical event of decay is *not* observed to occur in nature, and the beta decay of ^{60}Co nuclei displays a distinct preference for one particular handedness.

PROBLEM 12.7 In Fig. 12.2 we demonstrate that a mirror reflection is not a symmetry operation for the ^{60}Co decay event. Show that the results of the experiment also disqualify parity symmetry; i.e. inversion is not a good symmetry operation.

PROBLEM 12.8 An electron is released from near the equator of a ^{60}Co nucleus spinning in a magnetic field; show that electromagnetic forces do not account for the results of the experiment.

Such cases, where a mirror does not present a physically realizable situation, are relatively rare, but they do occur. The expected symmetry is broken by the so-called *weak interaction*, and a Hamiltonian which contains such weak nuclear forces can not be invariant to the parity operator. In Chapter 11 we spoke of reductions of symmetry due to geometric distortions, but involving symmetric (e.g. Coulombic) basic forces. Here it is the fundamental (weak) interaction itself which is asymmetric!

SYMMETRY AND CONSERVATION LAWS

Consider, on the other hand, a system for which parity is a good, as opposed to broken, symmetry operator. If our system possesses a centre of inversion and has no weak nuclear forces at work, then

$$[\hat{P}_p, \hat{\mathcal{H}}] = 0 \tag{12.4a}$$

From the definition of the time-displacement operator $\hat{T}(t_2, t_1)$, Eq. (5.11),

$$\hat{T}(t_2, t_1)\psi(t_1) = \psi(t_2) \tag{12.4b}$$

it follows by Eqs (5.16) and (12.4a) that

$$[\hat{P}_p, \hat{T}(t_2, t_1)] = 0 \tag{12.4c}$$

Since $\hat{P}_p$ and $\hat{\mathcal{H}}$ commute, the Hamiltonian eigenstate $|\phi_m(t_1)\rangle$ can also be an eigenvector to $\hat{P}_p$:

$$\hat{P}_p|\phi_m(t_1)\rangle = p_1|\phi_m(t_1)\rangle \tag{12.5a}$$

From Eqs (12.4) and (12.5a)

$$\hat{P}_p|\phi_m(t_2)\rangle = \hat{P}_p\hat{T}(t_2, t_1)|\phi_m(t_1)\rangle = \hat{T}(t_2, t_1)\hat{P}_p|\phi_m(t_1)\rangle$$
$$= p_1\hat{T}(t_2, t_1)|\phi_m(t_1)\rangle = p_1|\phi_m(t_2)\rangle \tag{12.5b}$$

which says that the parity eigenvalue does not change with the passage of time, regardless of what sort of events and interactions are possible under $\hat{\mathcal{H}}$. This is what is meant by saying that p_1 is a good quantum number for our system, and that parity is conserved.

This conservation argument made use only of the fact that $\hat{P}_p$ commutes with $\hat{\mathcal{H}}$, and is therefore quite general. Similar laws are found if the Hamiltonian is insensitive to the replacement of t by $-t$ (time reversal symmetry) or to the replacement of every particle by its anti-particle (charge conjugation symmetry).

PROBLEM 12.9 Show that the eigenvalues of the parity operator $\hat{P}_p$ are ± 1.

†PROBLEM 12.10 Show that

$$\hat{P}_p \hat{r} \hat{P}_p^{-1} = -\hat{r} \tag{12.6a}$$
$$\hat{P}_p \hat{p} \hat{P}_p^{-1} = -\hat{p} \tag{12.6b}$$
$$[\hat{P}_p, \hat{L}] = 0 \tag{12.6c}$$

The spin operator $\hat{\sigma}$ also commutes with $\hat{P}_p$

$$[\hat{P}_p, \hat{\sigma}] = 0 \tag{12.6d}$$

Show that the *helicity* operator $\hat{\sigma} \cdot \hat{p}$, obeys

$$\hat{P}_p \hat{\sigma} \cdot \hat{p} \hat{P}_p^{-1} = -\hat{\sigma} \cdot \hat{p} \tag{12.6e}$$

†PROBLEM 12.11 It is observed experimentally that the neutrino has only a negative helicity. Is this a proof that there exist events in nature for which parity is not conserved?

PROBLEM 12.12 Show that a system in a well-defined state of non-vanishing linear momentum cannot be in an eigenstate of the parity operator.

†PROBLEM 12.13 A paramagnetic material of cubic structure exhibits full cubic symmetry in the absense of an external magnetic field. How can such a system undergo a ferromagnetic phase transition‡, and acquire a preferred axis of magnetization, as it is cooled below the Curie temperature?

12.3 Charge Conjugation

The laws of physics are invariant to a displacement in space, if such a displacement is a symmetry operation. Thus we can carry out an experiment at point A, and then repeat the experiment at B, where magnetic fields, electric fields, effects of the rotation of the earth, altitude, etc., are the same; we should obtain exactly identical results at A and B, and any physical laws deduced from our observations at A should be valid at B. And as we saw, associated with invariance of a system (and of the laws of physics) to displacement in space, there is a conservation law: the conversation of linear momentum.

The above does not constitute an isolated case; for all the known conservation laws, corresponding symmetry operators have also been found. Where conservation laws are valid (unlike, say, the non-conservation of parity in beta emission) the laws of nature are invariant to the associated symmetry operations. The development of every branch of physics (elementary particle theory in particular) has rested heavily upon exploring the properties of these conservation laws and symmetry operations.

Consider, for example, Coulomb's law, $F = q_1 q_2 / r^2$, which is invariant to 'charge conjugation', or a replacement of particles 1 and 2 by their antiparticles. The transformation $q_1 \to -q_1$, $q_2 \to -q_2$ leaves the law unchanged; this says, among other things, that through the use of Coulomb's law, it is possible to ascertain only the relative, and not the absolute, signs of the two charges. The same considerations apply to particles interacting through magnetic forces (note that if one changes the sign of a moving charge, the direction of the magnetic field it produces is reversed).

†Bransden, Evans, and Major[3], p. 230.
‡Landau and Lifshitz[25], *Statistical Mechanics*, p. 433, discuss changes of symmetry which accompany second order phase transitions.

SYMMETRY AND CONSERVATION LAWS

We might consequently be led to expect that if one observed an event involving any sort of interaction between charged particles, then a similar event involving the corresponding antiparticles should also be physically realizable; study of the two kinds of occurances, moreover, should not allow one to afix an absolute sign to the charge of any of the constituents.

The 'long-lived neutral kaon (K meson)', K_L^0, can decay into a pion, an electron, and neutrino through two channels:

$$K_L^0 \to e^+ + \pi^- + \nu$$
$$K_L^0 \to e^- + \pi^+ + \bar{\nu} \tag{12.7a}$$

One channel may be obtained from the other by replacing each particle by its antiparticle. By our above arguments, we expect the two rates of decay to be the same; otherwise, the preferential emission of e^+, say, by K_L^0 would allow us to define the sign of its charge unambiguously. The ratio of decay rates, however, is measured experimentally to be

$$\frac{K_L^0 \to e^+ + \pi^- + \nu}{K_L^0 \to e^- + \pi^+ + \bar{\nu}} = 1 \cdot 00315 \pm 0 \cdot 0003 \tag{12.7b}$$

Charge conjugation symmetry is broken here again by the weak nuclear forces.

†PROBLEM 12.14 What is conserved as a consequence of charge conjugation symmetry (when it holds)? Note that *conservation of charge* appears *always* to be valid; no event is known in which the net charge of the universe is increased or decreased.

12.4 Other Symmetries

We shall close this chapter by mentioning that there are a number of other symmetries and associated conservation laws of importance. For example, the Dirac equation, a general quantum mechanical equation to which Schroedinger's is a non-relativistic limit, is placed in 'covariant' form precisely so as to appear invariant to a Lorentz transformation; thus observers in inertial frames will record results of an experiment which are independent of their (constant) velocity relative to one another. Several other symmetries of current research interest are introduced in the next two problems.

PROBLEM 12.15 Imagine three systems composed, respectively, of two neutrons, a neutron and a proton, and two protons. Suppose the measured binding energies of the first two are the same, and e^2/r_0 greater than that of the third, where e^2/r_0 is the electrostatic interaction energy of the two protons. Can the interchange of a neutron and a proton be considered a symmetry operator (i.e.

neutron and proton are 'identical', just as two electrons are identical) which happens to be broken by the electromagnetic interaction? This problem leads to the study of 'isotopic spin' in nuclei and other particles.

PROBLEM 12.16 Are the laws of physics invariant to the sizes of the bodies involved? That is, do they display 'scaling' symmetry? The ways in which the laws of physics scale with respect to energy (as opposed to size) are of current interest in particle physics.

*PROBLEM 12.17 Which symmetry operations can be placed in the form $\hat{P}_R = \exp\{-i\hat{A} \cdot \hat{a}/\hbar\}$, where $\hat{A}$ is the quantum mechanical operator representing a constant of the motion (conserved quantity) and $\hat{a}$ is the associated generalized displacement? Is $\langle A \rangle$ conserved if and only if $\hat{R}$ is a symmetry operation, such that $[\hat{\mathcal{H}}, \hat{P}_R] = [\hat{\mathcal{H}}, \hat{A}] = 0$? Do $\hat{A}$ and $\hat{a}$ in general commute? See Problem 10.5.

In addition to the conserved entities we have already mentioned, there exist a number of others which govern the properties of the 'elementary' particles. It is not easy to grasp intuitively these symmetries and laws, as they have no simple analogues in every day experience, and they can be found only through examination of the patterns of behaviour of the particles themselves. The list of symmetries and conservation laws is long and seems to be growing every day—strangeness, SU(2), baryon number, μ-lepton number hyper-charge, and most recently, charm, colour, flavour and gentleness ... and where it all may lead is far from obvious. The time is clearly ripe for some inspired young Mendeleevs to put the pieces in their properly appointed order.

And that, perhaps, is where you come in!

12.5 Summary

1. Both classically and quantum mechanically, the invariance of a system to displacements in space means that (a) linear momentum is conserved in the system, and (b) all the laws of nature are invariant to the displacement. Similar conservation laws are associated with invariance to rotations, the passage of time, and, more generally, Lorentz transformations.

2. Conservation laws, however, are not inviolate. The breaking of parity, charge conjugation, and other symmetries by the weak nuclear force is manifest in beta-decay and elsewhere.

CHAPTER 13

Epilogue

A question which seems to nag at the back of the minds of many students and practitioners of physics is: 'Group theory and all that is fine, but what's the *really basic*, physically fundamental link between the possible degeneracies of a system and its geometric symmetry?' This book presents the simplest answer I have been able to find; in summary:
1. Any physical object is invariant to certain specific real-space geometric transformations $\{\hat{R}\}$, and complete information concerning these symmetries can be recorded in terms of the object's symmetry group structure. Through the $\{\hat{P}_R\}$ link, various sets of functions of $\bar{r}$ may be used to generate representations of $\{\hat{R}\}$; examples of such representations are the 'trivial' A_1 representation, and any isomorphic representation.

With proper alignment of function space, it becomes clear that *some* subsets of functions transform only among themselves under $\{\hat{P}_R\}$; equivalently, they inhabit mutually orthogonal invariant subspaces, and are capable of producing a representation of $\{\hat{R}\}$ in which all matrices have the same block-diagonal form. The block structure is preserved under matrix multiplication, and consequently one can separate each matrix (and thus the entire representation) into non-interacting irreducible representations. Any symmetry group will give rise to a unique† and characteristic set of possible irreducible representations, and to each irreducible representation there correspond families of generating functions of well-defined symmetry.

A listing of all the non-equivalent irreducible representations is essentially the same as a statement of the group's multiplication table; more than anything else, it tells us what sorts of distinct

†Up to similarity transformations.

symmetries may be displayed by the sundry functions associated with the system.

2. A quantum system displays a well-defined symmetry group $\{\hat{R}\}$ in real space which leaves the Hamiltonian invariant

$$\mathcal{H}(\hat{R}\bar{r}) = \mathcal{H}(\bar{r}) \tag{13.1}$$

Its energy eigenstates $\{\psi_i(\bar{r})\}$, for which

$$\mathcal{H}\psi_i(\bar{r}) = E_i\psi_i(\bar{r}) \tag{13.2}$$

display (like any functions whatsoever) *some* sort of transformational behaviour in function space

$$\psi_i(\hat{R}^{-1}\bar{r}) = \sum_j \Gamma_{ji}(\hat{R})\psi_j(\bar{r}) \tag{13.3a}$$

Or, if we define the operator $\hat{P}_R$ by means of $\hat{P}_R\psi(\hat{R}\bar{r}) = \psi(\bar{r})$, then

$$\hat{P}_R\psi_i(\bar{r}) = \sum_j \Gamma_{ji}(\hat{R})\psi_j(\bar{r}) \tag{13.3b}$$

The link between symmetries in real space and those in Hilbert space is that since $\mathcal{H}(\hat{R}\bar{r}) = \mathcal{H}(\bar{r})$, the functions $\{\hat{P}_R\psi_i(\bar{r})\}$ for all $\hat{R}$ must be degenerate; that is, the only states which can mix under the $\langle\hat{P}_R\rangle$ are those belonging to the same energy eigenvalue E_ν. And we can now classify together such states, and label them meaningfully with the quantum number ν. In the language of vector spaces, an orthonormal set $\{\psi_j^{(\nu)}(\bar{r})\}$ for which

$$\hat{P}_R\psi_i^{(\nu)}(\bar{r}) = \sum_j \Gamma_{ji}^{(\nu)}(\hat{R})\psi_j^{(\nu)}(\bar{r})\}$$

span a subspace invariant with respect to the $\{\hat{P}_R\}$. This is exactly equivalent to saying, but in the language of group representation theory, that the $\{\psi_i^{(\nu)}(\bar{r})\}$ generate the νth irreducible representation of $\{\hat{R}\}$.

3. Thus the invariant subspaces of eigenvectors of $\mathcal{H}$ are one and the same as the invariant subspaces of functions of well-defined symmetry which generate the irreducible representations of the $\{\hat{R}\}$ for which $\mathcal{H}(\hat{R}\bar{r}) = \mathcal{H}(\bar{r})$. This is shown in diagramatic form in Fig. 13.1.

The connection between real space and function space is still rather abstract, but none-the-less invaluable for two reasons. First, it provides us with a way of relating the physical shape of a system to the possible general shapes of its states in Hilbert space; this alone would fully justify the pursuit of group representation theory. But secondly, the orthogonality properties of functions belonging to different irreducible representations, manifest in the rules for finding vanishing matrix elements, makes the theory of

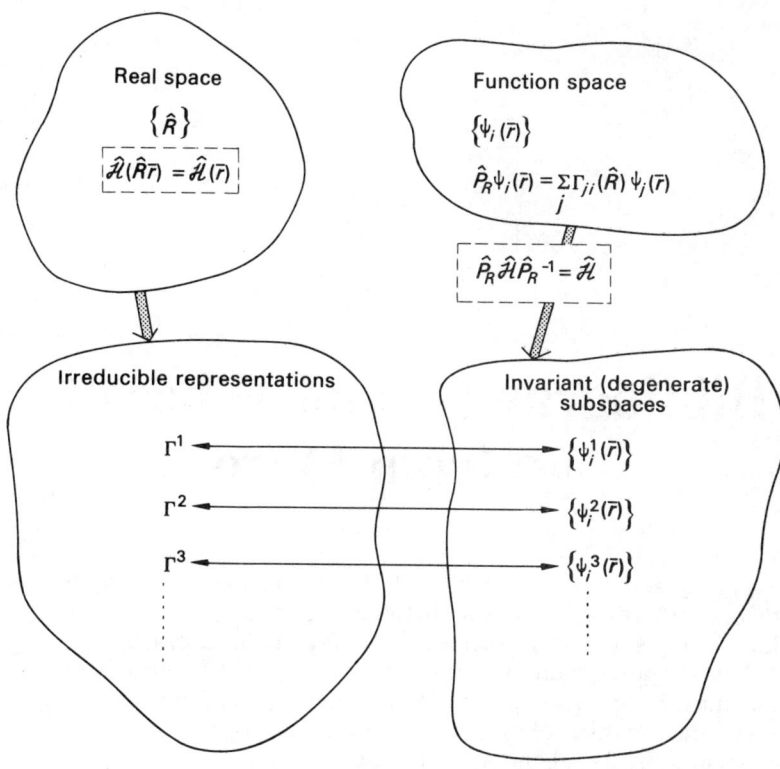

Figure 13.1.

great use in everyday calculations as well. It is to such diverse and subtle applications that a number of the books in the bibliography direct their attention.

Bibliography, and Where to go from Here

You may wish to continue your studies of group representation theory and/or the connection between symmetry and conservation laws, and a short list of suggested reading might be of help.

Before approaching the more advanced facets of group theory one must have a good grounding in quantum mechanics. Although there are a number of excellent texts on the subject, a few which are especially useful and appropriate are those of Feynman[14] (Vol. III), Merzbacher,[31] Landau and Lifshitz,[26] Davydov,[8] Dirac,[10] Kaempffer,[23] and, for the mathematically inclined, Fano,[13] Jauch,[20] and Jordan[21].

As for the applications of group representation theory to quantum mechanics, Tinkham[39] and recently published books by Joshi[22] and Kaplan[24] should be examined as general-coverage texts. While the latter two stress the mathematical background, Tinkham delves deeply into applications of atomic systems, as do Edmonds,[11] Rose,[34] and Brink and Satchler[5]. Heine[19] covers much the same ground and discusses the use of group theory in nuclear and relativistic physics. Schonland[35] and Bishop[2] are less rigorous, but consider in detail the calculation of molecular orbitals, molecular vibrations, and the spectra of many-electron systems. Wigner's classic[41] is rewarding but difficult and perhaps should be read after one of the others.

Ballhausen[1] and Lax[27] concentrate on solid-state problems, as do the recent advanced review by Cracknell[7] and text by Streitwolf[37]. An interesting and readable article which nicely illustrates the breadth of group theoretic arguments is Zak's[42] discussion of 'kq' electronic states in solids.

With regard to symmetry principles and conservation laws, Emmerson's[12] and Lipkin's[28] short books are excellent, and Chapter (1-52) of the Feynman lectures[14] is a must. Also have a look at Brereton's[4] article.

Finally, several texts dealing primarily with mathematical methods offer interesting approaches to the use of group theory. Two of the best of these are in Mathews and Walker[29] and in Ziman.[43]

Bibliography

1. BALLHAUSEN, C. J., *Introduction to Ligand Field Theory*, McGraw-Hill (1962)
2. BISHOP, D. M., *Group Theory and Chemistry*, Oxford (1973)
3. BRANSDEN, B. H., EVANS, D. and MAJOR, J. V., *The Fundamental Particles*, Van Nostrand Reinhold (1973)
4. BRERETON, M. G., "Symmetry in Physics" *Physics Bulletin* **25**, 95 (1974)
5. BRINK, D. M. and SATCHLER, G. R., *Angular Momentum*, 2nd edn, Oxford (1968)
6. CLARK, H., *A First Course in Quantum Mechanics*, Van Nostrand Reinhold (1974)
7. CRACKNELL, A. P., "Group Theory in Solid State Physics" *Advances in Physics*, **23**, 673 (1974)
8. DAVYDOV, A. S., *Quantum Mechanics* Pergamon (1965)
9. DENNERY, P. and KRZYWICKI, A., *Mathematics for Physicists*, Harper & Row (1967)
10. DIRAC, P. A. M., *The Principles of Quantum Mechanics*, 4th edn, Oxford (1958)
11. EDMONDS, A. R., *Angular Momentum in Quantum Mechanics*, Princeton (1957)
12. EMMERSON, J. McL., *Symmetry Principles in Particle Physics*, Oxford (1972)
13. FANO, G., *Mathematical Methods of Quantum Mechanics*, McGraw-Hill (1971)
14. FEYNMAN, R. P., *Lectures on Physics, Vol. I and III* Addison-Wesley (1963)
15. GEHRING, G. A. and GEHRING, K. A., "Co-operative Jahn-Teller Effects" *Reports on Progress in Physics*, **38**, 1 (1975)
16. GOLDSTEIN, H., *Classical Mechanics*, Addison-Wesley (1950)
17. HALMOS, P. R., *Finite Dimensional Vector Spaces*, Van Nostrand Reinhold (1958)
18. HAMERMESH, M., *Group Theory and its Application to Physical Problems*, Addison-Wesley (1962)
19. HEINE, V., *Group Theory in Quantum Mechanics*, Pergamon (1960)

20. JAUCH, J. M., *Foundations of Quantum Mechanics*, Addison-Wesley (1968)
21. JORDAN, T. F., *Linear Operators for Quantum Mechanics*, Wiley (1969)
22. JOSHI, A. W., *Elements of Group Theory for Physicists*, Halsted (1973)
23. KAEMPFFER, F. A., *Concepts in Quantum Mechanics*, Academic (1965)
24. KAPLAN, I. G., *Symmetry of Many-Electron Systems*, Academic (1975)
25. LANDAU, L. D. and LIFSHITZ, E. M., *Statistical Physics*, 2nd edn, Pergamon (1969)
26. LANDAU, L. D. and LIFSHITZ, E. M., *Quantum Mechanics*, 2nd edn, Pergamon (1965)
27. LAX, M., *Symmetry Principles in Solid State and Molecular Physics*, Wiley (1974)
28. LIPKIN, H. J., *Lie Groups for Pedestrians*, North-Holland (1965)
29. MATHEWS, J. and WALKER, R. L., *Mathematical Methods of Physics*, 2nd edn, Benjamin (1970)
30. MCWEENY, R., *Symmetry: An Introduction to Group Theory and its Applications*, Pergamon (1963)
31. MERZBACHER, E., *Quantum Mechanics*, 2nd edn, Wiley (1970)
32. MESSIAH, A., *Quantum Mechanics I and II*, North-Holland (1963)
33. NYE, J. F., *Physical Properties of Crystals*, Oxford (1957)
34. ROSE, M. E., *Elementary Theory of Angular Momentum*, Wiley (1957)
35. SCHONLAND, D. S., *Molecular Symmetry*, Van Nostrand Reinhold (1965)
36. SLICHTER, C. P., *Principles of Magnetic Resonance*, Harper & Row (1963)
37. STREITWOLF, H. W., *Group Theory in Solid State Physics*, MacDonald (1971)
38. STURGE, M. D., 'The Jahn-Teller effect in Solids', *Solid State Physics*, Vol. 20, eds. F. Seitz, D. Turnbull and H. Ehrenreich, Academic (1967)
39. TINKHAM, M., *Group Theory and Quantum Mechanics*, McGraw-Hill (1964)
40. WEYL, H., *Symmetry*, Princeton (1952)
41. WIGNER, E. P., *Group Theory*, Academic (1959)
42. ZAK, J., 'The kq Representation in the Dynamics of Electrons in Solids' *Solid State Physics*, Vol. 27, eds. H. Ehrenreich, D. Turnbull and F. Seitz, Academic (1972)
43. ZIMAN, J. M., *Elements of Advanced Quantum Theory*, Cambridge (1969)

List of Symbols

$\bar{r}$	position vector, 3
f_+, f_-	even, odd function, 3
$\hat{}$	denotes operator, 4
$\hat{R}_p$	real-space inversion (parity) operator, 4, 75
$\hat{P}_p$	function-space inversion (parity) operator, 4, 75, 96, 231
$\hat{M}_h$	'horizontal' mirror operator, 5, 117
$\hat{R}, \hat{R}_\phi$	real space rotation operator; generally, but not necessarily, a symmetry operator, 5, 9, 45, 104
$\hat{M}_I, \hat{M}_\phi$	real space 'vertical' mirror operator; generally, but not necessarily, a symmetry operator, 6
$\{\psi_i\}, i=1,\ldots,6$	orientational state of C_{3v}, 6–8
C_n	symmetry group of system with n-fold axis of rotation, 9
C_{nv}	symmetry group of system with n-fold axis of rotation and n 'vertical' mirror planes, 9
$[\ ]$	commutator, 9
$\{|i\rangle\}$	set of orthonormal basis vectors, 13
$(v_1, v_2, \ldots)$	representation of vector $\bar{v}$, 13
$(v_1, v_2, \ldots v_m)$	m-tuple, 15
δ_{ij}	Kronecker delta, 18
$|u\rangle, \langle v|$	Dirac bra-ket notation, 20
$\langle u|v\rangle$	scalar product, 20
$\hat{I}$	closure operator, 21

242 LIST OF SYMBOLS

$=$	denotes matrix, 23
$C, C_{ji} \equiv \langle j\|i'\rangle$	transformation matrix, 23, 29
$\bar{\bar{A}}, A_{ji} \equiv \langle j\|\hat{A}\|i\rangle$	representation of operator $\hat{A}$ generated by $\{\|i\rangle\}$ basis, 27
$\hat{A}^{-1}$	inverse of $\hat{A}$, 29
$\hat{A}^+$	Hermitian conjugate of $\hat{A}$, 31
$\hat{H} = \hat{H}^+$	Hermitian operator, 32
$\hat{U}^{-1} = \hat{U}^+$	Unitary operator, 32
$\{a_n\}, \{\|a_n\rangle\}$	eigen-spectrum of $\hat{A}$, 34
$\langle r\|f\rangle = f(\bar{r})$	representation of $\|f\rangle$ in the continuous $\{\|r\rangle\}$ basis system, 38, 94, 103
$\langle r\|r'\rangle = \delta(\bar{r}-\bar{r}')$	Dirac delta, 38
$\hat{r}, \hat{x}$	position operator, 40, 104, 150
$\bar{\bar{R}}, \bar{\bar{R}}_\phi$	matrix representation of real-space rotation operator, generated by $\{\|1\rangle, \|2\rangle, \|3\rangle\}$, 45
$\bar{\bar{\Gamma}}(\hat{R})$	function space representation of $\hat{P}_R$, 53, 77, 102
A_1, A_2, B, E, T	irreducible representations of point groups, 56, 138, 213, 219
Γ	representation, not necessarily irreducible, 56
$\|\psi\rangle, \psi(\bar{r})$	general quantum state, 62
$\hat{T}(t_2, t_1)$	time development operator, 67, 231
$\mathcal{H}$	Hamiltonian, 68, 149
$\{\|\psi_n\rangle\}, \{E_n\}$	eigen-spectrum of $\mathcal{H}$, 69
$Y_l^m(\theta, \phi)$	spherical harmonic, 81
$\|s\rangle, \|p\rangle, \|d\rangle$	atomic orbital, 82
$\hat{S}, \hat{S}_3, \hat{I}, \hat{I}_3$	electron and nuclear spin operator, 85, 191
$\mathcal{H}^e\|\alpha_e\rangle = g_e\beta_e H\hat{S}_3\|\alpha_e\rangle$	electron Zeeman energy eigen-equation, 85
a	contact hyperfine constant, 86
$\hat{R}_a, \hat{P}_a = \exp(-i\hat{p}a/\hbar)$	generator of finite spatial displacements, 98, 183, 228
$\hat{P}_{R_\phi} = \exp(-i\hat{J}\cdot\bar{n}\phi/\hbar)$	generator of finite angular displacements, 100, 187, 193

LIST OF SYMBOLS 243

$\bar{\bar{\Gamma}}^{(\psi)}(\hat{R})$	representation of $\hat{P}_R$ generated by the function-space basis set $\{|\psi_i(\bar{r})\rangle\}$, 102, 53, 77
$\bar{\bar{\Gamma}}^{(d)}(\hat{R}_\phi)$, $\bar{\bar{\Gamma}}^{(d)}(\hat{M}_\phi)$	d-state representations, 107
D_{3h}	$C_{3v} \times \{\hat{I}, \hat{M}_h\}$, 117, 219
$[54312](X_1, X_2, X_3, X_4, X_5)$	permutations, 119, 163
$\hat{R}_c$	colour group operator, 120
c	number of classes, and number of non-equivalent irreducible representations, in a group, 135
N	order of a group, 135
n_v	dimensionality of vth irreducible representation of a group, 135
$\chi^{(\mu)}(\hat{R}) \equiv \sum_i \Gamma_{ii}^{(\mu)}(\hat{R})$	character, 142
$\hat{\Omega}_C$	Dirac character, 146, 171
$\psi_i^{(v)}(\bar{r})$	ith eigen vector of $\mathscr{H}$, belonging to vth energy level and vth irreducible representation of Group of the Schroedinger equation, 150
$\hat{\mathscr{P}}^{(\mu)}$, $\hat{\mathscr{P}}_{ji}^{(\mu)}$	projection operators, 159
$\{|Q_i\rangle\}$, $i=1,\ldots,6$	normal modes of C_{3v}, 172
$C_{\infty v}$	group of axial symmetry, 187
$\{\bar{\bar{\sigma}}_i\}$, $i=1, 2, 3$	Pauli spin matrices, 192
$\hat{R}_{4\pi}$	double group operator, 193
$D_{m'm}^{(l)}(\hat{R})$	element of representation of $\hat{P}_R$ generated by $\{Y_l^m, m = -l, -l+1, \ldots, l\}$, 197
$\{\hat{A}_l^{-l}, \hat{A}_l^{-l+1}, \ldots, \hat{A}_l^l\}$	irreducible spherical tensor operator of rank l, 197
$C(j_1 j_2 j_3; m_1 m_2 m_3)$	Clebsch-Gordan coefficient, 199
$\langle j_3 \| A_{j2} \| j_1 \rangle$	reduced matrix, 201
$\{Q_{E1}, Q_{E2}\}$, $\psi_e(\bar{r}_i, \bar{Q})\psi_n(\bar{Q})$, $E_e(\bar{Q})$, $V(Q_\alpha, \bar{r}_i)$	215–218
$Q_{JT}, E_{JT}, T_{JT}, X, \Theta$	218–223

Important Relationships

1 Vectors and Operators

(2.20) $\begin{cases} \hat{I} = \sum_i |i\rangle\langle i| \text{ closure (completeness)} \\ \langle i|j\rangle = \delta_{ij} \text{ orthonormality} \end{cases}$ 21, 23

(3.6) $\hat{A}|i\rangle = \sum_j A_{ji}|j\rangle$ 27

(3.8a) $u'_i = \sum_j A_{ij} u_j$ 28

(3.24) $\hat{A}|a_n\rangle = a_n|a_n\rangle$ 34

(4.7) $\bar{\bar{R}}_\phi = \begin{pmatrix} \cos\phi & -\sin\phi & 0 \\ \sin\phi & \cos\phi & 0 \\ 0 & 0 & 1 \end{pmatrix}$ real space rotation matrix   46

(4.14) $\bar{\bar{A}}' = \bar{\bar{R}}\,\bar{\bar{A}}\,\bar{\bar{R}}^{-1} = \bar{\bar{R}}\,\bar{\bar{A}}\,\bar{\bar{R}}^+$ similarity transformation   50

2 Symmetry Operators

(5.17) $i\hbar \dfrac{d}{dt}\langle \hat{A}\rangle = \langle [\hat{\mathscr{H}}, \hat{A}]\rangle + i\hbar \left\langle \dfrac{\partial A}{\partial t}\right\rangle$ 69

(5.27a) $[\hat{P}_R, \hat{\mathscr{H}}] = 0$ 76, 149
(5.27b) $\hat{\mathscr{H}}(\hat{R}\bar{r}) = \hat{\mathscr{H}}(\bar{r})$ } symmetry operator   76, 149
(5.33) $\hat{\mathscr{H}}(\hat{P}_R \psi_n) = E_n(\hat{P}_R \psi_n)$ 79, 150
(6.4a) $\hat{P}_R \psi(\hat{R}\bar{r}) = \psi(\bar{r})$ 96

(6.19a) $\hat{P}_R|\psi_i(\bar{r})\rangle = \sum_j |\psi_j(\bar{r})\rangle \Gamma_{ji}^{(\psi)}(\hat{R})$ }   102, 151
(6.19b) $\Gamma_{ji}^{(\psi)}(\hat{R}) = \langle \psi_j(\bar{r})|\hat{P}_R|\psi_i(\bar{r})\rangle$

3 Irreducible Representations

(8.3) $\sum_v^c n_v^2 = N$ 136

IMPORTANT RELATIONSHIPS

(8.6) $\begin{cases} \sum_{R}^{N} \Gamma_{ij}^{(\mu)}(\hat{R})^* \Gamma_{i'j'}^{(\nu)}(\hat{R}) = \dfrac{N}{n_\nu} \delta_{\mu\nu} \delta_{ii'} \delta_{jj'} \\ \text{Great Orthogonality Theorem} \end{cases}$ 141

(8.11c) $\begin{cases} a_\mu = \dfrac{1}{N} \sum_{R}^{N} \chi^{(\mu)}(\hat{R})^* \chi^{(\text{red})}(\hat{R}) \\ \text{if } \Gamma^{(\text{red})} = \sum_{\mu}^{c} a_\mu \Gamma^{(\mu)} \end{cases}$ 144

(9.9c) $\begin{cases} \hat{\mathscr{P}}^{(\mu)} = \dfrac{n_\nu}{N} \sum_{R}^{N} \chi^{(\mu)}(\hat{R})^* \hat{P}_R \\ \text{projection operator} \end{cases}$ 158

(10.26) $\hat{P}_R \hat{A}_l^m \hat{P}_R^+ = \sum_{m'=-l}^{l} D_{m'm}^{(l)}(\hat{R}) \hat{A}_l^{m'}$ 197

(10.32) $\begin{cases} \langle j_3 m_3 | \hat{A}_{j2}^{m2} | j_1 m_1 \rangle = C(j_1 j_2 j_3; m_1 m_2 m_3) \langle j_3 \| A_{j2} \| j_1 \rangle \\ \text{Wigner-Eckart Theorem} \end{cases}$ 201

$\begin{cases} \text{if } \Gamma^{(\nu)} \times \Gamma^{(\eta)} \text{ does not contain } \Gamma^{(\mu)}, \\ \text{then } \langle \psi^{(\mu)} | F^{(\nu)} | \psi^{(\eta)} \rangle = 0 \end{cases}$ 209

INDEX

active and passive view point, 47
adiabatic approximation, 215–218, 222
admixture coefficients, 33, 51, 79, 109
algebra, 146
amino acids, 229
ammonia molecule, 4
angular momentum, 81, 108, 186, 189, 195, 199
associativity, group property, 8, 15, 114
A_1, 56, 138, 204, 217, 220, 226
A_2, 136, 139, 210, 220, 226

baryon, 234
basis, 13, 16, 41, 63, 102, 150
 –continuous, 36, 103
 –change of, 13, 16, 22, 29, 41, 45, 50, 64, 70, 121, 127
Bethe, H., 193
bin, 118
Bloch function, Theorem, 183, 229
block diagonalization, 51, 55, 78, 109, 128, 133, 144, 235
Bohr magneton, nuclear magneton, 85
Born-Oppenheimer approximation, 215, 222
bra vector, 14, 20
B, B_1, 138, 219

Celebrated Theorem, 140
center of inversion, 3
character, 10, 142
 –Dirac, 146, 171
character table, 142
 –for C_{3v}, 144
 –for C_{3v} double group, 194
 –for 0, 213
class, 120, 134, 136
Clebsch-Gordan coefficients, 81, 199

closure, group property, 8, 14, 101, 114
closure relation, 19, 21, 32, 37, 45, 102, 110
cobalt 60, 229
color, 234 (see also color group)
commutation relation, 63, 76, 108, 114, 189, 192, 198
compatible observables, 62
complete basis set, 16, 19, 65, 70, 110
complete set of observables, 62
complex number, 9
configuration coordinate, 215
conjugate, 32, 121
constant of motion, conserved, 70, 228
conservation
 –of angular momentum, 229
 –of linear momentum, 99, 100, 228, 232
Curie temperature, 232
C_n, C_{nv}, 9
C_{3v}, 4, 54, 76, 110, 117, 120, 124, 144, 152, 155, 161, 167, 172, 177, 188, 193, 206, 209, 217, 220, 226
 –character table, 144
C_{4v}, 108
C_{6v}, 120
$C_{\infty v}$, 182, 187

degeneracy, 10, 78, 79, 90, 150, 152, 180, 184, 206, 212, 213, 222, 224, 226, 236
 –accidental, 152, 214
delta function
 –Dirac, 38
 –Kronecker, 18, 38
differential equation, 9, 35
dimension of vector space, 16
Dirac equation, 233

INDEX

Dirac notation, 19
displacements, 68, 96, 99, 106, 128, 187, 228
distributivity, 38
dual space, 17, 21, 30
dynamic variable, 33, 69, 74
D_{nh}, 117, 122, 219
d-state, 82, 106, 134, 196, 210, 213

earthworm, 1
eigen equation, 4, 33, 70
eigen vectors, eigen values, 34, 64
electron states, 10, 80, 94, 106, 110, 122, 165, 212, 220
energy band, 165
electronic conductivity tensor, 198
ellipse, 73
expectation value, 66, 106, 195
E, 56, 138, 217, 220

Fermi Golden Rule, 206
ferromagnetic transition, 232
field, electric and magnetic, 213
flavor, 234
fourier decomposition, 19
fourier transform, 36, 38
function
 –Bessel, 35
 –even, odd, 3, 75, 78, 159, 224
 –symmetry of, 3, 53, 75, 109, 137, 150–155, 163, 165, 195, 200, 210, 221, 226, 235
 –vector, 30, 35
functional, 30
f-state, 210

generator, 117
 –of finite spatial displacements, 99, 100, 128, 183, 228
 –of finite angular displacements, 100, 106, 108, 187, 193
 –of infinitesimal spatial displacements, 96
 –of infinitesimal temporal displacements, 68
gerade, ungerade, 155
graphite, 198
group, 8, 54, 114
 –abelian, commutating, 9, 79, 114, 122, 128, 139, 184
 –color, Shubnikov, 118, 120, 194
 –continuous, Lie, 117, 146, 154, 186
 –cyclic, 117
 –double, 193
 –generator, 117
 –isomorphic, 115

 –Lie, continuous, 117, 146, 154, 186
 –permutation, 84, 118, 122, 162
 –point, 114
 –product, 117
 –Schroedinger Equation, 149
 –Shubnikov, color, 118, 120, 194
 –space, 114
 –sub-, 117, 121
 –symmetric, 84, 118, 122, 162
 –symmetry, 9, 54, 114, 149, 236
Great Orthogonality Theorem, 141, 144, 188

Hamiltonian, 60, 68, 149, 160, 215, 218, 230
 –diagonalization of, 10, 70, 77, 84, 89, 134, 157, 179, 221, 224
 –representation of, 71, 86
 –spin, 86
harmonic oscillator, 172, 217
Hartree, Hartree-Foch, 84
Heisenberg Uncertainty Principle, 66
helicity, 232
Helmholtz equation, 35
Hermitian conjugate, 32
Hilbert space, 10, 62, 63, 103, 161, 236
homomorphism, 124, 136
Hooke, 172, 217
horizontal reflection, 4, 117
hydrogen, 80, 84, 166
hyperfine interaction energy, 86

identity, group property, 8, 15, 114
infra-red absorption, 222
integers, 9
integral, vanishing, 51, 91, 203, 209
invariance, 2, 14
invariant space, 29, 51, 109, 128, 133, 141, 150, 235
inverse, group property, 8, 15, 114
irreducible representation, 56, 102, 126, 133, 149, 156, 162, 235
isomer, 9
isomorphism, 54, 100, 105, 115, 124

Jahn-Teller effect, 177, 209, 218, 222, 225

K meson, kaon, 233
ket, 20
Kramer Theorem, 164, 218

Laplace equation, 35
LCAO, linear combination of atomic orbitals, 10, 166
level-crossing, 209, 213

level—complete set, 110, 150
linear displacements, 96, 99, 128, 228
linear independence of vectors, 15
linear molecule, 187, 219
linear vector space, 15
Lorentz transform, 229

magnetic dipole, 85
magnetic material, 120
magnetic resonance, 165, 222
matrix, 9, 21, 27, 45, 54
 —Pauli, 192
 —reduced, 201
 —transformation, 23, 41, 64, 71
Mexican hat, 222, 225
molecular vibrations, 172, 177, 217
MO, molecular orbitals, 161, 165
multiplication table, 114, 235
mu meson, muon, 234
m-tuple, 15, 94

Neumann Principle, 198
neutrino, 232
neutron, 222, 234
Newton, I., 177, 228
normalized vector, 17
normal modes, 41, 172, 177, 215

observable, 62
optical absorption, 222
operator, 25
 —angular momentum, 81, 108, 186, 189, 195, 199
 —antilinear, 20, 165
 —Cartesian tensor, 196
 —commutating, 15, 63, 74, 80, 86, 146, 150
 —composition, product of, 6, 26, 114
 —equivalent, 26
 —exponential, 99, 100, 187
 —Hamiltonian, 68, 149
 —Hermitian, 32, 40, 65, 69, 80, 99, 100
 —inverse, 8, 15, 29, 114
 —irreducible spherical tensor, 194, 197, 198
 —linear, 20, 26
 —parity, 4, 75, 78, 96, 230–232
 —position, 40, 104, 150
 —product, composition of, 6, 26, 114
 —projection, 21, 146, 157, 204
 —quantum mechanical, 69
 —reflection, 6, 47
 —representation, 27, 44, 102, 151
 —rotation, 6, 32, 45, 121
 —scalar, 195
 —sum of, 26

 —symmetry, 1–4, 8, 74, 111, 133, 149
 —tensor, 194–198
 —time-development, 67, 231
 —time-reversal, 165
 —unitary, 32, 48, 80, 99, 100
 —vector, 195
ordered set, 13
order of group, 114, 135
orthogonality, 17, 23
O, O_h, 122, 213
 —character table, 213
 —double group, 193

parity, 4, 75, 153, 229
Pauli Exclusion Principle, 84, 163
Pauli spin matrices, 192
periodic boundary conditions, 183
perturbations, 10, 34, 89, 205, 208, 212, 221, 225
phase transitions, 232
pi meson, pion, 233
pressure, 165
principal axis, 73, 88, 134
product space, direct; product state, 84, 86, 159, 199, 204, 206, 217
protein, 229
proton, 85, 234
pseudo-Jahn-Teller effect, 223
pure state, 62, 64
Principle of Superposition of States, 26, 63
p-state, 82, 91, 110, 122, 126, 196, 212, 219

quantum number, 62, 110, 155, 231, 236

relativity, 68, 233
representation
 —d-state, 106, 109, 127, 134, 156, 189, 210
 —equivalent, 126, 134
 —faithful, 124
 —f-state, 210
 —group, 10
 —Hamiltonian, 71, 86
 —homomorphic, 124, 136
 —irreducible, 56, 102, 126, 133, 149, 156, 162, 235
 —isomorphic, 124, 134
 —matrix, 27, 54
 —of operator, 27, 45, 102, 151
 —p-state, 109, 122, 127, 134, 156, 189, 210
 —reduction of, 56, 130, 143, 189
 —regular, 139
 —s-state, 124, 127, 156, 189, 210

–true, 124
–vector, 13, 16, 27, 38, 41, 94, 103

scalar, scalar field, 14
scalar product, 17, 20, 36
Schur's lemma, 142
Schroedinger Equation, 67, 78, 164, 172
secular determinant, equation, 71, 89, 221
similarity transformation, 50, 121, 127
Slater determinant, 163
S_N, S_n, 118, 122, 162
spherical harmonics, 35, 81, 106, 186, 188, 197, 199, 210
–real valued, 82
spin, 85, 120, 191, 231
–isotopic, 234
Stern-Gerlach experiment, 85
strangeness, 234
string, vibrating, 39–43
subgroup, 117, 121
subspace, 23, 29, 51, 78
SU(2), 234
symmetry, 1–4, 8
–axial, 154, 186
–broken, 231
–charge conjugation, 231
–exchange, 119, 162
–of functions, 3, 53, 75, 109, 137, 150–155, 163, 165, 195, 200, 210, 221, 226, 235
–full rotational, 80, 154, 156, 188, 210
–group, 9, 54, 114, 149, 236
–inversion, 3, 9, 75, 155, 224
–mirror, 2, 4

–octahedral, 122, 213
–operator, 2, 74, 111, 133, 149
–reduction of, 209
–reflection, rotation, 2, 6
–scale, 234
–time-reversal, 164, 231
s-state, 82, 196

tensor, 194–198
trace, 142
transition, electronic, 159, 206, 208
triangle, 2, 4–8, 41
T, T_1, 138, 213

ungerade, gerade, 155

vector, 3, 10, 13, 14, 62, 103
–axial, polar, 45, 36, 137
–column, row, 17, 21, 157
–component, 13
–eigen-, 34, 64
–field, 14
–model of atom, 189, 200
vibrational state, 10, 172, 177, 217

weak interaction, 230, 233
wheel, 51
Wigner Theorem, 149, 157, 162, 172, 177, 183, 209, 212, 219
Wigner-Eckart Theorem, 198, 221

X-ray scattering, 222

Zeeman effect, 85
10 D_q, 213